Natural Obsessions

NATALIE ANGIER has written copiously on science for *Time, Discover*, and many other periodicals, including the *New York Times*. Her contributions to that newspaper, which won her a Pulitzer Prize, were published in a volume entitled *The Beauty of the Beastly: New Views on the Nature of Life*. She has won numerous awards, including the Lewis Thomas Prize and the AAAS Science Journalism Award. Her most recent book is *Woman: An Intimate Geography*, a celebration of the female body. She lives with her family in Takoma Park, Maryland.

Natural Obsessions

STRIVING TO UNLOCK THE DEEPEST SECRETS OF THE CANCER CELL

Natalie Angier

A MARINER BOOK
Houghton Mifflin Company
BOSTON NEW YORK

First Mariner Books edition 1999

Library of Congress Cataloging-in-Publication Data is available.
ISBN 0-395-92472-3 (pbk.)

Printed in the United States of America
QUM 10 9 8 7 6 5 4 3 2 1

TO MY MOTHER

AND THE MEMORY

OF MY FATHER

"Then," she asked, "what *is* the matter?"

"Why it has a crack."

It sounded, on his lips, so sharp, it had such an authority, that she almost started, while her colour rose at the word. . . . "You answer for it without having looked?"

"I did look. I saw the object itself. It told its story. No wonder it's cheap."

"But it's exquisite," Charlotte, as if with an interest in it now made even tenderer and stranger, found herself moved to insist.

"Of course it's exquisite. That's the danger."

— Henry James, *The Golden Bowl*

Acknowledgments

This book would not have been possible without the utmost consideration, patience, and good humor on the part of Robert Weinberg, David Baltimore, and Michael Wigler. They made my life easier as I made theirs more difficult, and for that unbalanced exchange I can only express my deepest gratitude.

I would also like to thank the many scientists who took the time to talk with me, not all of whom appear in the narrative. I am especially grateful to James Watson of Cold Spring Harbor Laboratory; Salvador Luria and Phillip Sharp of MIT; Howard Temin of the University of Wisconsin; Renato Dulbecco, Tony Hunter, Inder Verma, Bart Sefton, and Walter Eckhart of the Salk Institute for Biological Studies; Gerald Fink and Richard Mulligan of the Whitehead Institute for Biomedical Research; Philip Leder, Charles Stiles, and Geoffrey Cooper of Harvard Medical School; Raymond Erikson of Harvard University; John Cairns of the Harvard School of Public Health; Bruce Ames, Steven Martin, Peter Duesberg, and Ira Herskowitz of the University of California, Berkeley; Michael Bishop and Harold Varmus of the University of California, San Francisco; Mariano Barbacid of the Frederick Cancer Research Facility; Hidesaburo Hanafusa of the Rockefeller University; Arthur Levinson of Genentech; William Hayward, Richard Rifkind, and Lloyd Old of Memorial Sloan-Kettering; James Feramisco of Cold Spring Harbor; Frederick Alt and Richard Axel of Columbia University; Michael Waterfield of the Imperial Cancer Research Fund Laboratories; Sheldon Penman, John Buchanan, and Jonathan King of MIT; Stuart Aaronson of the National Cancer Institute; James Broach of Princeton University; Thaddeus Dryja of Massachusetts Eye and Ear Hospital; Brenda Lee Gallie and Robert Phillips of the University of Toronto; Webster Cavenee of the Montreal Branch of the Ludwig Institute for Cancer Research; Wen-Hwa Lee of the University of California at San Diego; Alfred Knudson of Fox Chase Cancer Center; and Robert Ellsworth, David Kitchin, and David Abramson of New York Hospital–Cornell Medical Center.

The spirit of any great research laboratory rests with the young scientists who ply their trade at the bench. To all who shared that spirit with me, my heartiest thanks; this book is for you.

In the course of my research, many people tutored me in molecular biology, but I would like to express a special debt to my original tutor, James Lupski.

Thanks to Clifford Tabin for his detailed review of the manuscript. He probably regretted halfway through having agreed to the chore, but he remained unflinchingly generous to the end.

I could never have written the book without the professional guidance and moral support of Peter Davison and the editors of Houghton Mifflin. If a book is a kind of living organism, my text would have remained protoplasmic had they not intervened.

Finally, thanks to my family and friends for their encouragement, their lost sleep, their forgiveness, and above all their love.

Contents

Introduction to the Mariner Edition x
Foreword / *by Lewis Thomas, M.D.* xx
Prologue: Building Blocks 1

1 / The Optimist 11
2 / The Family 44
3 / The Blackest Magic 70
4 / Pyrrhic Victories 96
5 / No Reason to Stay in Science 117
6 / Cooperation and Collapse 141
7 / Desert Days 170
8 / Down the Garden Pathway 203
9 / Something Borrowed, Something Blue 237
10 / The Alternate Pathway 269
11 / Pink Cadillac 304
12 / Superfecta Key 324
13 / Through the Looking Glass, and What May Be Found There 348

Epilogue 368
Appendix: Major Publications Cited in the Text 375
Index 377

Introduction to the Mariner Edition

IF I HAD TO DRAW the leaves on my family tree, I'd give a lot of them the jagged, angry silhouette of a cancer cell. I've lost more relatives to cancer than to anything else, and some of them at a fairly young age. My father died at fifty-one, when a malignant melanoma tumor spread to his brain. My maternal grandfather also died at fifty-one, of pancreatic cancer. My father's father died of colon cancer. Great-aunts, great-uncles, and cousins have died of cancer. So when I mull over the likeliest storyline for my own death, which I do to self-indulgent excess, I start with the assumption that it will be from cancer, and then entertain myself by wondering, "When?" and "What kind?"

Natural Obsessions is a book about the search for the molecular origins of cancer. It is a book about the nature of basic research and the blood-sweat-and-bones scaffolding of two highly competitive research laboratories: Robert Weinberg's group at the Whitehead Institute of MIT, and Michael Wigler's team at Cold Spring Harbor Laboratory on Long Island. It is about how scientists think, and how they feel, and how they behave. It is about the rush of ecstasy that comes when an experiment works, the virulent paralysis that follows failure, and the many stretches of confusion and ambiguity in between. It is about how scientists are as human as the rest of us, only smarter and with less attractive footwear.

What the book most emphatically is *not* about is the search for a cure for cancer. As I explain in the first chapter, most basic scientists do not search for cures. They ask how the cell grows or stops growing. They ask about dominant cancer genes and tumor-suppressor genes, signaling pathways and membrane ruffling. They don't like anybody's mentioning the phrase "cure for cancer." It makes them nervous. Are taxpayers getting impatient? they worry. Will the government cut off our grant money and make us look for a real job?

Basic researchers don't like talking about cures for cancer, and neither, I admit, do I. I'm scared of cancer, and I fret hypochondriacally about every headache or new freckle. I would love to imagine that the people whom I profile in *Natural Obsessions*, or any of the thousands of others in the field of oncogene, or cancer gene, research, will soon make spectacular breakthroughs with immediate bedside applications. I would love to imagine that scientists are on the verge of conquering cancer, yet I can't say I'm confident that they are. They may be, or they may not be. Nobody knows. In the eleven years since this book first appeared, scientists have made relatively little progress in applying the fruits of basic research to the treatment of cancer. They have not come up with any magic bullets; they haven't even found the right gun yet. It makes sense to hope that a firm understanding of the genes and proteins responsible for cancerous transformation will yield more effective therapies. Right now, oncologists rely on the standard treatment troika that they relied on have for decades: surgery, radiation, and chemotherapy—or, as the blunt tongues among them put it, slash, burn, and poison. Chemotherapy and radiation are quite good at killing cancer cells, but they are also notoriously good at killing normally dividing cells, which is why the therapies lead to side effects such as intense nausea, hair loss, and immune suppression, and why the treatments can't always be used in doses high enough to destroy every last tumor cell. If researchers could design drugs that target mutant genes or abnormal proteins found only in malignant cells, they theoretically could destroy those cells while leaving normal tissue unharmed.

Researchers have identified a large number of genetic mutations that are specific to cancer cells. One of them affects the so-called *ras* gene, a major molecular figure in this book and a subject of ongoing research among scientists, pharmaceutical companies, and biotechnology firms. The normal *ras* gene operates in the body to help orchestrate healthy cell division. But when the gene in one cell mutates, it can help instigate the growth of a malignancy. The mutant *ras* gene is a nasty character, found in about 25 percent of all human tumors. It would be magnificent to have a drug that homes in on the mutant *ras* gene, or on the aberrant protein that a mutant *ras* gene specifies. Researchers have worked mightily to develop an anti-*ras* drug. So far, they don't have one. For all the hype surrounding the biotechnology business, and despite the inherent logic behind the targeted approach to cancer therapy, designing drugs is extremely difficult. Even drugs that initially look selective often end up with a suite of unexpected side effects, and any drug that is going to be powerful enough to destroy or disable a full-blown cancer, when millions of malignant cells are disseminated throughout the body, is likely to have a few harrowing surprises up its side chains.

Progress in this area is always a matter of one step forward, one slap in the face. In Chapter 9, I describe the Weinberg lab's work on an oncogene called *neu,* which has since been given a double name, *Her-2/neu.* Recently, Dennis Slamon, of the University of California at Los Angeles, designed a drug called herceptin, a monoclonal antibody that interferes with the *Her-2/neu* gene in tumor cells. Slamon has fought, heroically and monomaniacally, to develop the drug and bring it to clinical trials, and early results suggest that it shows promise for the treatment of advanced breast cancer, when the *Her-2/neu* gene often is hideously hyperactive and has amplified its numbers to grotesque proportions. But herceptin cannot be called a cure; in some cases it can prevent recurrence or lead to remission, but in others its use appears to increase the risk that stray breast cancer cells will spread to the brain, beyond any therapy's reach.

Similarly, in Chapter 8 I describe the excitement over preliminary work with a compound called angiogenesis factor. As scientists were learning at the time, a tumor must grow new blood vessels if it is to increase in size beyond a millimeter or so. The tumor needs the new blood vessels to feed it oxygen and nutrients; it needs a blood supply to expand its aggressive reach and eventually to seed the body with metastases. A tumor that cannot grow new blood vessels will stop proliferating and essentially suffocate to death. From the beginning, the process of vessel growth, or angiogenesis, looked like a beautiful target for cancer therapy. If angiogenesis could be blocked, tumors in theory could be killed at a very early stage in their development, long before they posed a danger to the patient. By the mid-1980s, Judah Folkman and Michael Klagsbrun, of Harvard Medical School, had isolated angiogenesis factor, a protein that tumor cells (and some normal cells) will release to stimulate the local proliferation of blood vessels. The scientists began working on ways to interrupt the activity of that stimulatory factor. They and others have since created a number of angiogenesis inhibitors, which are at varying stages of clinical trials and in some cases look quite heartening. Even here, though, nobody is predicting that angiogenesis factors will be *the* cure for cancer. Instead, as Dr. Folkman has emphasized, such inhibitors will likely find their greatest utility as adjuncts to standard therapy—slash, burn, and poison.

We can expect speedier progress in designer diagnosis than we do from magic-bullet therapeutics. For example, researchers at Johns Hopkins University have shown that they can detect mutant *ras* genes in colon cells shed in stool samples from patients with colon cancer. If the scientists can improve the sensitivity of the technique, then it could be used to screen stool samples in people who are risk for colon cancer and perhaps detect a *ras* mutation early on, when the lesion is still a precancerous polyp and easily scraped out. But even here, the reality is messier than the theory. It won't be enough to screen stool samples for *ras* mutations. Many colon cancers show no evidence of *ras* mutations. Instead, an effective diagnostic technique will require DNA probes to root out every possible genetic mutation that could contribute to malignant growth, and we are not there yet—not for colon cancer, not for any other cancer.

So I don't like talking about cures. As far as I can tell, the assessment of Michael Bishop, of the University of California at San Francisco, which I quote in Chapter 1, is still correct. "There's been far too much hype in this business, too much cocksureness," he said. "Anybody who walks around and says that we've got this problem almost licked is a fool, a knave, or both."

Yet if we reconfigure what we view as "the problem" and ask not "When are we going to cure cancer?" but instead "What have we learned by studying cancer?" then the answer has to be: an extraordinary amount. It is the second question rather than the first with which my book ultimately is concerned. Through studying the cancer cell, that fluttering little leaf of death, we paradoxically have learned about the roots of life. The cancer cell is the distilled cell, doing what all cells are designed to do, which is proliferate; and because the cancer cell does its job so tirelessly and ostentatiously, it can be analyzed. In dissecting the cancer cell, we get a handle on the healthy cell, its organization, rhythm, and aesthetics, and what it takes to live and to perpetuate life. We have learned much, and we are learning more by the day—or by the night, when many scientists do their best work. This is a spectacular enterprise, this exploration into the rudiments of the cell. It deserves our applause, our communal pride, and our tax dollars. If we were given our oversized frontal lobe to understand nature and the universe, then we have a duty, a moral imperative, to study the cell, to scratch away at its pitiless complexity until it squeals. We are obliged to keep exploiting the cancer cell, even as we have trouble taming it.

And we're lucky that biologists love to do this sort of work, because it is grueling, often tedious, badly paid, and anonymous. It is practically impossible to become a scientific celebrity. Look at poor Harold Varmus, who appears in this book and who has subsequently won a Nobel Prize with Michael Bishop for their work on oncogenes. He also went on to become the director of the National Institutes of Health, one of the most prominent scientific posts imaginable. No matter. On hearing that Varmus would be giving the commencement address at Harvard University in 1996, the editors of the student newspaper, the *Harvard Crimson,* expressed their indignation. "Dr. Who?" was the title of their editorial. Sure, the guy may have a Nobel Prize, and he may be the head of the biggest research organization on the planet, but who the hell *is* he? Couldn't the university officials have invited somebody more "glamorous"? the paper asked. Undeterred by the rudeness of the Harvard brats, Varmus gave the address. He is, after all, accustomed to rudeness. Scientists are trained to question every bit of data they see and to speak their mind, however roughly. Such skills can spill over into their social interactions, as in, "You've gained weight, haven't you?" or "Did you know you have a pimple on your face?"

Scientists are not high-minded martyrs, not by a long shot of any heigh-ho silver bullet. They may rarely gain widespread public fame, but their egos are large and in need of chronic stoking, which they seek from their peers. Scientists compete fiercely for professional acclaim; they want to be known and respected by those who know better, who understand what they do. They also want to do what they want to do. They want to follow their whims and instincts, they want to solve puzzles, and they want to learn how things work. In this book, I follow researchers on a series of those profound and whimsical quests. The field of oncogene research is fast-moving, and some of what I describe is a bit dated, particularly the technical aspects of how experiments are done. The art of gene cloning is orders of magnitude easier today than it was a decade ago, in part as a result of the Human Genome Project, the federally financed campaign to identify and spell out every one of the three billion subunits of human DNA. A number of procedures that once had to be painstakingly carried out by hand can now be performed with off-the-shelf kits. The use of computer databanks to analyze, compare, and categorize genetic information has exploded.

Yet much of the progress in oncogene research over the past few years has built on, rather than superseded, the early and revolutionary discoveries from the Weinberg, Wigler, and other laboratories. For example, one insight from the pioneering days that remains salient is that the cell is a perpetual waffler. It is poised to grow and must be told not to; and it is inclined toward slothfulness, toward stasis, and must be strong-armed to divide. Each cell is outfitted with scores of genes that can impel a cell to divide, and perhaps an equal number that tell it to stop dividing. The system works spectacularly well most of the time. As Weinberg points out in his 1998 book, *One Renegade Cell,* a human being experiences about 10 million billion cell divisions over a seventy-year lifespan, and those divisions are almost always orderly, precise, and effective. The immune system produces fresh T- and B-cells; the lining of the stomach is replenished with new protective epithelial tissue. Keeping the growth steady but predictable are two categories of genes, and both are capable of becoming oncogenes when mutated. One class includes genes like the *ras* gene, the dominant oncogenes, which normally receive growth signals from the bloodstream and translate those signals into the act of cell division. In aberrant form, such genes become hyperactive bullies, prompting cell fission even in the absence of growth signals from the outside. The other class of genes consists of the growth-stanchers, called tumor-suppressor genes. These genes ordinarily operate to keep cell growth in check, and when they are mutated into worthlessness the cell loses an important brake on its growth and rumbles toward cancer. In *Natural Obsessions,* I describe the breathtaking and bitterly contentious race to isolate the first of these tumor-suppressor genes, the retinoblastoma gene, which is involved in childhood cancer of the eye. Many other tumor-suppressor genes have since been cloned, most famously the breast cancer gene called BRCA-1. The concept that dominant oncogenes and tumor-suppressor genes are co-executives in the life and growth of both healthy tissue and cancer remains a central tenet of the field.

As much as I'd like to be able to boast that my original text says it all, or at least vaguely gestures in the right direction, I'm not that foolish a knave. There have been a number of important new insights into the cell's personality, which deserve mention here. Among the most profound is the realization that cells die the way they divide: by design. Phalanxes of scientists have converged on the study of programmed cell death, which also goes by the name of apoptosis, a coin termed from the Greek word for "falling from," as leaves fall from a tree or, as one balding scientist put it, as hair falls from the scalp. Biologists have determined that under some circumstances, a cell essentially "decides" to die for the good of the body around it, and initiates a sequence of molecular events to literally blow itself apart. Many scientists now believe that disorders of the cell's suicide program are responsible for a host of diseases, especially cancers. They have discovered genes that help initiate the cell's self-destruct sequence, and they have shown that when these genes become mutated, cells continue to live long after they should have expired, and malignancies such as lymphomas and leukemias can result. In other words, cancer sometimes is a matter of too little cell death rather than too much cell growth.

Conversely, a disordering of the body's apoptosis mechanism sometimes can lead to too much cell suicide too soon, again with catastrophic results. The massive death of brain tissue seen in Alzheimer's patients may be the result of a neuronal chain reaction, with the untimely self-destruction of one brain cell initiating a frenzy of copycat suicides. Likewise, a communal, aberrant spasm of apoptosis, a kind of Jonestown downing of the poisoned Kool-Aid, may explain why in some AIDS patients, even immune cells that are not infected by the human immunodeficiency virus suddenly start dying.

The conceptual link between cell growth and cell death, between knowing when to divide and when to die, extends beyond genes to the very architecture of the chromosomes, the viscous strands on which the genes are arrayed. Another extraordinary finding in fundamental cancer research is of the importance of chromosome tips to cell proliferation. Human cells have twenty-three pairs of chromosomes enclosed in their nuclear hearts, and at each end of each chromosome are protective structures known as telomeres. They are like the plastic tips on shoelaces, which prevent the laces from fraying. They are also like timepieces, keeping track of a cell's age and telling a cell, Sorry, your warranty has just expired. Under normal circumstances, the telomeres get slightly shorter each time a cell divides; and when the cell has divided a set number of times—generally between fifty and one hundred, depending on the cell type—the telomeres become too short to sustain chromosome integrity, and the chromosomes begin wobbling. Somehow—we don't know how—that instability sets the cell's suicide program in motion. Telomeres dictate the lifespan of individual cells, and they play a role in the death of the entire human being as well. If you look at the cells of an infant, you'll see that her telomeres are much longer than the telomeres of her grandmother. The infant has many rounds of cell division before her, while the older woman . . . well, let's not get too specific about the chromosomal ticks and tocks that remain.

Not all cells in the body are subject to the unidirectional shortening of the telomeres. It turns out that in some cells, an enzyme called telomerase is active, and telomerase helps restore the telomeres, builds them back up again. The enzyme is active in stem cells that give rise to a man's sperm, for example, and in the primordial cells of the immune system. All cells of the body have the capacity to flick on telomerase, for they needed the enzyme during fetal growth and the gene responsible for the enzyme remains in place in every nucleus; but in the vast majority of body cells after birth, telomerase is switched off and stays dormant, so the telomeres gradually shrink and cells die graciously when their time is through. But wouldn't you know it, some cancer cells snub the rules of decorum and flick on their telomerase enzyme once again, with the result that their telomeres are replenished no matter how many times the wicked cells divide. The cells have mastered the trick of immortality. It is a stupid, selfish gesture, and a futile one, for the disease will end up killing the host and and the tumor cells along with it; but cancer cells are no better at long-term planning than your average American corporation is.

Scientists who study telomeres and telomerase now see the chromosome tips and the enzyme that repairs them as tempting sites for two types of therapeutic intervention. On the one hand, telomerase might prove useful, in a controlled fashion, for treating diseases of premature cell death, as happens in heart attacks and strokes. On the other hand, a number of investigators are striving to develop telomerase blockers, to shut off the enzyme in cancer cells and in theory remove the malignancy's private, pestilent fountain of youth. Whether these blockers will work is not yet known, for some cancer cells have shown an ability to continue dividing indefinitely without the restorative aid of telomerase. Nevertheless, the take-home message of the telomere story is sad and clear: for life to continue, there must always be death, and the portrait of Dorian Gray, that seeker of eternal youth and immortality, bears the malign mark of a cancer cell.

Yet good scientists do not dwell on difficulties and setbacks. They don't lapse into gloom. They are too disciplined for that, and they are too curious, and so they keep following their noses. The title of Chapter 1, "The Optimist," is a description of Weinberg, and he is still optimistic about the current progress and future payoffs of basic research. He also has expressed the opinion—one I heartily agree with—that the best way to reduce cancer mortality is to prevent the disease in the first place, by wiping tobacco off the cultural landscape, by cutting down on animal fat in the diet, by gaining a more sophisticated understanding of the components of fruits and vegetables and of why some populations suffer higher rates of various cancers than others do. Weinberg continues to be a leader in the field of oncogenes. His laboratory is among the most productive in the business; papers from his young scientists appear regularly in the journals, on the subjects of old faithfuls like dominant oncogenes and tumor-suppressor genes and of trendier fare like telomerase and apoptosis. He doesn't dwell on disappointments of any sort. I recklessly predicted in my prologue that "when the Nobel Prize is parceled out for oncogenes, he will almost surely be a recipient," but I was wrong. The Nobel committee awarded the prize for oncogene research in 1989 to Harold Varmus and Michael Bishop, so unless Weinberg makes another really big breakthrough—a cure for cancer at last!—I'm not going to win the Cassandra Prize for Excellence in Forecasting. No matter. Weinberg always said that the Bishop-Varmus discovery—that cancer-causing viruses pick up their malignant genes from the cells they have infected—was the watershed event in his discipline, more revolutionary, he insisted, than anything he'd ever done.

Whether or not the oncogene vocation will yield any more Nobel Prizes, it is still vibrant and seductive. Most of the characters in *Natural Obsessions* are still a part of it, and most of them are thriving. Many were young and junior when I met them, graduate students and postdoctoral fellows, and they were scrambling to make a name and future for themselves, and sometimes they sparred with each other over rights to experiments and authorship, and sometimes they fought with their supervisors, for as I describe in the book, they recognized that their professional needs did not always harmonize with those of their renowned leaders. Now many of them are lab leaders themselves, at top-tier institutions. To name a few: Clifford Tabin is at Harvard Medical School, David Stern at Yale, and Paolo Dotto at Massachusetts General Hospital; Mien-Chie Hung runs a large group at the University of Texas M. D. Anderson Cancer Center in Houston; Cori Bargmann and Rudy Grosschedl both ended up at the University of California at San Francisco; Stephen Friend is with the Fred Hutchinson Cancer Center in Seattle, where he does basic research and also devotes himself to new drug discovery. Many of the European scientists have gone back home, René Bernards to the Netherlands Cancer Institute, Carmen Birchmeier to the Max Delbruck Center for Molecular Medicine in Berlin. Now the once uppity juniors are themselves griping about their demanding subordinates, and spending more time administering, writing grants, and attending conferences than they do running gels or clicking pipettes. Almost none of the book's cast has abandoned science altogether, even though it is harder than ever to wrest money from the government. When they have strayed from the fold, it hasn't been terribly far; Alan Schechter, for example, a postdoctoral fellow in Weinberg's lab, decided to leave science for medicine and became a dermatologist.

That the scientists have stuck with science underscores my most enduring impression of the lot: that they love science, really love it. They love science with their heads the way most of us love people with our hearts. They love it at 3:43 in the morning, they love it on Sabbaths and holidays, they love it when they should be so bored they're starting to drool. Their latest results may look like the smears on a preschooler's placemat, but still they scan the results for a scrap of sense and get excited and want to do more science. In the many years I've been writing about science, I've had my gripes about its practitioners. Science is like Paris, I've thought irritably: it would be a great place if it weren't for the Parisians. Yet always I have admired and envied scientists for the depth with which they love their calling. In this book, I try to give a sense of that love. The names and technical details of the science may change, but the love, at least, keeps on burning.

Foreword

BY LEWIS THOMAS, M.D.

THIS IS A BOOK about the way biomedical science is done these days, based on close scrutiny of a scientific community by a young observer who came in from outside and settled down to watch and listen, living within the community almost like an anthropologist attached to a remote tribe. And, at times, an exceedingly wild tribe.

To an old hand, a survivor from earlier generations of investigators, the changes in the way science must be conducted these days come as a surprise and, at times, shock. The underlying drive, pushing the enterprise along, is of course the same: it is the insatiable curiosity of the scientists to find out how nature works, at deeper and deeper levels. There is also the hope, just as always, that pieces of the new information, whatever, may sooner or later be put to use for betterment of the human condition. It is important to note, however, that that hope has never been the driving engine; it is just there, in the back of the minds of the workers. The steadier, irresistible push is the plain wish to find out, and whenever possible, to be the first to find out.

There is, I suppose, a way of going about work of this kind that can be called the scientific method, but I have never been quite clear in my mind about what this means. "Method" has the sound of an orderly, preordained, step-by-step process; one maneuver leading sensibly and logically to the next; a beginning, a middle, and then an ending. I do not believe it really works that way most of the time, and surely it doesn't in the firsthand anecdotes that abound in this book. As Natalie Angier saw over and over again in the Whitehead Institute laboratories, the work usually began with what seemed at the time a good hypothesis, a very bright idea, sometimes emerging in the middle of an excited conversation about something else. Next came the making up of a story about how nature might go about doing this or that thing if nature had an intelligence like that of the investigator. If the idea became a sufficient stimulus, the response was the starting up of work on a new line. But then, more often than not, the step-by-step process began to come apart because of what almost always seemed a piece of luck, good or bad, for the scientist: something unpredicted and surprising turned up, forcing the work to veer off in a different direction. Surprise is what scientists live for, and the ability to capitalize on moments of surprise, plus the gift, amounting to something rather like good taste, of distinguishing an important surprise from a trivial one, are the marks of a good investigator. The very best ones revel in surprise, dance in the presence of astonishment. Others, less gifted, cannot endure bewilderment and find other ways to make a living.

These aspects of research have not changed, but other things have become very different from the old days. And, although the work is moving faster than ever before, with more new discoveries popping up each week in the headlines of the science sections of the newspapers and newsmagazines, some of the changes are not for the better. Or anyway, I feel certain, not for the better in the long run ahead.

Doing science on hard biological problems has to be the greatest fun in the world, or it can't be done at all. I have some apprehensions, after reading this book and looking around at the scene in other, comparable institutions, that some of the real fun may be draining away from the game. It has always been intensely competitive, especially when the stakes for new information become as high as they are today in cancer biology, but the competition seems now to be nearing the point of intellectual ferocity, and I take this as a bad sign. Partly, this is due to the sheer size of the enterprise; there are now more young scientists launched in the early stages of research careers, and they face a pyramidal hierarchy with a sharply limited number of tenure posts or secure jobs in industry. The frenetic pace is also due in part to the incredible power of the new instruments employed in day-to-day biomedical research; questions that used to take months or years to contemplate can now be answered within days, sometimes minutes. But this has caused a new change in the way laboratories are organized and run, with much larger teams of young investigators than ever before, most of them graduate and postdoctoral students, each assigned to a very small piece of the problem and each, naturally enough, driven to regard success with that small piece as a matter of life and death.

Another worry: money has become a more central matter in science than used to be the case. There is more of it, all told, thanks to the institutional marvel of the National Institutes of Health, but still seemingly not enough to go round, and clearly not enough to provide decent wages for the scientists who are now the important figures, both the absolutely indispensable and the most junior of the lot living on derisory stipends. Something new has to be done about this asymmetry. I don't know what, or how it should be done, but it is there as a nagging worry, raising questions about the future recruitment of young scientists in this country.

Finally, the essential role of pure gossip in the furtherance of science may be coming into some jeopardy. If there is any single influence that will take the life out of research, it will be secrecy and enforced confidentiality. The network of science, nicely illustrated in this book by the international and transnational collaborations now the style for molecular biology, works only because the people involved in research are telling each other everything they know, out in the lobbies at international congresses, in nearby bars and diners, and by spontaneous long-distance telephone calls. Telling the world (the scientific world, that is, not at all the press) everything you know, including the unprecedented observation made yesterday in your own laboratory, is a large part of the fun of doing science. I am worried that something may be happening to interfere with this high privilege.

Prologue: Building Blocks

THE LATE APRIL DAY in southern New Hampshire couldn't decide if it was winter or spring. Clouds marbled the sky in thin white fingers arcing from northeast to southwest; the spindly trees were still leafless and uncertain, like awkward teenagers waiting to see what their bodies would do next. We had gathered at Robert Weinberg's house, set amidst a hundred wooded acres in Rindge, about seventy miles from Boston. We wore blue jeans and sneakers or work boots; I'd made the mistake of wearing a cotton knit sweater, which Weinberg insisted I trade for one of his plaid flannel shirts. Embarrassed, I attempted to decline the offer. "Look, I'm not telling you this for my benefit," he said. "It's a nice sweater, but one drop of cement and that's the end of it." Bob has a knack for saying things that are very persuasive yet are worded to leave the decision entirely up to you. I changed my sweater.

Weinberg had designed and built his weekend house by himself. Or not entirely by himself. Five days a week he is a principal investigator at the Whitehead Institute for Biomedical Research in Cambridge, a privately financed basic research center affiliated with the Massachusetts Institute of Technology. He is studying oncogenes, the brief coils of DNA that twist and prod a cell toward cancer. The discovery of these genes and their role in human malignancy has been one of the most dramatic events in biology in the late twentieth century. Bob Weinberg is numbered among the Magellans or Neil Armstrongs of the field; when the Nobel Prize is parceled out for oncogenes, he will almost surely be a recipient. And like other important scientists, Weinberg has a big lab, comprising some fifteen postdoctoral fellows, graduate students, and technicians. But unlike most scientists, Weinberg does more than research with his lab members: he has them help him build his house. Whenever he needs to have a floor laid or a wall erected, he simply throws a party. People are always glad to come, to spend a day working outdoors with their hands and muscles, rather than indoors with their hands and heads.

That Saturday, it was a concrete-pouring party. Weinberg's two children, Aron and Leah Rosa (called Rosie), were growing bigger, and he'd decided it was time to build an extension to his single-room cabin. We were going to pour the concrete for the foundation of that addition. As we awaited supplies, Bob outlined the plan of attack. He wore paint-stained overalls and the funny fat-soled shoes that he needs to support his flat feet; the wind ruffled his thin hair, which he grows long on one side to cover up the balding patches. Yet he spoke with the booming authority you expect at the podium or in the classroom, peppering his explanations with "Do you follow me?" and "Everybody got that so far?" Bob may sometimes seem harried or distracted or the picture of the absent-minded professor, but details of the foundation to his house were worked out down to the last bolt and bit of guy wire.

When the pickup truck finally arrived, rumbling and tossing and skipping along the stray logs and gullies of the driveway, work proceeded quickly. Constance Cepko, a postdoctoral fellow in another lab at the Whitehead, stood in the back of the truck, all strength and no-nonsense, shoveling concrete from the troughs and plopping it into wheelbarrows. Luis Parada, a handsome and muscular graduate student in Bob's group and a former actor from Colombia, swiftly navigated the laden wheelbarrows up treacherous inclines, where fellow Weinberg acolyte Cornelia Bargmann, a Southern wraith who may well be the brightest student at MIT, helped slough it into the foundation holes. Whatever she does, Cori does flawlessly, and her concrete never spilled around the edges of the postholes, the way mine sometimes did. Her husband, Michael Finney, another Southern wraith and a graduate student at an MIT lab kitty-corner from the Whitehead, was equally adept; I felt that between the two of them they could have tooled together Chartres cathedral in about nine days.

Not all were so indispensably precise. René Bernards, a tall, blond, and elegant-faced postdoc from Holland, did his anxious share of shoveling and stirring concrete, but his left arm had yet to recover from a skiing accident, and he moved as awkwardly as a sparrow with a broken wing. Snezna Rogelj, a beautiful postdoc from Yugoslavia, worried repeatedly about René's zealous efforts, fussing and scolding like a counselor at a sleep-away camp. Sometimes she would stoop down and gather little Rosie Weinberg into her arms, and then walk around the compound bouncing the child against her. Later, when I saw a photograph of Snezna and Rose taken by Luis's girl friend, I was reminded of Flemish paintings of voluptuous Madonnas, their hips languorously thrust out to support a plump and sacred cargo.

Bob Weinberg doesn't think that his house building has much to do with his science. And at first I agreed. I want to believe that people can be many different things at different times, changing their moods and personae as easily as they change shampoos or computer diskettes. But when I came up to New Hampshire again, three weeks later, I began to see what the house really meant to him. I always thought that building anything bigger than a kitchen cabinet would take months, and that you could come back from week to week seeing no change more perceptible than a new roof gutter or a virgin sprinkling of sawdust. Now, however, there was an entire floor for the addition, and wall trusses, set up in interlocking V's. Weinberg and two strapping New Hampshire neighbors had been working for fourteen hours that day, pounding and grunting. I was mildly shocked at how ornate it was becoming.

"It looks like a château-in-progress," I said.

Weinberg yawned noisily. "I'm exhausted," he said, and he repeated two more times that they had had been toiling for fourteen hours.

Several days later, his wife, Amy, a small, slim, dark-haired woman, showed me a wood model that Bob had made of the addition. The roof was actually going to be two roofs, intersecting to form a pattern of twin gables on all four sides. I thought of Robert A. M. Stern and Michael Graves and all the other postmodernists who culled evocative scraps and corners from history and just barely avoided making a mess.

"We once visited this old medieval dining hall in Bennington," said Amy. "The room must have been over a hundred feet long. Then Bob found out that the posts supporting the ceiling over this enormous space were the same width as those he used to build our one-room cabin." Amy is shy. She stands with her arms crossed over her chest, her thick black hair streaming down her shoulders. "I guess you could call it overkill."

Just as Bob constructs models for his house, so he draws models to illustrate his ideas about the quirks and methods of DNA. During group meetings with his lab mates, his chalk will trace messy circles representing the cycle of a dividing cell, with lines slashed through at various points to pick up the putative activation of this or that gene. But though his house is baroque, his plans for the mammalian cell are almost wistfully simple. Bob Weinberg believes that there are only two or three or several ways that a cell can receive its signal to divide, that beneath the tangles and brambles of 50,000 genes in the average human cell, a few quiet dictators are running the show. More important, he thinks that you needn't map the position and function of every last bureaucratic enzyme — a trembling, miserable task — to get at those key genes and their corresponding proteins.

"I believe that nature is ultimately organized on very simple principles," he told me one afternoon as we sat in his office, surrounded by a diminutive jungle of house plants, the fronds of which occasionally strayed near my hair or cheek. "The same gene that must be activated to signal a cell to divide is likely to be the gene that is activated after a sperm has penetrated an egg. I have no proof of that yet, but I have a conviction." He wagged his finger at the word *conviction*. "Sometimes, when one of my students is writing a paper and trying to make it more and more complicated, I say, 'You shouldn't be trying to obscure your point. You should be trying to say something as simply as possible.'"

Bob Weinberg has some reason for his faith in simplicity. The discovery of oncogenes in the late 1970s was one of those rare events that punch through the snarl of ignorance and — at least for a little while — make everything seem easy. Until that small set of genes was plucked from the chaos of a tumor cell, scientists knew virtually nothing about the biochemistry of cancer. Nothing. They could look at a malignant cell under a microscope and see that it wasn't flat and happy, as a normal cell is, but straining away from the dish into a jagged abstract pattern, like a rug made from an animal hide. They could tug the cancer cell apart and find gross chromosomal changes in its nucleus, exaggerated quantities of hormones and related growth proteins oozing from its surface, warps and gulches along its membrane, and any number of ugly deviations from health. Yet they couldn't say which, if any, of these changes were important — that is, which were the primary events that triggered the cancer and which were the consequences of transformation. For all its flamboyant aberrations, a tumor cell remained a black box. Small wonder that the treatment of the disease had progressed little beyond the surgeon's scalpel and a desperate assortment of poisons.

The science of oncogenes offered the first real hope for understanding. In a few febrile years, researchers revealed that the beginnings of cancer lay not in a wholesale rewiring of the cell, but in a subtle alteration of a fistful of key genes among the human quota of DNA. Under normal circumstances, such genes play a vital, growth-related role in all or most tissues of the body. In some tissues, the genes may set up the rounds of simple division, helping skin cells to proliferate into a scab around a wound, or allowing the immune system to send out a host of antibodies to assail an invading pathogen. In other tissues, the genes may prod an infant cell toward its mature destiny as a muscle cell or neuron or liver cell. Whatever their assigned tasks, the genes that scientists have designated *oncogenes* share a common characteristic: they are vulnerable to mutations. And once mutated, the genes contribute to the birth of a tumor. That's why the genes are called oncogenes; onco is from the Greek *onkos*, meaning mass. Some scientists prefer to say "proto-oncogene" when referring to the healthy progenitor of a cancer gene, but most biologists rather imprecisely say "oncogene" for any gene that is prone to becoming tumorigenic. Nevertheless, it's important to keep in mind that our cells possess oncogenes not because some nasty natural or supernatural force placed them there to keep our population in check, but because the body requires the genes to grow. To date, about four dozen different oncogenes have been identified in human cells, and all are assumed to be necessary to survival. Only when the genes are mutated do they become agents of death.

Indeed, so indispensable to life are the normal versions of oncogenes that humans are not their sole possessors. Biologists have discovered that these genes are preserved virtually intact across the phylogenetic spectrum: in cats, in cows, in rats, in roaches, in bluefish, even in yeast. Oncogenes have been around for more than half a billion years — which, to an evolutionary biologist, can mean only that the genes are doing something pretty basic. Cancer may be the tithe that Westerners pay for their advanced society (in less developed nations, most people die of parasite-caused illnesses, malnutrition, or infectious diseases), but the genes that participate in cancer are Anasazi genes, the Ancient Ones.

All told, oncogenes are a big, sprawling deal; it would be difficult to overstate the importance of their discovery. Through studying oncogenes, scientists may learn what trips the switch to cancer and figure out some way to switch it off. What's of more embracing value, we may at last get to know the private heart of our own cells, the ones we tote around and expect to work, with only the slimmest notion of how they do it.

The field of oncogenes, then, is doubly blessed: on the one hand, it promises practical and important spin-offs in the advancement of cancer treatment; on the other, it already has yielded a lot of intimate details about what makes a cell a cell. Scientists like large problems, and they have tramped to this one by the battalion. There is no major or modest university in the United States or abroad that isn't in some way involved in the analysis of oncogenes. And leading the pack through the years of the field's most explosive growth has marched the short, disheveled figure of Bob Weinberg.

When I first approached Weinberg to ask whether I could write a book about his lab, he was already quite famous, in the *People* magazine definition of the word. Abundant articles have featured him, television crews have cluttered his corridors, and journalists have phoned from Australia, Kuwait, and Bangor, Maine. Yet he wasn't blasé about my request, nor did he mutter, "Not you, too." Instead, he invited me to lunch at a nearby seafood restaurant, and as he mopped squibbles of fish chowder from his thick Victorian mustache, he asked for more details.

The book I described to him would be about oncogenes: their history, their importance, what they are, how they work, their role in human cancer. A standard approach to the subject might have been a survey course of the field, with cameo appearances by all the players and frantic gropes for clever similes about genes and proteins that would nudge the reader awake. But not only would such a book be boring; it would be out of date before it was even edited, let alone on the stands. If many of the smartest minds in science are working on something, how could I possibly keep up? True, the field had reached a point where many conclusions could be drawn and separate pieces threaded together, but that seemed more like a job for a molecular biologist or a scholar than for a journalist.

I wanted to do something else. I figured that if I couldn't capture the here-and-nowness of the science, maybe I could succeed with the scientists. After all, I'd long believed that the problem with most popular science writing, my own included, was its emphasis on the gee-whizardry of science, the "spectacular" or "revolutionary" discoveries that seem to spring to life parthenogenetically, with little or no evident effort behind them. There's another side to science, of course: the fits, the starts, the false leads, the sailing hopes, the chalky tedium. I believe that scientists really do live in a moat-girdled world of their own, far more so than, say, advertising executives or sportswear designers. Not only do they have their technical language, indecipherable to the layperson, but they have their own way of thinking, of looking at problems, of dealing with each other and with the rest of us. The way science gets done is unlike the way any other business of life is conducted.

That, I told Weinberg, was what I wanted to capture: *doing* science, rather than done science. How do scientists think? How do they conceptualize a problem and then erect an experiment, including an elaborate bulwark of controls, to prove it? Where do they get their ideas from? What are the many things that can go wrong, and how does it feel when something goes right? What motivates a person to spend twelve, sixteen hours a day in a fluorescent-lit laboratory, hands thrust under a vacuum hood squirting serum into a petri dish or gently slicing through the shaved backside of a pink-eyed mouse? What are the dynamics between the senior scientists who orchestrate a lab and the young post-doctoral fellows and graduate students who actually do the work? Are the politics of science any more savage than those of a corporation? Is the competition keener?

"It sounds to me," said Weinberg, "as though you want to use oncogenes as an excuse to write about life in a modern molecular biology laboratory."

"That's right," I said, pleased that I hadn't bungled my point too badly. "But I didn't make my choice at random. I think that oncogenes are one of the most exciting things to happen in a long time. Don't you agree?"

He spread his hands apart into a pair of wings, indicating, "That's what I do, isn't it?"

"But why my lab?" he pressed. "There are many people in oncogenes now who are doing work as good as, or even better, than ours."

"You've been the most consistently successful scientist in the field," I said. "Besides, I had to choose somebody. If I tried to write about two dozen different labs, I'd make a real mess of it." I told him that I would eventually spend time with another oncogene specialist, although I had yet to decide who might provide the best contrast. Later, when I had learned more about the internal hierarchy of the business, I would settle on an obvious and faintly melodramatic choice: Michael Wigler of Cold Spring Harbor Laboratory on Long Island, a relentlessly brilliant scientist and one of Weinberg's rivals for supremacy in the field.

It didn't take long for Weinberg to agree to my proposal, with the provision that David Baltimore, the director of the Whitehead Institute and a Nobel laureate, also consent. I talked to Baltimore and gave him a batch of articles I'd written as a staff member of *Time* and *Discover* magazines, together with a number of free-lance pieces. Baltimore granted me permission to come and go as I wished, and he even arranged for me to be given a magnetic card that would get me into the institute after hours. I think my prior affiliation had a lot to do with his robust acceptance. Unlike most intellectuals, Baltimore is fond of *Time*: in 1974, a year before he received his Nobel Prize and the consequent adoration of the Western world, *Time* had featured him among the two hundred rising leaders in America.

For a total of six or seven months, over a period of a year and a half, I would trail molecular biologists through their lengthy days and longer nights in the lab, struggling to keep up with their schedule even when my eyes were sore and blood-laced from such pungent chemicals in the air as ammonia and chloroform. I liked being at the lab, particularly at night, after the senior staff had gone home and the young postdocs and graduate students ruled the halls, cranking up their radios and breaking regularly for conversation. Apart from watching the scientists at their work and occasionally making them botch an experiment by asking an ill-timed question, I tagged along to whatever ancillary activities they might engage in — conferences, seminars, meetings, beer hours, picnics, dog races, jogging, meals. Especially meals. Young scientists, I found, are like Italians: they turn meals into events, and I got to know the restaurants of Cambridge and Boston more intimately than I ever did the dining spots of my home town of New York.

Through it all I attempted to mimic the scientists' thinking and to see biological problems as they did. I tried to whittle large and intimidating puzzles into manageable pieces and to figure out how I could design an experiment to deal with a given piece if I were doing the work. Occasionally I tried to interpret results on my own. The scientists would show me the splotchy x-ray photographs of DNA that had been carved up and probed with radioactive markers, and they'd tell me what they thought was going on. Then I'd look at the next x-ray, or autoradiograph, and attempt to make out patterns myself. If I was feeling particularly hubristic, I'd declare to the scientists that I disagreed with their analysis, or that I thought they were seeing what they wanted to see rather than what was on the autoradiograph. Sometimes they'd laugh at my efforts, and admittedly I never did learn to "think like a scientist." Molecular biologists spend seven to twelve years in training, and they spend the rest of their lives struggling to keep abreast of their field.

But what I did do, by narrowing my scope and struggling to consider problems in cautious, baby-step fashion, was to sharpen the focus of my book. I realized that I couldn't possibly draw the paradigm of a Scientist or of a Laboratory. The best I could do was convey the concerns of a limited group of people: Bob Weinberg himself; the researchers who helped lift Weinberg from his early obscurity to the first stages of his fame; and the researchers I watched in action at the Whitehead Institute, who labor under the burden of Weinberg's now enormous reputation and yet who, miraculously, continue to push the field forward. I attempted to put the concerns of the Weinberg scientists in a larger perspective, both by spending time at the competing Wigler lab and by interviewing scores of other molecular biologists around the country. Nevertheless, what emerged was a nonfiction version of a drawing room novel. I have written on a small scale, and from a very personal perspective, about a small segment of a very big chase. I don't pretend to have limned any universals.

In writing a story scaled to the drawing room, I've placed no more emphasis on the celebrated scientists who run the labs than on their still uncelebrated protégés. Senior scientists receive much of the professional and public acclaim for any discoveries made in their labs; their names are cited by the Nobel committee, and they are the ones who are quoted in the press. That kind of glory is an accepted feature of science, and even the younger scientists who may feel slighted don't grouse audibly. Nevertheless, seniors and juniors alike know that great ideas bubble up from below as often as they trickle down from above. A portrait of the Whitehead or Cold Spring Harbor lab that stressed the glory and genius of Robert Weinberg or Michael Wigler would be false and sterile. As David Baltimore told his benefactor, Edwin Whitehead, "Young scientists are the lifeblood of basic research, and my institute cannot survive without them." In a sense, the young scientists are the lifeblood of my book as well.

One last point about the personal nature of the book. As exhilarating as my scientific foray was, the months I spent in the laboratories, particularly at the Whitehead Institute, proved to be the most difficult of my life, and I don't try to gloss over that. I committed stupid mistakes in dealing with the researchers; thinking about those makes me wince even now. The scientists are, almost without exception, among the brightest people I've ever met, in any profession, but I got angry when that brightness made them too hard on each other and on me. More than once I felt that it would be easier to be a war correspondent in Tripoli or Johannesburg. Yet by the time my stay was through, I deeply respected and even loved Bob Weinberg, Mike Wigler, and their teams; they had become my friends. They taught me to appreciate the lush beauty of doing science, and I think that if I had it to do over again, I'd become a molecular biologist.

1 The Optimist

If this is the best of possible worlds,
what then are the others?
— Voltaire, *Candide*

BOB WEINBERG has yet to win the most coveted scientific laurel of all, but already he is as lavishly decorated as the Joint Chiefs of Staff. Two thirds of his curriculum vitae are taken up by a list of prizes. He's been honored by august societies of his peers, including the National Academy of Sciences, the American Association for the Advancement of Science, and the American Cancer Society. He's received awards from hospitals, state departments of health, European academies. And he's also the man industry turns to when it feels flush and philanthropic. In May 1987, Weinberg won the bounteous General Motors Cancer Research Prize; it was $100,000 to spend outright and another $30,000 to organize a conference on the science of oncogenes. He's been the beneficiary of awards from Bristol-Myers, U.S. Steel, the Armand Hammer Foundation.

Bob Weinberg is not a physician. He has never diagnosed a cancer patient, prescribed a course of chemotherapy, interpreted a mammogram, or tottered through oncology rounds. Yet most of his awards come from companies and nonprofit groups that believe he is the best hope in the protracted crusade against cancer. The General Motors prize, for example, is given annually to those doctors and scientists "who have made important contributions to the understanding, treatment and prevention of cancer." Corporate prize committees honor Weinberg not because he's a superlative scientist who is helping to decode the scrambled language of a living cell, but because they think his work, and the work of others like him, may someday spawn a cure for the second biggest killer of the Western world. And Bob Weinberg believes his benefactors just may be right.

Molecular biologists who work with oncogenes almost never talk about cures for cancer. When pressed, they'll talk around the subject, but they're pressed so often by nonscientists that their talk is usually tinged with weariness. I asked each biologist I interviewed variations on the question "How long before lab results may translate into clinical treatment?"

"I'll tell you the same thing I tell everybody else," said Howard Temin, a Nobel Prize–winning biologist at the University of Wisconsin in Madison. "I'm not a psychic, and I don't like trying to predict the future."

"Asking scientists to understand cancer is asking quite a lot of them," said James D. Watson, Nobel laureate in physics, co-discoverer of the structure of DNA, and director of Cold Spring Harbor Laboratory. "You're asking them to tell you what makes a cell a cell. I believe the problem is tractable, but you must understand the depth of the problem."

"I have no idea when we'll know enough to develop anything that's clinically applicable, and I don't know who's going to do it," said David Baltimore. "It's not a high priority in my thinking. I'm happy to work on the model systems we're working on."

"Who knows if we're close to finding one big cure for cancer?" said Richard Rifkind, director of research at New York's Memorial Sloan-Kettering Center, where many experimental cancer treatments are tested. "We won't know until we have it, and I resent those people who rise to the question and make postulations. To demand an instant cure and immediate applications is disappointing as a goal for society. It may be understandable — we don't want ourselves or our loved ones to die of the disease — but in a way it's childish. We all have to die someday."

"If you're giving me money, I'll talk about cures. Since you're not, I won't. Talking about cures is absolutely offensive to me. In our work, we never think about such things even for a second," said Mariano Barbacid, a molecular biologist at the Frederick Cancer Research Facility in Maryland. "But there are some people who do think they're looking for a cure. They're not basic researchers; they're in a sort of applied field, medical research. And any time I've gone to one of their meetings, I see how poor the quality of their work is. They really don't understand what's going on inside the cell. They'll try any kind of desperate thing. That's what happens when you look for 'cures.'"

"You can't do experiments to see what causes cancer," said Harold Varmus, a biologist at the University of California, San Francisco. "It's not an accessible problem, and it's not the sort of thing scientists can afford to do. You've got to live, you've got to eat, you've got to keep your postdocs happy. Everything you do can't be risky."

In skirting the issue of a cancer cure, oncogene scientists can sound cool, offhanded, and even disingenuous. After all, they're working with the genes that they think are the key to human tumors. They've *chosen* to study oncogenes, and they've chosen by the pack: their field is one of the most densely populated specialties in molecular biology. Surely it's not a coincidence that a subject related to human cancer has seduced so many basic researchers. Surely all those scientists must think about the medical implications of their work. If they're not altruists, they must at least fantasize about the glory that discovering a cure would bring them. To which basic researchers reply that of course the medical angle occasionally crosses their mind — but they don't obsess about it. And what good would it serve anybody, they say, if they were to talk about cures?

The scientists' apparent caginess is complicated — and largely justified. To begin with, they don't want to bolster false hopes in the estimated 5.9 million people around the world who contract cancer each year; the public has too often been disappointed in the past. "I've been keeping a file for about ten years now," said Robert Hoffman, a biologist at the University of California, San Diego. "You know what I call it? I call it my 'Answers to Cancer' file. It's got everything that's been touted over the years as *the* breakthrough in understanding human cancer. It's got clips from newspapers, announcements from conferences, papers from journals. Every one of those answers has turned out to be a crock."

"There's been far too much hype in this business, too much cocksureness," said Michael Bishop of UC, San Francisco. "Anybody who walks around and says that we've got this problem almost licked is a fool, a knave, or both."

"The first time news of our work began being talked about in the popular press, I realized how dangerous it was," Barbacid said. "I started receiving all these letters. One man wrote to me and said that his wife was dying, and could I please cure the point mutation in her cancer? I was really sad about that. What was I supposed to say to him? How could I explain to him that we're just a bunch of scientists working with DNA and petri dishes?"

Biologists resist speculating about clinical applications because they're a little defensive. The public has too often been disappointed by any number of claims that a cure for cancer is just around the bend, or that science has already made great strides in cancer treatment. University researchers are almost entirely dependent on government funds to support their labs, and they worry that the public's expectations and demands for results may affect the flow of grants. To scientists confronted with the question "When are you going to cure cancer?" there's a threatening subtext: "Why are you guys taking so long, and why should taxpayers keep giving you money?" Scientists complain that only a third of the projects that peer review committees deem worthy of financing are awarded government grants, but in fact molecular biologists have fared reasonably well in an era of penury and retrogression. In 1987, the National Institutes of Health — the largest source of funds for academic science — spent $1,402,660,000 on cancer-related research. That's up 42 percent from 1983, significantly greater than the combined inflation rate of 14.1 percent for the same period. (By contrast, the money earmarked by the federal government for the social sciences has been slashed more than 75 percent since 1981.) Nevertheless, there are ominous signs of official disgruntlement over the pace of cancer research. Richard Nixon declared a national War on Cancer in 1971, an act modeled after Lyndon Johnson's War on Poverty; after almost twenty years and approximately $6.5 billion in NIH outlays, tumors are still with us. In 1987, the U.S. General Accounting Office released a report lambasting the National Cancer Institute (a division of NIH) for exaggerating progress in cancer treatments. The subcommittee charged that many of the statistics bandied about by NCI to demonstrate improved survival rates are misleading: they may be a consequence of doctors' ability to diagnose cancer early rather than evidence of better treatments once a tumor has been detected. If a malignancy today can be diagnosed two years earlier than it was in the 1970s, and the afflicted person now lives six years after the initial diagnosis instead of four, the patient's prospects can hardly be said to have improved. Yet by the standard definition of a cancer "cure" — a person in remission *five* years after diagnosis — that same doomed patient, who once would have been included in the mortality figures, currently falls under the rubric of survivor, thereby artificially inflating the encouraging statistics.

Many cancer researchers counter such criticisms with a recitation of their demonstrable successes. Less than a generation ago, Hodgkin's disease, a lymphoma that afflicts primarily children and young adults, was 80 percent fatal; today 85 to 90 percent of all Hodgkin's patients are cured of the disease. A similar statistical reversal holds true for acute lymphocytic leukemia, another childhood cancer. Testicular cancer, Burkitt's lymphoma, hereditary eye cancer, malignancy of fetal membrane tissue — all have yielded to innovations in chemotherapy and radiation technology and are now 80 to 90 percent curable. But the experts grudgingly and unhappily confess that their miracle stories are limited to the rarer malignancies. "Here at Sloan-Kettering, we manage to send about fifty percent of our patients home, cured of their cancer," said Richard Rifkind. "And we have some very sick people coming through here; patients are only referred to us when more traditional therapies have failed. But the other half of our patients don't make it. And that half is a hard half to accept, because it includes many cases of the three big cancer killers: breast, lung, and colon. When it comes to those common cancers, there's not much we can do beyond primary surgery, and the real survival rates have hardly changed since the sixties and seventies." Nobody wants to dwell on failure, but so far science has failed to eradicate the major cancers. And even though basic researchers are not working directly on the treatment of human cancer, they can't help being chastened by the apparent intractability of the disease.

The real reason that basic researchers don't discuss cures, however, is that medical problems are not what motivate them. If they'd wanted to study human disease, they would have become doctors. Instead, they chose pure research. Motivations for doing science vary from person to person, but a statement I heard from at least a dozen researchers was "I like to solve puzzles." For molecular biologists, the living cell is the most challenging puzzle in the toy store. And cancer, as Watson noted, happens to be a problem intrinsic to the meaning of a cell. Cancer is the cell speaking in hyperbole. So, yes, it's true, say researchers, that the oncogene field has extrascientific appeal: any discoveries related to cancer attract attention and money, and scientists appreciate both as much as the rest of us. But the more profound value of oncogenes, biologists argue, in their utility in solving the puzzle of the cell. "I think all of us would say honestly that it's the normal processes of the cell that are our real concern," said David Baltimore. "And the thing that's always given us a handle on any normal biological process is the way it can go wrong." Cancer is the cell gone wrong, and oncogenes are genes gone wrong. The disease that we find so loathsome makes scientists' work ironically easier. When they study malignancy, molecular biologists are not grappling with a human tragedy; they're studying biochemical pathways, the interplay of proteins, the cycle of DNA division, the structure of the plasma membrane. They're studying the cell, and they're solving puzzles. "Biology has this terrible problem of being reductionist," said Renato Dulbecco, a Nobel laureate and Distinguished Research Professor at the Salk Institute in La Jolla. "But reductionism is the only way to get anything done. You have to break down a problem into its components. Without reductionism, we would never have learned anything about enzymes or DNA replication or the cell itself."

In pursuing abstractions and components, biologists often go far afield from human concerns. They'll ignore the patient whose body is riddled with metastases, and focus on one isolated enzyme from one fractionate of that patient's neoplastic tissue. They'll use whatever biological "system" is amenable to manipulation, however unlike us the organism may seem. They'll study yeast cells or translucent worms called nematodes or newts or fruit flies — or rats and mice. They refuse to promise that their work will ever have any "relevance," but they will say one thing: if past experience holds true, their arcane basic research is the soundest and finally the most selfish investment society can make. Biologists confronted with "Why should we support you?" can always swing back with the platinum word *spin-off*. And once they get going on spin-offs, they're as unstoppable as singing whales. All that we know about DNA and the genetic code stems from work performed on an obscure and insignificant virus that infects bacteria. The reason scientists have so quickly determined the basic machinations of the AIDS virus is that they've been studying similar pathogens for over three quarters of a century, ever since a young biologist named Peyton Rous decided, in 1910, that chicken tumors might have something to say about human disease. The methods that genetic counselors now employ to screen for hereditary diseases are the result of earlier studies of yeast chromosomes. Michael Brown and Joseph Goldstein, winners of the 1985 Nobel Prize in Medicine, spent twelve years investigating a single protein on the surface of animal body cells: a membrane receptor that ferries packets of cholesterol into the cytoplasm. Their work was of the most basic and abstruse variety, yet already it has helped save the life of one little girl stricken with severe cholesterol disorder, and it may eventually lead to improved therapies for coronary heart disease. The same serendipitous outcome, say scientists, is likely to occur for cancer.

Again and again molecular biologists told me that the only way they can be useful to cancer patients is to be given the financial and professional license to follow their instincts; not to be commanded "Find a cure for cancer," but to be encouraged "Do whatever you have to do — whatever you want to do — to understand the cell."

"The public is entitled to worry about spin-offs; they're paying us, and cancer is a scourge," said Baltimore. "But I'm not at all ashamed to say to them that we're doing the best we can in their interests, even though it may not look it. It's in their interests for me to do what I want to do. In studying problems about how cells grow and differentiate, I feel that I'm doing exactly the sort of thing that I ought to be doing to study cancer, even though I'm not really studying cancer."

The giving of license isn't restricted to dollars. It means not holding scientists to timetables or production schedules. ("You can't have breakthroughs on command," said Phillip A. Sharp, director of the Center for Cancer Research at MIT.) It means allowing scientists to judge the merits of one another's work by an elaborate peer review system that precludes outside interference and suggestions. It means trusting that what basic researchers do will eventually accrue to our interests. In a sense, we are at the researchers' mercy. We need their obsession with puzzles, their elitism, their willful flippancy. The study of oncogenes may not be the best hope of banishing cancer; it may be the only hope.

Which is why it's such a relief to talk with Weinberg. He has an infectious, carbonated optimism about the future of oncogenetics. Bob Weinberg is the Doris Day of molecular biology.

It's not that his motivations for being in cancer research are especially lofty. Weinberg shares with his colleagues a general aversion to questions of medical applications. "My reflex has not been automatically to go lusting after human tumors in order to make my work socially relevant," he said. "I've never thought to myself, Ah, yes, if I work on human cancer cells and human oncogenes that will make my work more 'humane.'" He claims to be as cerebral as the next scientist. His mother died of cancer in 1971, but it was not mournful resolve that drove him to oncogene research. His personal motive for working in the field is, he says, the "sheer delight" he takes in seeking intellectual paradigms. "It's not that I believe it would be a pollution of the mind to think about curing cancer, or that it would be compromising my ideals," he said. "It's not that I'm ideologically committed to avoiding a direct interest in cancer. It's just that that's not what fascinates me.

"My central passion, which I realize more and more is consistent in my life, is to look at a very complex, multifactorial, muddled system and try to extract out of it a small number of simple truths. I'm really interested in how one can explain the actions of a small number of genes in creating the complexity of a cancer cell. What excites me are those moments when I'm able to see, quite clearly and quite suddenly, how two apparently unrelated cellular events may be intimately intertwined.

"That's what I really like. I really like those occasional moments when I pace back and forth in my living room and all of a sudden I have an idea. Or I can't go to sleep at night because I'm obsessed with some problem, some simple insight into a connection between events in the cell. That's the high of it. But that sort of high is quite a few steps removed from a direct and emotional involvement in human cancer."

In concert with his colleagues, Weinberg insists that his ability to divorce the problem of oncogenes from the dilemma of the disease is one of his greatest strengths. "I believe that someone whose stated goal is to cure cancer will never have any chance to do so," he said. "You have to approach it as a desire to understand how cells work. That motivation is really essential for being able to function effectively as a researcher. It's necessarily secondary whether our work will affect clinical cancer. We can't let that be the driving force, because if we did, I believe that our effectiveness would be rapidly reduced to zero.

"I don't know how to respond to those who say I should be more involved in human cancer — and there are people, a few scientists among them, who make that claim," he said. "I can say, however, that my interests, as well as the interests of hundreds of other molecular biologists, may ultimately prove to be of more benefit to the solution of cancer than the concerns of those who have daily and ongoing and passionate involvement in the clinical treatment of the disease. The way the American people — whoever they are — should think of this is that the best way to get science done is to let people follow their own passions, and to have the trust, which is often vindicated, that the truths uncovered by people like me are going to work in their great interest."

And work they probably shall, says Weinberg. Where Weinberg differs from some of his colleagues is in his occasional willingness to play seer. As firmly as he believes in the fundamental simplicity of the cell, he believes that the oncogene field is on the brink of big breakthroughs: Answers to Cancer that will not end up in a waste basket. He doesn't talk about how daunting the puzzle of cancer is; he talks about it as though it were practically solved. And he expresses his beliefs with a rare conviction. "I think that by 1990, we're really going to understand what goes wrong metabolically with cells," he said. "I have no doubt that much of the work now under way — not only in my lab, of course, but in hundreds of others around the country — will have real relevance to the understanding of the transformations that induce cancer. I don't have a sliver of doubt about that. I can't tell you what the relevance will be, but of the importance I have no doubt. Zero doubt."

Weinberg is not only willing to go out on a limb in his prognostications; he'll dangle from twigs. Many molecular biologists who may agree with him that researchers will soon understand the metabolic basis of cancer are not so sure that the understanding will translate into a cure. Weinberg predicts that lab results will have direct and fairly rapid clinical significance. "I think there are going to be some big changes in the next twenty years in our approach to treating cancer," he said. "Even by the year 2000. Again, I can't tell you what the changes will be. But I'm confident they'll happen."

There are several examples of work from his and other labs that already point to potential bedside applications. One oncogene his group helped discover seems to be a primary participant in human colon carcinoma, one of the three big cancer killers. The finding suggests that the gene soon could serve as a marker, allowing doctors to diagnose colon cancer and perhaps treat it at an early stage. Another oncogene studied in his lab has been found to be dangerously overactive in certain common types of breast tumors, again raising hopes that the gene will be invaluable to diagnosticians. A third oncogene cloned under his directorship appears to be the key to eye and bone cancer; through analyzing the protein produced by the gene, scientists may be able to devise therapeutic drugs that prevent the malignancies from even beginning. Other labs have detected DNA segments that are consistently mutated in lung cancer, childhood brain cancer, and adult leukemias. Whenever scientists spot consistent patterns, they feel confident they're on the right trail to comprehending the genesis and progression of the disorder. "The more we learn," said Weinberg, "the more I feel that my faith in simplicity will be vindicated."

I loved talking with Weinberg about cancer. Whenever I repeated to him any of the reservations I'd heard from other scientists, he always had a heartening rebuttal.

"What do you think of the idea," I said, "that pharmaceutical companies already have tried every chemical known to man against cancer cells, so that it doesn't matter what you guys find out about the biochemistry of cancer — it won't result in a drug that hasn't already been tested?"

"Nonsense," he replied. "What was the figure I read the other day? That some fifty thousand chemicals have been tried? Well, the fact of the matter is that the number of organic chemicals that could be synthesized well exceeds the number of stars in the universe. Therefore, the chances of having highly specific and targeted chemotherapeutic agents are excellent."

I raised the argument that any therapies resulting from oncogene analysis could be no more effective than the current "cures" for cancer. Standard radiation treatment and chemotherapeutic drugs work by attacking all growing cells in the body — healthy as well as cancerous. That indiscriminate destructiveness explains the many traumatic side effects of cancer therapy: the rapidly dividing hair cells are killed, and the hair falls out; the growing cells that line and protect the stomach are destroyed, resulting in racking nausea; the cells of the immune system are blasted, and the patient contracts a welter of secondary infections. If the oncogenes that Weinberg and other molecular biologists study are as fundamental to cells as he said, how then could their activity be nullified without knocking off normal body cells as well? "Won't we be just where we are today — restricted to cancer treatments that are worse than the disease?" I asked.

Weinberg replied that there are subtle molecular differences between the normal genes and the oncogenes that had been mutated to cause cancer. Drugs could be designed to attack only the mutant forms of the genes, he said, leaving healthy cells unscathed. "Your question illustrates the need for us to understand oncogenes in precise molecular detail," he concluded. "It argues strongly in favor of our approach."

"Some people say that the importance of oncogenes to human cancer has been blown way out of proportion," I said.

"Oh, I'm well aware of that," he said. "There are some who prefer to continue with their phenomenological approach. They just want to look at the cancer cell as an entity and try to understand its overall biology without worrying about the molecular basis of cancer." Those are the researchers, he explained, who will study a cancer cell and describe its appearance and behavior, ignoring the genes that contributed to the malignant traits. "They strongly resist and resent and even derogate molecular biology in very explicit terms. They resent the attention that has been showered on us. In my view, though, the phenomenological approach has not yielded much over the past decade, and I'm cynical about its ever getting to any fundamental truths about the causes of cancer."

By contrast, he said, when it comes to understanding the genetic mechanisms that underlie the biology of a tumor cell, "the difference between now and ten years ago is like night and day. We now have a conceptual framework in which to understand the disease."

"So you're completely convinced that oncogenes are relevant to human cancer?" I asked.

"Of *course* they're relevant," he said. "They're relevant, they're real. They grow more relevant virtually by the hour."

Bob Weinberg believes that molecular biologists are on the brink of understanding the biochemical basis of cancer. He believes that cancer, the second biggest cause of death in the Western world, will be tamed in our lifetime. I told him that he was one of the most optimistic people I'd ever spoken to.

"Yes, I'm an optimist," he said. "It's in my nature to be optimistic. But I don't base my optimism on blind faith. I'm not a foolhardy optimist. The facts have borne out my optimism in the past, and I see no reason why they shouldn't bear me out in the future. I don't see any reason for despair."

No wonder Weinberg prefers basic research to medicine. Oncology wards today remain ghettoes of pain, where the simple job of living becomes an exercise in heroism, and where cancer triumphs once for every two times it is vanquished. In the United States, where cancer treatments are more sophisticated than they are anywhere else on the planet, twelve hundred people nonetheless die of the illness each day. That's one cancer death every 1.2 minutes.

But in Weinberg's lab, the scientists remain aloof from the "daily and ongoing and passionate involvement" in human cancer. Instead, they carry out the equally passionate if less poignant rituals of doing basic research: Weinberg in his role as lab impresario, his staff scientists in their role as experimentalists. They may be optimistic, pessimistic, or fatalistic about where their work is going. They're more caught up with themselves and each other than they are with "humanity." Yet whether they'll admit it or not, the subtext of their research is hope. They may not all agree with Weinberg that the cancer cell is simple — yet they keep solving puzzles.

• • •

Bob Weinberg's lab has been responsible for a number of the conceptual advances in oncogene research, and Bob Weinberg has been photographed and filmed for dozens of magazines, newspapers, and TV documentaries. Photographers don't like taking pictures of a person seated in his office, so they'll ask Weinberg to do something that looks scientific. They'll hustle him into the laboratory, temporarily displacing a researcher from his or her work station to give Weinberg room to pose. They'll ask him to pick up a test tube, an x-ray film, or a beaker filled with bright vermilion culture serum. Weinberg's young scientists snort derisively when they see pictures of their furrow-browed leader pretending to do benchwork, as lab duty is called. The last time Weinberg performed a full-blown scientific experiment was around 1976.

Weinberg is a typical director of an academic laboratory. He does not wear a white coat, squint through a microscope, or, strictly speaking, make discoveries. Although some lab directors insist on doing benchwork to the end of their careers, they're in the minority. Most, like Weinberg, leave experimental details to their subordinates. Benchwork requires intense concentration and continuity. An experimentalist cannot afford the repeated disruptions of a jangling telephone. And Weinberg, like other principal investigators of his stature, lives for the telephone. Once, after he had given me an hour-long interview in his office, I came out and counted the pink message slips on his secretary's desk. Sixteen people awaited his return call, some of them well-known scientists.

Weinberg also lives for the road. Between 20 and 25 percent of his time is spent at meetings and seminars, to give lectures, hear lectures, or to skip the scheduled talks altogether and schmooze. In a four-week period in 1987, for example, Weinberg traveled to Berkeley, California; Cold Spring Harbor, New York; St. Louis; Detroit; Washington; and Rumson, New Jersey. Over a six-month period, he's likely to attend conferences in France, West Germany, England, and Australia. He claims to turn down invitations to scientific forums about twice as often as he accepts; if he wanted to, he could be peripatetic 365 days a year. "But as much as I enjoy traveling, I can't overdo it," he said. "I'm not up to it physically. I suppose that's the price of galloping middle age."

Weinberg spends all that time on the phone and in transit because he must keep up with any development that may relate to his own work. Among the many misleading images of scientists is that of the lonely, dour researcher who never ventures from the lab. In truth, scientists are professional socialites. If they're not gregarious by nature, they learn to be by necessity: scientists get most of their ideas from other scientists. Only rarely does science lurch forward by nocturnal eurekas. More often, science is a process of accretion and synthesis. A scientist hears what somebody else is doing, relates that information to a recent report in a scientific journal, blends in a few personal ideas, and decides to do an experiment that is not radically different from the work that antedated it. That may sound like copycat gamesmanship, but refinements on past results can have profound implications. And the more disparate bits of data that scientists can synthesize, the more likely they are to be the ones who push the barriers forward. Scientists who aren't tapped into the larger community of peers risk becoming provincial, repeating old experiments, or, worst of all, pursuing a line of published research that their savvier colleagues already know to be wrong. The people in Weinberg's lab try nobly to stay abreast of their field, but they have to spend so much time at the bench — and molecular biology is sprinting forward so quickly — that they can learn the latest only in hints and murmurs. It is up to Weinberg to be the reliable repository of scuttlebutt, and he doesn't miss much. He returns from meetings bearing spiral notebooks filled with the details of the seminars he attended and the patter he exchanged with his vast network of friends and friendly competitors. His notes on his travels are written in careful, legible script, and are organized as though by the Dewey decimal system, to ensure that his scientists can review his reporting. On occasion, they actually do.

Weinberg must also spend time writing grant proposals to raise money for his lab. Although corporations seem eager to shower him with unsolicited gifts, an award like the one from General Motors cannot begin to cover the cost of running a large research team — and besides, that's not how prizes generally are spent. The expenses of Weinberg's lab amount to well over $1 million a year. Approximately $500,000 is earmarked for the direct needs of the lab, including equipment, such as beakers, test tubes, and pipettes; biological reagents, such as enzymes, radioactive compounds, cells, and cell serum; lab animals and the cost of caring for them; Weinberg's travels; the salaries of Weinberg's secretary and lab assistants; and long-distance telephone calls.

The remainder of the grant is the designated "overhead cost," and is put, with other monies, into the substantial cache that supports the operations of Weinberg's institute: lighting, heating, and cleaning the building; disposing of radioactive and other hazardous wastes generated by experiments; computer, library, reception, and public relations services. All university labs have staggering overhead costs, and professors are expected to raise funds to cover a fraction of them. Thus, when Weinberg submits a grant to NIH, the grants office at the Whitehead Institute requests a significant percentage more than Weinberg needs to keep his own people going.

Weinberg's grant proposals, like those of any respectable scientist, consist of equal parts explication, propaganda, and farce. He generally requests money for five- to seven-year intervals, outlining the theme of his lab members' investigations, their past results, what they hope to discover, and why it will cost so much to discover it. Weinberg, of course, has no intention of adhering to his stated projects; he can't predict what the lab will be doing from month to month, let alone five years in advance. But the peers who review his proposal must be convinced that his overall approach is reasonable, and Weinberg writes forcefully enough to make most of his ideas sound reasonable. Whenever he wants his lab to embark on a project that he fears will be considered far-fetched, he just omits any mention of it in the proposal and then diverts the funds he receives on the basis of his more conservative propositions. Indeed, in the early days of his career, the work in his lab bore scant relation to the work he supposedly was getting paid for; he knew that his schemes, if revealed, wouldn't earn him a token for the Boston subway. Today, Weinberg is luckier than most scientists. Because he has been so successful, his applications for government money generally are met with little fuss or calls for rewrites, which leaves him proportionately more time to dedicate to the business of doing science — more time to pace around his living room and obsess over ideas, more time to talk on the phone and travel. It's a self-regenerating cycle: Weinberg doesn't have to agonize over raising money; he can dedicate his efforts to making sure that his lab succeeds; his lab succeeds; he receives more money.

But Weinberg's talents at gossiping, synthesizing, and fund raising would be meaningless if he didn't have a great lab behind him. That point seems obvious, yet putting together a superlative team of scientists is anything but. Weinberg prides himself on his ability to choose excellent people. Once, when he was talking about the caliber of his staff, he started banging on the table so hard that as I later listened to the tape of the interview, I could hardly make out his words for all the rat-a-tat-tatting. "To the extent I've had success in science, it stems in small part from my own scientific intuitions and in larger part from my success in having good people work for me," he said. "It's important to be prophetic about a person's future performance, and I at least have the illusion that I can read human nature. It sounds trivial, it sounds like a minor aspect of functioning as a scientist, but having intuitive abilities about people is as important as having ideas. Because if you end up with a laboratory populated with scientists who in one way or another don't function well, you can have the best ideas in the world, and they'll all be seeds thrown on sterile ground." Weinberg's casting skills are in constant demand: turnover at an academic lab is guaranteed. Graduate students spend between three and six years working under a director's aegis. Once they receive their Ph.D., they must leave the lab to continue their scientific training elsewhere as postdoctoral fellows. Postdocs work with a senior scientist for about three or four years, and then they move on again, either to establish their own university lab or to join a biotechnology firm. With an average of fifteen to eighteen young scientists, Weinberg's lab isn't huge. Michael Bishop has about twice the number of people working for him; Leroy Hood of Cal Tech, three times the staff. Nonetheless, Weinberg frets continually over hiring dilemmas. He receives about seventy-five to a hundred letters a year from Ph.D.'s around the world who want to study with him as postdocs. Because he's famous, the quality of the applicants is high, and Weinberg told me that he wishes he had room for every one of them. "They knock on the door," he said, "and I'm tempted to say 'Come on in.'" But he has to make choices, and that means he depends on his "nose." "I do a lot of espionage before I accept somebody into my lab," he said. It's not enough that the person has published papers in *Nature* and *Cell* as a graduate student; impressive publications merely provide a solid base line. More revealing is what a thesis adviser says about the applicant's imagination, mental suppleness, work habits, personality. Or Weinberg will introduce the applicant to the members of his lab and later solicit their opinions on his or her overall demeanor. "We met this one guy who seemed, at first, to be a superb scientist," said Cori Bargmann, a Weinberg graduate student. "Then it came to light that he had really shat all over the people in his old lab. He'd do things like leave a centrifuge running even when he wasn't using it, just so that he'd be sure he'd have access to it if he needed it. That's not the sort of person you want to have working next to you."

"One of the most important things to me is that the person be nice," said Weinberg. "On a number of occasions I've turned away extraordinarily good people because they were pains in the ass. My science may have suffered as a result, but at least my health and well-being have been preserved." In short, Weinberg seeks scientists who are like junior Bob Weinbergs: creative, flexible, and reasonably pleasant. One day, while standing outside Weinberg's office waiting to interview him, I overheard him discussing a postdoctoral applicant with a lab director at the University of Wisconsin. The director obviously was praising his graduate student's intelligence and accomplishments, but Weinberg wanted to know whether the applicant "evinced any creative spark." And was he cooperative? Did he ever fight with his lab mates? What are his flaws, his warts? "When you say that Jon is 'hard on himself,'" said Weinberg, "do you mean he has unreasonably high expectations? Is he likely to get discouraged or demoralized too easily?" The Wisconsin director must have maintained his glittering appraisal, because six months later Jon Horowitz showed up at Weinberg's lab.

The constituency of Weinberg's team has changed markedly in the past ten years. Throughout the 1970s, he had many MIT graduate students working with him. By mid 1987, there was only one Ph.D. candidate in his lab. The shift has not been of Weinberg's design. Professors do not pick graduate students; graduate students pick professors. After completing a year of course work, an MIT graduate student petitions a professor for the privilege of training in the scientist's lab. And lately, students have not been petitioning Weinberg as frequently as they once did. Graduate students need assurance that they'll be working on an experiment that will evolve into a thesis project. The field of oncogenes has become so crammed and competitive that there are no straightforward or easy experiments left to do. Students who come to study in the Weinberg lab commit themselves to Weinberg's research specialty, and they risk getting placed on a technically difficult and open-ended project that wends on for years without producing thesis-worthy results. Nor do novice scientists necessarily relish the challenge of sharpening their skills in a rabidly high-powered field. "Weinberg's lab has had grad students who ended up working on the same things that Genentech was going after," said Luis Parada. "Is it fair that one graduate student should be competing against the biggest biotechnology company in the country? Who wants to get scooped by Genentech?" Weinberg told me how sorry he was that students no longer asked to join his lab. "Graduate students have been some of the greatest stars in my laboratory," he said. "So I'm very upset about not having more of them around. I'm upset, and I tell people about it. But weeping and wanting it and beating my chest isn't going to help things. I can't blame myself for the way my field is perceived."

The glut of qualified postdocs amply compensates for the graduate student shortage. And postdocs not only don't mind the prickly competition of oncogenes; they welcome it. Postdocs have only a few years in which to distinguish themselves from the gaggle of other protoscientists. The quickest way to gain recognition would be to make a major discovery at one of the top labs in one of the most visible fields in molecular biology. Hence, postdoctoral fellows flock to Weinberg's lab from across the United States, Europe, Asia, South America, the Middle East. They're trained in disciplines as diverse as immunology, protein chemistry, virology, bacteriology, medicine. They arrive well aware of the pressures of their new undertaking and of Weinberg's reputation. They may be "nice people," but they don't come to Weinberg's lab for the fraternity, the picnics, the construction parties, or even the Friday afternoon beer hours that are an MIT tradition. They come because they hope that Weinberg's magic will dust off on them.

Whether staffed by graduate students, postdocs, or some mixture of the two, an academic lab in no way resembles a corporation. Weinberg's main source of labor is either free or very cheap. Postdocs normally must provide their own income by applying for fellowships from the government or from a charitable organization such as the American Cancer Society. (Postdoctoral labor, in fact, is sometimes better than free; in addition to covering the researcher's annual salary of about $18,000, some U.S. fellowships include a "bench fee" of several thousand dollars a year, which helps defray Weinberg's lab expenses.) Graduate students are paid an annual stipend of about $7000, either by Weinberg or by a federal agency like the National Institutes of Health. The only salaries that must always come out of Weinberg's grant money are those of his secretary and the lab technicians, who assist the graduate students and postdocs with their experiments. And university lab technicians, however talented and educated, are notoriously underpaid.

Accompanying the young scientists' financial independence is their professional autonomy. Weinberg would not and really cannot fire researchers who he thinks are doing a poor job. He can attempt to hasten their departure from the lab, but once he's agreed to take them on, he's stuck with them to the end of their graduate schooling or postdoctoral fellowship. Nor can he flatly tell his people what to do. He can make suggestions; he can sweet-talk; he can question the experiments they would prefer to work on. He can and will do all of the above. But he cannot say, "Do this or else." Threats and demands run counter to the custom of academic freedom; they would ultimately undermine Weinberg himself. Bright scientists do not want to work for a grand vizier, and imaginative scientists who are saddled with an undesirable project do not work well. Weinberg needs his staff to talk back to him. He needs their sauciness, their arrogance, and their ideas. As I discovered, their impertinence has been behind many of his successes.

The most outstanding feature of an academic lab like Weinberg's is that it fulfills two seemingly contradictory functions. As a part of a university, it is a center for learning. Weinberg thinks of himself first and foremost as a teacher who trains his investigators how to do science. But it is not enough that Weinberg's acolytes master the known; they simultaneously must create the new. Weinberg's lab is also a research lab, and the greenest students may be thrown into projects of supreme moment. They know almost nothing, but suddenly they're expected to disclose a fundamental truth about the cell. While they're learning, they're discovering new things for others to learn. That's one reason why the distinction between mentor and protégé in a university lab is so easily blurred. It's hard for scientists who, with their own hands, have brought to light even the smallest truth from the canyon of ignorance to retain perspective or modesty. After all, they're getting *results*. "Common sense might tell you that an academic lab would be more efficient if it were run like a private biotechnology firm, with a hierarchy and predetermined goals," said Phillip Sharp. "But I'm here to teach my people, not lord it over them. And letting them figure things out for themselves is finally all for the best. It's from the wildness and woolliness of academic research that most of the major breakthroughs arise."

That Weinberg's students are themselves teachers helps to keep the lab humming. Weinberg frequently is out of town and often rings up to see how people are doing, but the Weinberg scientists aren't working for him; they're working for themselves. They're working for results. The delectable pleasure of results makes benchwork, the dailiness of science, bearable. And the dailiness of doing science does demand bearing.

The spectacular innovations in recombinant DNA technology introduced in the early 1970s, and subsequently refined beyond all expectations, have transformed molecular biology into one of the "high-tech" fields that supposedly presage the future economic and professional base of Western society. The Weinberg scientists exploit the latest advances in experimental protocols, and there are many machines in the lab that impressively wink and flash red digital readouts. An automatic cell counter can calculate exactly how many millions of cells are in a given milliliter of solution. An automatic oligonucleotide synthesizer will serve up prefabricated pieces of spliced DNA. The scientists have access to computer data banks that provide information on several thousand genes that other scientists have cloned from dozens of creatures.

Yet for the most part molecular biology remains an old-fashioned, labor-intensive industry. The scientists use much of the same equipment that biologists have relied on for decades: petri dishes; test tubes; beakers; centrifuges, which spin a test tube at anywhere from 1000 to 10,000 rpm; optical microscopes; tanks of liquid nitrogen that, when tapped, will spew out a vapor of nitrogen at minus 150 degrees for freezing cells in preparation for storage; vacuum hoods that keep air impurities out and noxious experimental byproducts in. A molecular biologist almost never is seen without his or her pipette, a dagger-shaped device that operates essentially like an eyedropper, except that solutions are sucked up through a nozzle by the push of a button at one end rather than by the squeeze of a rubber bulb. The materials from which it's made have changed, and its calibrations have become more precise (today's version allows the measurement of volumes as small as a millionth of a liter), but the basic principle of a pipette has been around since Pasteur's day and probably before.

Even the recombinant DNA techniques that truly are new look old. To analyze individual genes, the Weinberg scientists spend many hours preparing various types of blots, which are just pieces of paper speckled with the treated components of genes. There is the Southern blot, named after the Scottish biologist Edward Southern, who devised the technique. A Southern blot is a slip of paper smeared with diced DNA. Expanding on Southern's work, biologists developed a similar protocol for transferring purified RNA onto paper. Because RNA is the molecular antithesis of DNA, the technique was cutely named the Northern blot. Representing the next point on the compass is the Western blot, a sheet paisleyed with proteins precipitated from shattered cells. (There is, thank goodness, no Eastern blot, a term that too easily would lend itself to bad puns.) The Weinberg researchers produce scores of Southern, Northern, and Western blots every month, following a precise series of several dozen steps for each blot. Preparing a blot requires attention and care, but the procedure is routine, and the end product looks about as glamorous as a tablecloth after a good meal.

Much benchwork is routine. The Weinberg scientists grow millions of animal cells in tissue culture. Twice a week they have to feed the cells a nutritious serum of calf's blood, bicarbonate of soda, vitamins, and hormones, a chore that one postdoc compared with watering plants. When the cells replicate in their dishes to maximum density, they must be "split" — divided into multiple dishes so that the tissue can continue growing. To separate genetic components according to size, the scientists prepare gels; this entails pouring a liquid preparation into a plastic frame and allowing the liquid to solidify into a lavender-tinted slab of jelly. The most challenging part about running a gel is making sure that nothing dribbles out of the sides of the frame. If the researchers are working with proteins or RNA, which disintegrate rapidly at room temperature, they don warm sweaters or jackets and endure a few hours in the cold room, an air-locked compartment kept at 4 degrees centigrade. If they're working with bacteria, they retreat into the warm room, a stuffy space, maintained at 37 degrees centigrade, that smells like an overused sauna. To manipulate recombinant viruses that have the potential of infecting humans, the scientists swaddle themselves in plastic coats and hats and proceed through two air-locked doors into the P-3 room, where biologically hazardous experiments are performed. For any one project, a researcher may have to go from the bench to the warm room to the quarters where the lab animals are housed, then over to the incubators and back to the bench, darting quickly enough to prevent the multitudinous biochemical reactions from overcooking. The scientists frequently misplace their lab gear or accuse each other of filching a prized gel frame. More often than not, an experiment will fail, and all the steps have to be repeated. "I find benchwork incredibly boring," said Clifford Tabin, a former Weinberg graduate student who now runs a lab at the Massachusetts General Hospital. "Here you've had this long and expensive education, and you're spending most of your time either looking for equipment or labeling tubes."

Not all scientists agree with that assessment. A Weinberg postdoc, David Stern, insisted that he loves doing benchwork. "It's sexy," he said, waggling his rubber-gloved fingers. "Think of all the physical stimulation it gives us." He picked up a tiny plastic tube and thrust it into the opening of a centrifuge, which chirpily vibrated both the tube and his hand. "Don't you think that must feel good?" he said. "I'll never give this stuff up." And benchwork can be beautiful to watch. Biologists often compare doing an experiment with cooking: you follow a recipe, pipetting in 20 nanograms of ATP (adenosine triphophate), 3 milligrams of protein, a splash of distilled water. But while observing scientists at work, I was reminded less of chefs than I was of medieval guildsmen — the goldsmiths and ivory carvers and manuscript illuminators. The scientists ply their craft with nimbleness and delicacy; their fingers are long and sinuous; their spines curve over their workbench in perfect quarter moons. Sometimes they whisper to themselves, as though singing a throaty harmony to the clickety-clicking of their pipettes. Benchwork may be routine, but it needs highly skilled performers. The few times I attempted to do experiments, my wrist grew sore from gripping the pipette too tightly and clumsily, and once I spilled a radioactive solution all over the lab sink and floor.

Even the biologists who treasure their hours at the bench, however, agree that the real peaks of joy occur at either end of the experiment. Before beginning a project, the researchers must design its scope and set its parameters. They must determine which detail of the muddled cell they want to learn about and what experimental conditions are likely to clarify that detail while eliminating as much dross as possible. Should the scientist begin with rat cells or human cells? Is the gene under investigation likely to be more active in skin tissue or in lung tissue? Are there any DNA probes around that may resemble the desired gene? How long should the test cells incubate before their RNA is extracted and sprinkled onto a Northern blot? Weinberg may help his researchers develop the larger framework of a project, but because he's not always familiar with current experimental techniques, his junior scientists must turn theory into blueprint on their own. And though there are lab manuals outlining standard protocols, the key to a successful experiment lies in how cleverly the parts and protocols are combined and modified. After all, anybody can sketch a dress: two sleeves, a skirt, a bodice, a back, buttons or a zipper. But not everybody can design a number that looks like a Norma Kamali or a Christian Dior. The adjective that scientists use to describe a well-wrought experiment is "elegant," which means not only that it worked, but that it worked with style.

More joyful still than the conception of an experiment is its conclusion. Getting results is so much fun that I hadn't been in the lab two weeks before I was living for the fruits of others' labor. I couldn't wait to see the outcome of an experiment. The researchers also anticipate their results well in advance, jabbering so longingly about what they'll know in X number of weeks or days or hours that when the red light on the darkroom door indicates that a scientist at last is developing an autoradiograph, others may stand outside to catch the rushes. Sometimes the results are disastrous, unmistakable evidence that the researcher has botched the entire experiment. Less often, the results conform precisely to what the scientist has hoped for, in which case that person will strut around waving the film in everybody's face. "Look at this beautiful 2-D gel," said Stern, showing me a clear plastic sheet dotted with clusters of proteins. "It was worth every single minute of it." By far the most common outcome of an experiment, though, is ambiguity. The results differ from the scientist's prediction, yet the differences are tantalizing. Something is going on in the cell that requires interpretation. The scientists go back to doing what they like to do best: solving a puzzle.

The puzzles that intermittently amuse and taunt the Weinberg lab are puzzles of cancer genes — of the oncogenes that transform living cells from compliancy to lunacy. Oncogenes are specific to cancer, but they claim as their progenitors normal genes that perform necessary tasks in healthy tissue. Thus, the Weinberg scientists are studying the same things that concern every other molecular biologist — genes. From their perspective, researchers can understand a cancer cell only by examining the genes within and by considering a given gene in all of its manifestations: as DNA, as RNA, and as protein.

Before I arrived at the Weinberg lab, I didn't quite believe in DNA. I believed in it intellectually, of course. I believed that a molecule of deoxyribonucleic acid exists in each cell of my body, as it does in every cell of the two trillion tons of biomass that make up what we call life on earth. I knew that without the magnificent autonomy of DNA, its ability to create self from self, we would all be so much carbon and molten quartz, cyanide-tinctured water snaking through slags of clay, vaporous pillows of methane, sulfur dioxide, and ammonia. We would, in short, be as we were four billion years ago, when our planet was young and senseless. I knew that DNA was the best synonym for *life*.

I tentatively understood some biochemistry of DNA, which I'd learned in part by studying the gaily colored ball-and-stick models of the double helix that perch atop many biologists' desks. I knew that DNA consists of four alternating subunits of nucleotide bases — adenine, thymine, cytosine, and guanine — held together vertically by a firm backbone of sugar and phosphate; that one twisting ribbon of sugar and nucleotides faces a complementary ribbon of sugar and nucleotides; and that the nucleic acid units of the opposing strands are joined at their waists by hydrogen bonds — bonds between hydrogen and oxygen atoms. I'd learned that a hydrogen bond is both specific and weak: specific because an adenine subunit can usually be bonded to a thymine, and a cytosine to a guanine; weak because when the double helix of DNA needs to separate, either to replicate itself in preparation for the synthesis of a new cell or to expose a piece of itself for the propagation of a protein, the hydrogen bonds linking the nucleotides are easily dissolved by water.

I even knew a few amazing facts about DNA. Ripley's fact number one: A single portion of DNA is so tightly coiled into the nucleus of a human cell that it measures only a millionth of a meter in diameter, but if that DNA were unsprung and extended into a straight string, it would be as tall as a four-year-old child (and if the collected DNA molecules from a person were stretched out end to end, a scientist told me later, they'd reach to the far side of the moon and back, with plenty of slack to tie all the knots). Another fact: Under the proper conditions, DNA is immortal. Scientists have taken the DNA from frozen remnants of woolly mammoths — which perished thirty thousand years ago — cleaved the DNA apart, inserted the pieces into gene-cloning vehicles, and prodded the DNA to reanimation in bacterial cells: through genetic engineering technology, and relying on the placid durability of DNA, biologists have in essence raised the dead. Final fact: DNA is so monotonous, it's subtle. Some human genes and rat genes are about 75 percent identical. The DNA of a person differs from that of a chimpanzee by less than 1 percent. Some may argue that human beings and their primate cousins are 99 percent identical on more than a molecular level, but it must be admitted that a chimpanzee has never been known to clone a gene. Whatever physiological vagaries constitute the difference between the human and the simian brain can be encapsulated in a very modest fraction of DNA.

So I came to the Whitehead with a rational respect for the intricacies of DNA. I could look at my left arm and see the way the fine hairs all sweep across the skin in a single direction, and I'd know that the DNA of every skin cell had commanded each individual hair to curve clockwise as it grew, rather than counterclockwise or up in the air. I could go for a long, painful run, and be thankful afterward that the DNA of my pituitary cells saw fit to bathe my brain in soothing, opiating endorphins. Yet I didn't really believe any of it. DNA seemed to me to be a sort of religion, requiring the vigilance of faith. From the way DNA is talked about, in textbooks or in schools, it sounds indeed like the Yahweh of the later Old Testament, existing beyond the realms of man in some ill-defined netherspace. DNA is the immutable Other.

It took only a single event at the Whitehead to convert me. I *saw* DNA. From that moment on, I understood why molecular biologists are so enraptured by their subject. DNA was among the most dazzling substances I'd ever beheld. If you could melt the sun and catch it in your palms, you might have an idea of the beauty of naked DNA.

I was sitting on a stool in the lab one Saturday afternoon, scribbling notes and thinking about going out for lunch, when Ruth Wilson, a technician, came running over to me. Her light gray eyes looked lighter still with glee. "Have you ever seen DNA come out of solution?" she asked me. I said no. "Well, hurry up, then," she said. "It's ready." Not knowing what to expect, I hopped off my stool and quickly followed her to her workbench.

A little metal pot of what looked like ordinary tap water was sitting on her desk. Ruth explained to me what she had done. Long interested in plant genetics, she was searching for evidence of oncogenes in plant cells. She knew that many plants were vulnerable to so-called tumor galls, which resemble genuine tumors, but she didn't know whether that meant plants possessed the same sort of cancer-prone genes found in humans. To answer the question, she had grown the cells of several species of plants in tissue culture. When the cultured cells had replicated to a sufficient density in the petri dishes, she'd ground the cells into a pulp, employing a protocol that progressively separates the flotsam of the cytoplasm — the cell's equivalent of the egg white — from the prized nucleic acids in the yolklike nucleus. She'd added a strong dose of detergent to burst the cells apart and a variety of enzymes to degrade the cell walls and the proteins within. Then she'd spooned phenol acid into the test tube solution of pulverized plant cells, spun the tube in a centrifuge, and allowed the solution to settle, an operation that left the degraded proteins, fats, and membranes trapped in a layer of phenol at the bottom of the test tube while the nucleic acids floated in water at the top. She'd extracted the water and transferred it to her little metal pot. But the nucleic acids included RNA, DNA's chemical associate, as well as DNA. Ruth needed to treat the nucleic acids with a compound that would agglutinate the DNA into a single mass and leave the smaller molecules of RNA behind to dissolve. That was what she was about to do. She poured a solution of ethanol into the pot, and we waited. The water turned foggy before my eyes and almost seemed to be burbling.

Sixty seconds or so later the RNA had dissipated, the DNA had congealed, and it was ready to "come out of solution." Ruth poked a long glass rod into the liquid and stirred. When she pulled the rod out, a strand of something shimmering, viscous, and bouncy swagged from the end of it. I thought of eggdrop soup, and I thought of semen. The thick strand glowed and quivered as Ruth pulled it ever farther from the solution.

"That's it," she announced.

"That's DNA?" I asked, incredulous.

"That's it," she repeated, a Humpty-Dumpty smile of pleasure on her face.

She stuck the rod and the hanging strand back into the solution, stirred, and eased the rod out again, this time hauling forth a thicker, bouncier hank of DNA. She twirled it as though manipulating a forkful of fettucine. The DNA was simultaneously translucent and opaque, clear as crystal and cloudy as sleep. I hadn't known what to expect, but this elegant changeling necklace was not it.

"I can't believe it," I said. "It's so *clean*." I thought that if I grabbed the DNA, I could roll it in my hands, twist it, pinch it, shape it into little animals, even drop it on the floor, and still it would jiggle merrily, holding life so firmly that it was, on its own, alive. I might have thought I was being silly, reacting to purified DNA so strongly, if Ruth hadn't been equally captivated — though she'd seen the substance many times before. We both stood there grinning like girls making fudge as she yanked out the DNA of the plant cells and prepared to do her experiments. I think that watching DNA come out of solution should be required of every high school student. Then perhaps people would realize that DNA is not a set of balls and sticks or a diminutive staircase. Real DNA is like the sound you hear when you hold a conch shell to your ear. DNA is music.

Over the following weeks, I hungrily absorbed many of the finer details of DNA. I say *absorbed* because I was no longer learning from textbooks, but believing, like a witness, my faith continually replenished by the smallest experiments. Few of the experiments were as graphic as the isolation of DNA, but all entailed the miraculous transformation of the invisible into the seen: the Other into the We. Once, I picked up a small test tube that David Stern was working with. At the bottom was what looked like a drop of water. David told me that it was, in fact, distilled water, but that it contained a tiny quantity of DNA from rat cells.

"Doesn't that amaze you?" I asked.

"Doesn't what amaze me?" he said.

"This," I said, shaking the tube. "There's DNA in there, but you can't see it."

"Scientists are used to working with things they can't see," he replied, with the mixture of matter-of-factness and genial condescension that I'd come to associate with him. "That's been the case for a long time."

But I didn't agree with David Stern. Although scientists may start out with the unseen, their goal is to manipulate the invisible until they have a result that they can see, a result that will show up thick and inky-black on an autoradiograph, the x-ray photographs that are fundamental to molecular biology. They want a palpable and reproducible result that they can print in a scientific journal or flash onto a screen at a scientific meeting. They want hard data — not a chemical aswirl in a droplet of water. They want the trinity of molecular biology — DNA, RNA, and protein — and they want it in a format that makes sense.

The fragmentation of science was yet another reason that I learned to believe in DNA. Looked at whole, molecular biology is painfully difficult. Looked at in bits, it is lovely. An isolated bit of biology has an internal logic; you don't need to know everything to make sense of one thing or a few things. Which is a good thing for scientists. Quite frankly, they don't know very much, and they'll be the first to admit it. For instance, they don't know how many genes a human cell contains: Estimates range from 50,000 to 100,000. They don't know, in every detail, how a single human gene is expressed — turned into protein. They don't understand all the switches that flick a gene on, get it into high gear, and then shut it down again. They don't know how most newly synthesized proteins decide where to go in the cell or what to do once they get there.

They don't know why, given that every cell in the human body holds the same set of genes as every other cell, some cells become eye cells and other cells become liver cells. The don't-knows so far outstrip the do-knows that, though ignorance could fill a Borgesian library of the imagination, the entire field of molecular biology can still, even with the accelerating pace of research, be contained within a good, fat textbook. And yet, in the face of astounding ignorance, biology slouches on. One by one, genes are being cloned, and one by one they're being understood.

I remember what I thought a gene was before I learned molecular biology: something that I inherited from my parents and that keeps one or another part of my body working whether I will the working or not. I knew that because I have brown eyes, I had a "gene" for brown eyes, and that because my father had blue eyes, I also had a "gene" for blue eyes, which is silent in me but is capable of turning my baby's eyes blue should I marry a man with blue eyes. I had a vague sense of the connection between genes and chromosomes, genes and DNA, DNA and RNA, RNA and protein, but the sense was, like the strands of naked DNA that Ruth Wilson pulled out of solution, an undifferentiated mess.

Then I started learning molecular biology, and I started learning to think about genes in stages.

The first stage of a gene is not an event but an address: its structural position on the chromosome. In cells, the DNA doesn't sit in a pile, but is divvied up among the chromosomes, sausage-shaped links of nucleic acid and encasing, protective proteins. Human beings have twenty-three matched pairs of chromosomes, which contain all the genes in duplicate. When I said earlier that humans have about 50,000 genes, I meant that we have 50,000 different genes, each gene coming in two copies: one gene on one of the twenty-three chromosomes, one on the matching duplicate chromosome. The definition of the word *gene*, however, doesn't include the proteins that swirl around the chromosomes; it refers strictly to the nucleic acid of the DNA that is hidden inside them.

In essence, a gene is a distinct set of nucleotides, the subunits of nucleic acid — of DNA. Nucleotides are nothing but irregular baubles of atoms; they're molecules arrayed in three-dimensional space. The core of a nucleotide is a ring of nitrogen and carbon atoms. Surrounding the core are flounces of hydrogen, oxygen, and nitrogen atoms. The exact number and arrangement of the flounces are what differentiate guanine from adenine from cytosine from thymine; adenine and guanine, for example, are considerably bulkier than cytosine and thymine, and thymine has a little flange of hydrogen and carbon atoms that none of the other three nucleotides has. To mulish common sense, the atomic variations among the four nucleotides (hereafter abbreviated *a*, *t*, *g*, and *c*) seem ludicrously tiny, considering that the four must say everything worth saying. Yet when you start thinking about all the ways that *a*'s, *t*'s, *g*'s, and *c*'s can be stacked together along the tightly torqued double helix, you realize that DNA is not so much laconic as it is resourceful. In human beings, DNA is not concise at all. A single cell contains about three billion nucleotides: three billion *a*'s, *t*'s, *g*'s, and *c*'s chattering on to a deafening crescendo.

The exact pattern of nucleotides is what differentiates one gene from another and what signals the completion of one gene and the introduction of the next. All genes, in all organisms, begin with three particular nucleotides, the so-called start site: *a*, *t*, and *g*. And all genes end with one of three distinguishing triplets of nucleotides, the stop site: *t*, *a*, *a*; *t*, *a*, *g*; or *t*, *g*, *a*. In between start and stop, genes look very different: sometimes there may be ten *a*'s in a row or a repeated stutter of *t*, *g*, *c*, *t*, *g*, *c*, *t*, *g*, *c*. The average human gene extends for about 20,000 nucleotides, but there are short genes of fewer than 5000 nucleotides and colossal genes of over a million nucleotides.

On the most elementary level, then, a gene is the line-up of nucleotides from start to stop. But other important sequences precede the start site of a gene. There's the promoter, for example, the switch that turns the gene on and damps the gene down; and there's the enhancer, another switch that somehow switches on the promoter and throws the gene into overdrive. Promoters and enhancers, like most of any gene's sequence, differ from one gene to the next. Scientists don't yet know much about how promoters and enhancers work in mammalian cells; they know a lot more about what happens between the start and stop points of a gene. When biologists say in casual conversation that they have cloned a gene, be it the gene that produces color in the iris or the gene that synthesizes acetylcholine in brain cells, their isolated piece of DNA may or may not include the nucleotides of the promoter and enhancer elements; but the clone always encompasses the significant nucleotides from start to period.

DNA does not, however, do anything. It doesn't really color the eye or produce neurotransmitters that flicker across nerve synapses.

It lounges in the nucleus of the cell, curled up into chromosomes, a field of potential energy. It doesn't do; rather, it enfolds all doing. And before a gene can do, it must step across another field of potential, of nondoing. It must be transcribed into RNA.

The transcription of DNA into RNA is extraordinarily complex, and scientists have only begun to understand its many subtleties. The complexities arise because transcription, as the word implies, is the cell's version of writing. An RNA transcript is the gene written out into a working copy of itself, ready for circulation and execution; and writing, even in its molecular form, is inevitably a complex process. Transcription begins with a gentle shudder of the double helix. When a cellular gene is called on to shift from potential to action — say, when a brain cell requires a fresh supply of neurotransmitters — the portion of DNA that harbors the neurotransmitter gene begins to unwind. The tiny fraction of the chromosome where the gene sits pulls slightly apart: the hydrogen bonds that link the matched nucleotides together are broken, revealing a toothlike row of single bases that can now serve as a template. Only one of the two exposed rows of teeth is employed as a template at any given time; the facing row of nucleotides merely stands apart until transcription is complete.

The detached stretch of DNA initiates activity by a complex of transcriptional proteins that latches on to the promoter sequence that precedes the start site of the gene and begins the creation of RNA. Somehow the linking of the transcriptional proteins to the gene's promoter magnetizes the field of the template, attracting to the exposed nucleotides of DNA free subunits of RNA that are already floating around the nucleus.

The subunits of RNA are much like the subunits of DNA, with one notable exception. DNA is made up of *c*'s, *a*'s, *g*'s, and *t*'s. RNA uses the first three nucleotides, but in place of thymine it requires a uracil base. Uracil chemically resembles thymine, but it lacks a methyl group. These slight deficiencies don't prevent uracil from behaving like thymine in one important aspect, however: it can bond with adenine, just as thymine does along the double helix of DNA. Bonding is the key to the production of RNA. When those free-floating subunits of RNA drift over to the waiting template of DNA, they start pretending they're forming a new double helix. Wherever there's a companionless *g* subunit of DNA, a *c* nucleotide of RNA slips into place; where there's a *c*, a *g* is linked to it. If the DNA exposes an *a*, a *u* bonds to it. On and on the RNA micromolecules mate with the DNA template, until all bases, from the first nucleotide of the start site through the thousands of nucleotides of the body of the gene to the last nucleotide in the stop site, are outfitted with their RNA corollaries. And so the second phase of potential begins: an RNA transcript is born.

Twenty years ago, biologists believed that RNA was simple, a straightforward intermediary between gene and protein. And in bacteria, the organisms they studied at the time, RNA *is* prosaic. The transcript fashioned from DNA immediately moves on to be turned into protein, base for base. But in the 1970s, Phil Sharp of MIT discovered that the RNA of higher organisms must be thought about in phases. The first phase leads to the initial RNA transcript: the mirror-image sequence of the DNA template. Once complete, that transcript peels away from the template, allowing the exposed DNA to rebond with its facing strand of the double helix. As it lingers by the DNA that gave it form, the fresh RNA transcript represents a carefully crafted copy of an entire gene. It's the precise ribonucleic rendering of DNA.

Yet much of that exactitude and duplication goes to waste. The RNA of any organism more advanced than a bacterium is usually processed before it leaves the nucleus. Through a mechanism that is still a mystery, portions of the linear RNA transcript loop up into structures shaped like cowboy lariats. By the time the looping is through, the once straight transcript resembles a merry bow on a wedding gift. A complex of proteins then clips off the lariats, degrades them, and splices the unlooped segments of the transcript back together. The sequences of the transcript that are preserved are called exons: they are eventually *ex*pressed as protein. The segments that are sheared off and destroyed are called introns, because they're sequences that apparently *in*tervene in the serious business of the RNA. Hence, a good percentage of human DNA appears to be "junk," which is transcribed into RNA only to be discarded; in human cells, perhaps 90 percent or more of the average newly minted RNA transcript is thrown away right after it's produced. The spliced RNA transcript is known as the messenger RNA. It is the true ambassador of the gene, ready to journey from the nucleus into the cytoplasm and to swap becoming for being. It is ready to be translated into protein.

The word *protein* has lately acquired an unfairly negative connotation, probably because we still like it best when it comes in rich, greasy foods. In molecular biology, however, the word always evokes a positive image of dancing vitality. There are between 20,000 and 50,000 proteins in a human cell, and they do everything that must be done to keep the cell alive. Some proteins serve as the delicate beams and girders of the cell, lending it shape and elasticity. Some proteins poke through the membrane of the cell and communicate with the larger matrix of the surrounding tissue. Most proteins are enzymes: they react with other molecules in the cell, helping to break down bulky sugars into a usable energy source or deforming neighboring enzymes just enough to pass along a critical biochemical signal. Some enzymes are stationed in the nucleus, pinching the DNA when it's time for the cell to divide or gripping on to the promoter of a gene and setting off the construction of an RNA transcript. And some proteins are part of the ribosomes, little pear-shaped denizens of the cytoplasm that turn messenger RNA into new proteins.

When the messenger RNA trundles from the nucleus into the cytoplasm, a ribosome slides over and clasps the RNA at the start site. The ribosome moves along the message from left to right, reading the nucleotides in groups of threes. The ribosome doesn't really "read," of course; it interacts with the chemical twists and knobs of the bases, interpreting the sum configuration of three nucleotides and reacting accordingly. Each triplet of RNA nucleotides is called a codon, a code word for an amino acid. Just as nucleotides are the subunits of DNA and RNA, so amino acids are the subunits of a gene's endpoint, its protein. One codon signifies one amino acid. The triplet *g*, *c*, *a*, for example, is the molecular incantation for the amino acid alanine. If the ribosome, as it lurches along the messenger RNA, encounters the distinctive architecture of *g*, *c*, and *a*, it will grope around the cytoplasm for a spare residue of alanine (there are abundant loose amino acids in the cytoplasm); if it reads the codon *u*, *u*, *a*, it will rummage for a unit of leucine. There are twenty amino acids to choose from for the construction of a protein. The ribosome attaches one amino acid on top of another until it reaches the last base of the message stop site. At that point, the completed protein pulls free, its body bent and pleated into a shape that accommodates the sparring electric and chemical aspects of its individual amino acids. Some proteins are further modified after their sheets of amino acids have been folded. Other enzymes in the cell tack on frills of fat, sugar, or phosphates, and those extra side chains, in turn, alter the overall configuration of the neonatal protein.

Whether plain or fancy, the new protein eventually migrates to its appropriate position within the cell or is transported outside, and there it becomes the gene incarnate. It performs what before was only implied. The gene for eye color is a protein that affects the light-reflecting properties of the iris. The gene for lactase is the enzyme that breaks down sugar in milk. The gene that causes arm hairs to grow in one direction probably is a structural protein that stipples the outer membranes of follicle cells.

There is a lovely symmetry between DNA and the proteins that spring from its mouth. Not only does DNA make the proteins needed to replicate DNA and also make the proteins that blanket it safely within the nucleus; DNA makes the proteins that keep the very cell alive. DNA gives life to proteins, and without its proteins, DNA would die.

Like the subtlest of characters, like Lear or Orestes, a gene is a compendium of parts and more than its parts; a gene is open to many interpretations. A gene is its DNA, its RNA, and its protein. A gene is the exons that inform the ribosome on its pick of amino acids. A gene is its garbage, the introns that are destroyed after transcription. A gene is its promoter and its enhancers. A gene is the decorative side chains that aren't written into the gene proper.

I stress the multiple personalities of a gene because the Weinberg scientists, in their search for the genetic nature of cancer, look at oncogenes from every possible angle. All that I have said about the odyssey of a healthy gene, from energy held to energy released, can be applied to its malignant counterparts. Oncogenes are normal genes that have been mutagenically skewed, but oncogenes, too, must proceed stepwise from DNA to RNA to protein. Thus, the researchers consider oncogenes in the same piecemeal fashion with which they analyze normal genes; they seek anything, anything that will give them a handle on the tumor cell. A cancer cell, like a normal cell, possesses 50,000 genes, yet most biologists believe that only several of those genes are behind the malignant might of the cell. Scientists struggle to discriminate between the few oncogenes and the mass of innocent genes. They look at the DNA sequence of oncogenes, seeking evidence of mutations. They consider the RNA production of an oncogene. Is the gene transcribed into RNA too often or too quickly in tumor cells? Is the RNA synthesized at the wrong moment during the replication of the cell? Is the promoter out of kilter, unable to shut down the oncogene? They examine splicing: perhaps the initial transcript of the gene is being cut and pasted in the wrong places. As their greatest challenge, scientists examine the activity of the protein within the cell. It is proteins, after all, that twist and maul a normal cell into a cancer cell. The scientists attempt to understand what the protein of the oncogene is doing by tracing the Byzantine biochemical pathways that link the protein to a hundred or a thousand neighboring proteins. They test whether the protein is the correct size and shape, whether it's missing a side chain or outfitted with a side chain it's not supposed to have. They check to find whether the cancer protein stimulates the same enzymes that the analogous healthy protein addresses or prods enzymes it should ignore. They ask whether the oncogenic protein migrates to its proper location in the cell or wanders off to forbidden terrain. They ask any question that comes to mind. And when their ideas run dry, they retreat to the library, pick up the phone, pack their suitcase, or go to sleep, hoping that some dream will synthesize the pieces and shake them awake.

The overall theme of Weinberg's research is genes, cancer genes that explain the phenomenon of a cancer cell. Weinberg's obsession with genes and the parts of genes has served him well, and it isn't likely to desert him soon. I'm not a soothsayer, and it's foolish of me to make predictions, but I believe that when something really big breaks in cancer research, something that will benefit a substantial fraction of cancer patients, Bob Weinberg will, directly or indirectly, be involved with that breakthrough. He's intuitive, creative, always clever, and occasionally brilliant; the best young researchers in the world clamor to work with him. Weinberg is also one of the luckiest scientists alive. I know; I saw his luck in action, and I was flabbergasted. Just when it seemed that the great Weinberg laboratory had finally faded into mediocrity, just when it seemed, to quote one of the postdoctoral fellows, that "the golden age of Bob Weinberg was dead," the group hauled itself out of its lethargy and scored what could be one of its most significant coups ever. The researchers discovered a new kind of human cancer gene that nobody thought they would find. They pulled a Bob Weinberg.

As I pieced together the story of the Weinberg lab, from its early days to the time I spent at the Whitehead Institute, I realized that Weinberg's career has always followed a sine wave: up and down, triumph and slumber, parched confusion suddenly giving way to the lushest insights. Weinberg is blessed by the gods, but as a competitor of his pointed out, "Sure, the guy is lucky. But you make your luck, and Bob just keeps on making it."

2 The Family

I LEARNED about Bob Weinberg from his office, whether he was in it or not. We would talk in his office. He'd settle back in his squeaky, molasses-colored lounge chair and idly finger the fronds of his house plants, which completely blocked the view from his eastward windows. He talked about himself and the history of his lab. His memory was so bad that he'd forget months and even years at a stretch. I soon learned never to rely entirely on his account of a critical experiment, although his scattered recollections of the experiment were invariably revealing. He was often didactic, speaking slowly to make sure I followed him and punctuating his monologues with "Clear?" and "You follow me?" His preferred pronoun was "one," as in "One might suspect this" or "One would presume that." At first I worried that his formality was a distancing technique directed solely at me, but then I realized he talked that way to everybody. "Oh, I know all about the 'ones,'" said Anthony Komaroff, his best friend and the chief of general medicine at Brigham and Women's Hospital in Boston. "We can be hiking out in the wilderness and he'll start in with the 'ones.'" I also noticed that many people would tease Weinberg whenever he started sounding pompous, and he seemed to enjoy being teased. "That's one thing I like about Bob," said Cori Bargmann. "You don't have to laugh with him. You can laugh at him."

At some point during our conversations, Weinberg's secretary would inevitably interrupt to have him sign a batch of letters or answer a phone call in the outside office. The moment he left, I would leap up and stare at the carnival of photographs on his wall. I must have studied those pictures twenty-five times or more, but I never grew tired of playing voyeur: I'd always find something I hadn't noticed before, a detail or clue or a red herring that I imagined Weinberg had put there to amuse or mislead his audience. There were arty photographs of his young children. Aron and Leah Rosa had somber shadows cast across their faces, and they were playing musical instruments with expressions of concentration and discipline startling for a seven- and five-year-old. There was a picture of Weinberg's grandfather sitting beneath a wall hanging of an Indian deity. The old man looked so contemplative and so comfortably self-absorbed that, until I asked Weinberg to identify him, I figured he was a famous scientist or artist. There was a photograph of Weinberg's sister, Suzanne, taken at an angle that made her resemble Weinberg's wife, Amy. There was a photo of what I assumed to be Bob as a young man, with his sister, parents, and grandfather, but it wasn't easy to recognize Weinberg.

"Is this *you*?" I asked him.

"Yes, at age twenty-four, I think," he replied.

"You look incredibly dashing," I said.

"It's hard to believe," he said, "that I was once skinny and had a full head of hair."

"And you look so tall here, too," I said, gracelessly.

"I'm the giant in my family," said Weinberg, who's somewhere between five feet five and five-six. "My grandfather was five feet and a half inch, my father was maybe five-three or five-four, and my mother was also about five-three."

The photographs that I liked best were the group shots of the lab. It's impossible to go to a lab picnic, house-building party, softball game, birthday party, going-away party, or any gathering that includes Bob and two or more people without Weinberg at some point running to fetch his camera. "Picture time!" he yells, and everybody grunts but grins for the photographer. Looking at the product of Bob's passion is like watching the shifting blades of colored light that sliver the sea at sunset. A similar group of people shows up from picture to picture, but always two or three old faces have disappeared in one photo and new faces have taken their place in the next. Sometimes the group includes only postdocs and graduate students; at other times there are husbands, wives, boy friends, girl friends, babies on laps, or babies in strollers. The only constant is Bob Weinberg, who's usually up front; if for some reason he's at the back of the crowd, his neck is craned high to make him visible.

I think I found the group pictures so mesmerizing because the people in them always looked happy and connected, as though they were, for the moment anyway, the great family that Bob wants them to be. "I must spend sixty percent of my time trying to maintain a sense of harmony in the lab," he said. "I want people to get along, to cooperate. It's very difficult to do that, you know. It's hard enough to do that with your own family, but managing with twenty people is almost impossible. But really that's what my job is all about."

It's also what his heart is all about. If he had a choice, Bob Weinberg probably would like everybody to be related to him. Along with designing and building houses and taking pictures, Bob's favorite pastime is genealogy. He's traced his family tree back to 1675; he named his son after the first Weinberg, Aron, who was born in 1739. "Before him, there were no real surnames," he said. "We had only tribal names — the twelve tribes of Israel.

"I," he said, thrusting out his chest and wiggling in his seat in playful braggadocio, "am a member of the tribe of Levi."

Weinberg has filled in the smallest branches on his family tree; at last count, he had twenty-five hundred names written out. "It's very easy," he said. "The farther back you go, the broader the branches spread." He's identified illustrious and obscure ancestors on both sides of his family, among them Franz Boas, the great Columbia University anthropologist and an early champion of the nurture half of the nature versus nurture debate about human behavior; Alfred Flechtheim, who owned a famous gallery in Berlin in the 1930s and who helped introduce expressionism to the art world; and Ernest Wynder, who first alerted people to the connection between cigarettes and lung cancer. "A very unpleasant man," said Weinberg of his cousin Ernest. "His maiden name was Weinberg, but his family had changed it to Wynder when they came to the United States. I saw him once at a conference, and he tried to find all sorts of reasons why we weren't related. Perhaps he didn't want to confess to being Jewish. But I knew. He was a male Weinberg, and we had the same Y chromosome."

And, of course, there was Bob's great-grandfather Leffmann Abraham Weinberg. "Poor Leffmann," said Weinberg. "He was so short that he was called Männchen, which means Tiny Man in German. The only thing that Männchen was known for was his habit of drinking too much peppermint tea after the Sabbath and then walking around the perimeter of his horse corral, breaking wind.

"O Leffmann! If that's all the people remember about you, was life worth living?"

Wherever Weinberg travels, he finds a relative to visit. He has relatives in West Germany, Holland, England, South Africa, France, Argentina, Brazil, Bolivia, Japan, Israel, Mexico, and elsewhere. Sometimes he'll go to a conference of questionable importance simply because it's being held in a country where he knows he has a second cousin once removed whom he's never met before. After returning from a two-week meeting in Australia, he complained to me about the twenty-two-hour plane ride and the dull talks that he'd suffered through.

"But didn't you find any relatives in Australia?" I asked.

His face brightened. "As a matter of fact, I did," he said. "I saw my cousins in Perth." He then related the odyssey of his father's cousin, who had left Germany in 1933, moved first to Indonesia, and then to Singapore, where she met and married an Iraqi Jew. The couple fled to Australia when the Japanese captured Singapore, moved back to Singapore after the war, and then back for good to Australia. "It was their offspring I visited," said Weinberg. "It's always nicer to have some connection to a strange city." The migrations of the Weinberg family across the globe reminded me of maps of medieval Europe, with the lines drawn for the invasions of the Goths, Visigoths, Ostrogoths, Franks, and Saxons. Weinberg never gets his Visigoths and Ostrogoths confused.

As he travels, or when doing his genealogical research, Bob picks up scraps of foreign languages. He thinks of himself as an amateur polyglot, and it's his informal policy that all foreign students who come to his lab should be greeted in their native tongue. He's learned to say "Hello" and "How are you?" in Japanese, Chinese, Serbo-Croatian, Russian, Italian. When he welcomed the Budapest-born wife of one of his postdocs in Hungarian, she was so swayed by his flawless accent that she began jabbering back in Hungarian. Bob just stood there, smiling and nodding. "I'm very pleased to meet you" was the only Hungarian he knew.

Bob Weinberg's love of the family — or of the idea of the family — surely affects the dynamics of his lab. Not everybody who works for him thinks of Bob as a father figure; indeed, some would be deeply insulted by the suggestion. When I casually proposed to the postdocs Snezna Rogelj and René Bernards that maybe one reason they liked helping Bob build his house was a desire to please him, they almost expelled me from the Mazda we were driving in.

"That's *wrong*," René snarled from the back seat. "That's completely *wrong*. Is that your idea of an insight?"

"We do it because we enjoy doing it," said Snezna, "not to make a good impression on Bob or to soften him up."

I tried to explain that I hadn't meant they were toadying up to him — or "rimming" him, in the Weinberg lab's vernacular — but that they wanted to make him happy. It was too late: they thought I was imputing either infantile or manipulative motives to them. They're professional scientists, each over thirty, and consider themselves to be colleagues of Bob Weinberg rather than his subordinates. Which is mostly true. They're not Bob's children. And yet, at forty-five, Bob is a father figure. "Sure, we want his approval," said David Stern, "and it's very hard to do without it for long periods of time, which we all end up doing. You can't help feeling that you're being punished for something." Bob Weinberg's patriarchal personality seems as much a part of the total package as his immense mustache. "There are two Bob Weinbergs," said Cori Bargmann. "One is the formal European, the man who speaks in complete sentences and explains everything twice. The other is a guy who slaps you on the back and tries to be one of the gang. I feel a lot more comfortable around the formal European — it's so much more natural for him."

Enough people in the lab express a scrappy loyalty for Bob that the Weinberg group seems, at least to an outsider, like a family: a disputatious family, certainly, and a difficult family whose problems occasionally stray toward the Southern gothic. But it is Bob Weinberg's ability to domesticate the work place that makes the lab so intense. Weinberg once told me something that vaguely unnerved me. He said, "Leah Rosa is going to be a very interesting woman." It surprised me that he was already imagining his young child as an adult, which few parents do; the corollary of a child growing up is that the parent is growing old. But it surprised me even more that he would use the adjective "interesting." Creative scientists constantly describe problems as being either "interesting" or "not very interesting." If a result clarifies or substantiates previous work but by itself opens no new conceptual gateways, it may be important without being interesting. Bob Weinberg's science usually is interesting. He forces it to be interesting. And he wants his lab to be his family, not so much out of a desire for harmony and happiness and sunny smiles as from the knowledge that families are more interesting than a random assortment of people thrown together in a room.

Bob Weinberg will do anything to avoid being bored. If he has to, he'll adopt you, leaving you no choice but to follow in your father's footsteps and learn to beguile.

Bob Weinberg began training for his role as patriarch at a very early age. His parents had fled to Pittsburgh from Nazi Germany in the 1930s, bringing with them his mother's father. Surrounded by adults who were not yet assimilated into middle-class America, Bob spoke only German until he was four years old — a formal and precise German, at that. The first words he ever spoke, Weinberg told me, were "*Ach, Rosenthal. Kenn' ich gar nicht,*" which means, "Ah, Rosenthal. I don't know him at all." Young Bob was much closer to his grandfather than to his parents. "That's where I got my sense of humor from," said Weinberg on the day we stared together at the portrait of the solemn man beneath the solemn deity. "Such as it is." Suzanne, who's brassier and more jocular than her older brother, confirmed Bob's remarkable attachment to his *Opa*. "They spent hours together every day," she said. "They read, talked, did carpentry in the basement, or just sat around. Sometimes when you saw them together on the couch, they looked exactly alike, except that my grandfather's hair was a little grayer." Suzanne was jealous of her brother because he was a superb student, and her teachers would always compare the two. "They'd say, 'Are you really Bob Weinberg's *sister*?'" she said. "It was as though they were accusing me of imitating a saint."

Studious as he was, St. Robert evinced no greater love of science than of history, literature, or any other subject. Indeed, he had very little sense of what he wanted to be when he grew up. In high school, he flirted briefly with the idea of becoming a rabbi. "But when I was seventeen, my mother made a very salient point," he said. "She told me, If you want to be a rabbi, you'll have to have a congregation. And a congregation means a lot of Jews. She was right. We're a very fractious group." As an undergraduate at MIT, he majored in biology, more out of confusion than compulsion. "What was I supposed to do?" he said. "Become a physicist or an electrical engineer?" His grades were not extraordinary, but because he was known and liked by the department, MIT accepted him into its graduate program in biology.

Resigned to his destiny as a scientist, Weinberg began working harder and excelling commensurately. He studied RNA synthesis with Sheldon Penman and then decided to shift to the relatively young field of the molecular biology of tumor viruses. In the 1960s, virology was among the most fashionable of scientific specialties. Biologists believed that by analyzing the viruses that triggered cancer in chickens, mice, and rats, they might gain insight into the overall mechanism of tumorigenesis. Some biologists went so far as to claim that viruses were a major cause of human cancer, and that anything learned about viral malignancies in rodents could be applied directly to people — a theory that proved largely inaccurate. For his part, Weinberg didn't much care yet about cancer. He liked viruses because they were small and genetically well defined. "They represented, in essence, cloned genes," said Weinberg. "Understand that this was years before the actual cloning of genes through recombinant DNA techniques."

As a postdoctoral fellow, Weinberg spent one and a half years at the Weizmann Institute in Israel and the same amount of time in La Jolla at the Salk Institute, where David Baltimore, an old buddy from MIT, had once worked. Studying with Renato Dulbecco, Weinberg identified several of the RNA messages of the simian virus 40, a virus that causes cancer in monkeys. "It seems trivial now," said Weinberg, "but back then it had a big importance. Nobody had really done much work of that type before. Isolating messenger RNA was a royal pain in the ass."

Conceptually important though the work may have been, it didn't turn on Weinberg as an experimentalist, to judge from the funny stories I'd heard. Weinberg agreed that he never enjoyed benchwork very much. "I have flat feet," he said, "so it's hard for me to stand up that many hours, as it's necessary to do while one is performing experiments." But I'd heard worse than tales of fallen arches. I'd been told, for example, that he regularly spilled radioactive compounds all over the floor, and that he would leave his cells growing unattended for so long that they'd begin rising off their dishes like little magic carpets. When I visited Renato Dulbecco, I pressed him about his former postdoc, repeating the anecdotes.

"Was he really such a terrible benchworker?" I asked.

"Oh, no, I don't think so," said Dulbecco. "I recall him as being competent enough. Of course, I wasn't watching him every minute of the day."

I asked whether Weinberg had displayed any signs of creativity or native scientific instinct. Soft-spoken and stately, Dulbecco is not given to easy praise, and he hasn't revised his opinions of his student in light of Weinberg's subsequent successes. "I suppose he must have had it in him," said Dulbecco, "but I can't say that I ever consciously noticed his creativity. He seemed like a decent, hard-working scientist. Better than average, but not spectacularly so."

Dulbecco may not have had a chance to notice Bob's latent talents. "I was a postdoc with him for one and a half years," said Weinberg, as we sat at lunch in the Whitehead cafeteria. "And do you know how much time I spent talking science with him during those years? The total amount of time? Maybe an hour, an hour and a half." He turned to a couple of his scientists sitting beside him. "Just remember that the next time you complain about my not being around."

Midway through his postdoctoral career, Weinberg received a visit from Salvador Luria, a Nobel laureate who was then in the process of establishing MIT's Cancer Center. "You already know, of course," Luria told him, "that you'll be going to work there."

Weinberg replied that, on the contrary, he hadn't heard the first thing about it. He was honored by the offer, he said, but he hadn't even begun looking for a job yet.

Oh, well, think about it, said Luria.

Weinberg thought about it and grabbed at it. The Salk Institute for Biological Studies, Louis Kahn's stripped-down modernist Stonehenge overlooking the periwinkle sweetness of the Pacific Ocean, was too placid for him. He missed the neurotic energy of the Cambridge scientific community, its rambunctiousness and venom. He moved back east in 1971 to begin his independent career. He resolved to spice up his life. He got married.

Bob Weinberg met his first wife, Barbara, while visiting friends in Pittsburgh. She was his professional and spiritual antithesis, and contrast was what he wanted. He was a scientist; she was a brilliant opera choreographer and the daughter of the conductor of the Pittsburgh Opera orchestra. As a performance artist, she loved her late nights; he had a slight case of narcolepsy and had trouble enough staying awake through the day, let alone past 10:00 P.M. He avoided crises; she fomented them. "She was a gifted artist and artistically temperamental. I thought it would make an interesting match," he said. "But it turned out to be a little *too* interesting." Six months after the wedding, the couple divorced. "My family was worried that the suddenness of the breakup would upset me. They thought I'd take a nose dive after that," Weinberg recalled. "But my grandfather, in his wisdom, told me, 'Better an end with pain than pain without end.'"

Today, Weinberg never volunteers information about Barbara. He seems completely *married* to Amy, his wife since 1976, whom he met through her sister, Sally, a former technician of his; he seems like a man who was born married to one woman. I learned of his first, abbreviated union only when one of his postdocs mentioned it in passing. And Weinberg might not have agreed to discuss his former wife had he not been under the influence of two glasses of peppermint schnapps when I broached the subject. "The funny thing is that I just don't think about her. Ever," he said, looking almost quizzical. "In fact, when you brought it up, I had to stop for a moment and ask myself, '*Was* I married before?' Of course, I immediately realized that I was, but it seems like ancient history, another life."

The pain behind him, Weinberg focused his energies on chiseling his niche at the new Cancer Center at MIT. Salvador Luria had promised him that, though he initially would be working under David Baltimore, he would be given his own lab as soon as space and funds permitted. "Luria was true to his word," said Weinberg. "I never really worked for David. We never published any papers together."

Weinberg may not have published papers under Baltimore, but in some ways he felt intimidated by the presence of his more accomplished neighbor. Baltimore had skyrocketed to the upper stratosphere of scientific fame in 1970, when he and Howard Temin independently discovered an enzyme called reverse transcriptase. The enzyme is found in retroviruses, a class of viruses responsible for many types of animal tumors, including at least two rare forms of human leukemia-lymphoma; the most notorious retrovirus today is the AIDS virus. The genetic core of a retrovirus is built of RNA rather than the more standard DNA, but RNA is not what makes the pathogen so singular. Temin and Baltimore demonstrated that a retrovirus, once it has infected a cell, employs its reverse transcriptase enzyme to work backward, turning its RNA core into a strand of DNA. The revelation shocked many members of the scientific community because it seemed to defy the central dogma of molecular biology: DNA makes RNA makes protein. Biologists had believed that genetic information flowed in one direction, from DNA to RNA. But here were these viruses, skipping to their own rhythms, stitching RNA into DNA. In the press and among some of their peers, Baltimore and Temin were portrayed as giants, destroyers of the dogma. Five years later, they shared the Nobel Prize.

Bob Weinberg worked down the hall from a giant, and he of

course could not yet compete. All he'd done thus far was identify a few messenger RNA's of an obscure monkey virus. Nor did Weinberg help matters much with the research project he chose to launch in his new position at MIT. After the discovery of reverse transcriptase, Weinberg's monkey pathogen — a DNA-based virus — looked as compelling as last week's casserole. RNA viruses were the reigning fashion, so in 1972 Weinberg decided to trade the simian virus for the Moloney leukemia virus, a retrovirus that infects mice. In short, Weinberg entered a field dominated by none other than David Baltimore. Yet Weinberg felt that he had no choice. Everybody was studying retroviruses, and he didn't want to be intellectually isolated at MIT.

As he describes them in retrospect, Weinberg's first few years as an autonomous scientist were one "long shaggy dog story"; they seemed to have little point and less direction. Weinberg wasn't yet making any great leaps forward. He didn't have any brilliant ideas. He was just doing the same sorts of things many other molecular biologists were doing. He did them better than most, and he had a solid, upper-middle-level reputation, but he often felt himself to be, he said, "a half step behind." In his mild-mannered way, Weinberg worried about failing, and, less mildly, he worried about being bored.

"Bob fretted constantly about establishing a professional identity that was independent of David Baltimore's," said Tony Komaroff. "They were both at MIT, they were both studying retroviruses, and they'd been colleagues for so long that people tended to associate the two in their minds."

A good shaggy dog story, however, is a very particular kind of story. It drags; it meanders; it leads its listeners on tangents that fork off from tangents. But the ending is always worth the annotations. The ending is completely unexpected, a punch from behind. The husband sells his pocket watch to buy his wife combs at the same time the wife cuts her long, exquisite hair to buy him a watch chain. Rosebud is a child's sleigh. Bob Weinberg discovers the first human cancer gene.

Six years would pass before he realized it, but the best decision Bob Weinberg ever made was to step into his old friend's shadow.

Weinberg's lab started small. First, there was Bob himself. Then there was David Smotkin, an MIT graduate student with a crooked smile, a swarthy complexion, and a tendency toward philosophic melancholy; and Massimo Gianni, a tall, likable postdoc who wears Fellini-style glasses and retains an unblemished affection for his old boss. Until the mid 1970s, Weinberg continued to make occasional stabs at doing experiments, but he allowed administrative duties to interfere whenever possible — much to the relief of the people working under him. The three researchers struggled to find something offbeat to do with the mouse leukemia virus, something that would set them apart from the glut of virologists. They considered what was then understood about the replication cycle of a retrovirus. Biologists knew that a retrovirus is the purest sort of parasite, a mere packet of RNA and enzyme bundled in a coat of protein and sugar and utterly dependent on a host cell to replicate its genes, synthesize its proteins, and give it meaning. They knew that immediately after infecting a host cell, the virus rallies its imported molecules of reverse transcriptase to transcribe its RNA core into a double strand of DNA. That viral DNA then slithers right into the host chromosomes.

From its camouflaged position, the virus behaves just like a cellular gene, exploiting the host's translational and transcriptional equipment to propagate viral proteins, more reverse transcriptase, and new messages of viral RNA. To cap the viral cycle, RNA, enzyme, and protein coat are cobbled together into a baby retrovirus, which breaks free of the original host to infect surrounding cells. Viruses are extraordinarily fecund: a single parent virus can force the host to manufacture about a hundred infant viruses within half a day after infection. "They've really got survival down pat," said David Baltimore. "The life cycle of the retrovirus is one of the most elegant and astonishing things nature has devised."

As the co-discoverer of reverse transcriptase, Baltimore had a monopoly on the biochemistry of the enzyme and the details of how RNA is scribbled into DNA. Weinberg decided that he'd focus on the second part of the replication cycle, analyzing the new DNA molecule as it peeled off the retroviral template of RNA. No scientist had ever isolated the fresh viral DNA from infected cells, and Weinberg wanted to be the first. He wanted to see whether the transcribed DNA was the same size as the original RNA, whether any bits of the viral head and tail were left over after reverse transcription, and how long the new DNA molecule survived in the cytoplasm before crashing the host chromosomes as a sham gene.

Because he was the sole postdoc and thus the most senior junior member of the fledgling lab, Massimo Gianni was given charge of the project; Smotkin more or less tagged along and cleaned up after him. The experiments were conceptually simple but technically difficult. Gianni and Smotkin infected cultured mouse cells with the leukemia virus, allowing the virus time to infiltrate the cytoplasm and transcribe its RNA into DNA. They then burst the cells open with detergent, grabbing the viral DNA before it had a chance to maneuver into the host nucleus. Comparing the DNA to the original RNA template, they found that the RNA heart of the retrovirus was transcribed base for base into a double strand of DNA; none of the RNA went to waste.

The Weinberg lab also found that the DNA strand went through several tortuous steps before integration. It began as a linear string, then looped back on itself into a circle, and finally straightened out again to nest in the mouse DNA. "It's always hard to convey the importance of little things that, in retrospect, seem trivial," said Smotkin, "but identifying that viral DNA was really exciting at the time."

As exciting as structural questions were, Bob Weinberg's mind eventually wandered to broader problems. "We had found the unintegrated DNA, the object of our search, and that was pleasing," he said. "But I soon became interested in a question that is always of greater intrigue, that of function."

For intrigue, the group considered the impact of viral DNA on its grudging host. If the DNA looks very similar to its RNA parent, does it therefore act the same? Weinberg wanted to know whether the DNA copy of the retroviral RNA would have the same effect on the cells as the original virus, or whether there was something about the virus in its initial RNA state — a viral promoter, say — that was necessary for viral potency. One day, at a group meeting, Weinberg threw out an idea. He wondered whether there was any way his lab could skip using whole viruses and just add viral DNA to cultured cells. Now *that* would be a kinky experiment.

But his musings, he admitted, were idle. True, it was easy enough to get the viral DNA. All they had to do was repeat what they'd done before: infect a batch of cells with retroviruses, wait a few moments, and then pull out the newly transcribed DNA. But Weinberg didn't see how the lab could transfer that purified DNA into a second crop of cells. When the scientists spread active retroviruses across a dish of mouse cells, the viruses figured out the best method for traversing the cell membrane and wiggling into the nucleus. If the scientists were to try dripping isolated viral DNA onto the plates, said Weinberg, the DNA would just lie on top of the cells like an oil slick. There was no way the DNA, lacking the original bore of the retrovirus, could penetrate the cells and integrate into the rodent chromosomes.

Oh, yes, there was, interrupted David Smotkin. We could try DNA transfection.

Transfection? said Weinberg.

Black magic, Smotkin replied — and he was only half kidding. He ran back to his desk and pulled out a paper that had recently been published by two Dutch scientists, Frank Graham and Alex van der Eb. Returning to the conference room, Smotkin showed the paper to Weinberg. This is it, said Smotkin. This is our ticket. DNA transfection.

He explained what the Dutch experimenters had done. Through trial and error, they'd discovered a way to smuggle foreign genes into cultured cells and to get those genes expressed in their new home. They had isolated the genetic material of the adenovirus, a DNA virus that causes a variety of unpleasant respiratory diseases in humans. They'd mixed the viral DNA together with calcium phosphate in a test tube. The calcium phosphate had crystallized around the genes, creating little DNA bullets. The researchers had sprinkled the crystals onto cultured human kidney cells and incubated the cells at 98.6 degrees Fahrenheit for ten to fifteen days.

Now here's where the black magic part comes in, said Smotkin. For completely mysterious reasons, the calcium phosphate seemed to act as a lubricant, loosening the pores of the kidney cell membranes and allowing the DNA to slip inside. The foreign genes had integrated into the chromosomes of the kidney cells, precisely as they would have had the scientists infected the cells with live adenoviruses. What's more, the viral genes were thoroughly active; Graham and van der Eb could detect offspring adenoviruses streaming out from the cytoplasm of treated cells.

In other words, transfection has the same effect on cultured cells as infection. When a virus infects a cell (either in tissue culture or in the body), the virus burrows through the cell membrane, and the viral genes hop right into the host chromosomes. Once in place, the invading genes get switched on by the guileless machinery of the host cell, which transforms the viral DNA into viral proteins. Transfection has an identical net result, said Smotkin. It allows foreign genes entry into the host DNA of our cultured cells, where the genes then have access to the host's protein-making equipment. The only difference is that it's calcium phosphate that ushers in the alien genes, not the sneaky nose of a virus. And that is our trump card, he said. With transfection, we're not limited to putting viral genes into cells by infection. We can circumvent our dependence on pathogens and all their complexities, and force foreign, naked genes into cells with chemicals. Our lives will be much simpler if we can study genes without the messiness of viruses.

The biggest drawback of transfection, said Smotkin, was its terrible inefficiency. By Graham and van der Eb's calculations, only one in a million cells had responded to the calcium phosphate massage and absorbed the foreign genes. But that was a trifling difficulty. A single petri dish can hold about 500,000 cells and still give them room to grow. All the Weinberg scientists had to do was transfect twenty or forty plates of cells to get what they wanted.

Smotkin described what he wanted to try. He would slide the leukemia DNA into the cells through transfection. If the DNA molecules were as potent as their retroviral templates, the treated cells shouldn't be able to tell the difference between infection and transfection.

Weinberg skimmed over the Graham and van der Eb paper. It would never work, he said bluntly.

Why not? Smotkin argued. It worked for *them*.

Yeah, said Bob, but they were transfecting adenovirus DNA into cells. Adenoviruses are already DNA viruses. You're talking about transfecting DNA made from retroviral RNA. We barely know the first thing about retroviral DNA, let alone whether the DNA would survive calcium phosphate roughhousing. No, it would never work.

"It seemed highly unlikely to me that we could put large pieces of viral DNA into cells and get them expressed," Weinberg told me. "Sometimes I'm game for unlikely experiments, but this one looked like a waste of time."

"I guess it was a way-out idea," said Smotkin. "Bob didn't want me to bother, and he gave me other projects to work on. But I was tired of following Massimo around. I was getting very depressed, not having my own project, and I argued with Bob about that all the time." Their fights were often explosive. Weinberg would ask Smotkin to do something, and Smotkin would dismiss the idea as nonsense and stomp out of the room. Weinberg didn't mind his student's back talk; he'd run after Smotkin and try to convince him that he was right. But Smotkin had stopped listening to his mentor.

"When I got the idea for transfection, I just went ahead and did it," he said. "I pretended to do what he told me, but really I spent most of my time on transfection."

Because there was nobody at MIT versed in the craft of transfection, Smotkin taught himself by reading the published reports. "He was a very smart student," said Weinberg, "an immensely talented young man." Weinberg may well believe that of his first graduate student: on the day that transfection started working at the smart hands of David Smotkin, Bob Weinberg's Emancipation Proclamation effectively had been signed.

With Graham and van der Eb's papers spread out across his lab bench, Smotkin began applying the transfection protocol to the mouse retroviruses. He isolated viral DNA from cells that had been infected with the leukemia pathogen. He cut the DNA into pieces. He mixed the DNA with calcium phosphate just as Graham and van der Eb had done. Now what? The Dutch scientists had transfected their DNA into cultured human kidney cells. The Weinberg lab didn't have any human kidney cells. Smotkin rummaged through the lab freezer to see what sorts of cells were on hand. The only ones he could find were mouse cells. There were the fetal mouse cells that he and Gianni used for their infection experiments, and there was another type of mouse cell from the National Institutes of Health, an all-purpose line of skinlike cells called NIH 3T3 cells. Shrugging his shoulders, Smotkin pulled out the NIH cells. "That was the luckiest choice I ever made," he said. Much later, Smotkin would learn that the vast majority of cell types are xenophobes. Under no circumstances will most cells absorb and express foreign DNA. But a scattering of cell types are so hungry for company that they'll take any genes they're given. Human kidney cells are one gregarious variety; NIH 3T3 cells are another, and, in fact, a better variety. "It was fate," he said. "If I'd started with cells that weren't amenable to transfection, I probably would have dropped the whole project pretty quickly."

Smotkin applied his calcium-DNA crystals to several million mouse cells, stuck the dishes in the incubator, and prayed.

Two weeks later, he removed the cells from the incubator and began screening them for signs of active virus. If any cells had scooped up the transfected DNA, Smotkin should have been able to spot the same number of baby leukemia retroviruses bursting from the cell bodies as he had obtained when he infected the cells.

The first plate he checked was calm and clean. No viruses there.

Dish number two was the same. The cells looked exactly as they had prior to transfection. So too did the cells on the third dish. Smotkin wasn't worried. He had plenty of plates to go.

The fourth, nothing. The fifth, nothing. He screened numbers six and seven, and still he saw no viral particles. Now he *was* getting anxious. He hated the thought that Weinberg might be right about transfection after all.

Then he came to a plate that was bloated with viral particles, and he practically yelped in glee. Several of the cells in the population, he found, had propagated so many viral offspring that the new retroviruses had overwhelmed their surroundings. There was only one way those retroviruses could have been born. The cells had accepted the proffered calcium-coated DNA into their chromosomes and willingly kicked the viral genes into gear. Transfection had worked.

"When Bob saw that, there was no arguing with it," said Smotkin. "Transfection was a bit of hocus-pocus, but it was believable."

Over the next few months, Smotkin repeated the experiment many times, fiddling with the ratio of viral DNA to NIH 3T3 cells, length of incubation, serum conditions, and other cell culture variables. Again and again transfection worked, and Weinberg was delighted. Transfection was obviously a powerful tool, and he knew that only a scattering of labs had mastered the art of foisting alien DNA on cultured cells. As soon as a really interesting biological problem presented itself, Weinberg's lab would be in an excellent position to pounce.

Yet David Smotkin, who had inaugurated the new Weinberg specialty, grew increasingly frustrated and even disgusted with transfection. Although the procedure was reliable, it was far from reproducible. Under the same conditions and using the same proportion of DNA to recipient cells, Smotkin sometimes found his newborn leukemia viruses and sometimes didn't. He had to work with monstrous quantities of tissue culture, transfecting thirty, forty, fifty petri dishes for every experiment. The Cancer Center quarters were cramped, and Smotkin's plates had to compete for incubator space with the dishes of twenty or so scientists from three other labs. At times David would get so tired of doing the experiments that he'd take abnormally long vacations. He'd leave MIT for a month and wander off to Japan, come back to the lab for a month, and then depart again to go hiking out west. "Bob would grumble and groan about my taking vacations," said Smotkin, "but there were too many things that I wanted to do outside science. I had to leave the lab just to keep from getting too depressed, and he couldn't really stop me."

In spite of his frequent absences, Smotkin was exceptionally productive. His experiments proved that DNA transcribed from the viral RNA was functionally indistinguishable from the complete template and that no bits of viral packaging were necessary for retroviral reproduction. He showed that no matter what shape the DNA was in before transfection — whether it was straight, circular, or twisted into a supercoil — it operated like the most rambunctious of retroviruses. He'd amassed enough impressive data to begin writing his doctoral dissertation. And then he and Bob Weinberg came up with the sort of interesting biological problem worthy of Smotkin's skills.

Before David began his transfections, he and his boss had known that the mouse leukemia virus could not turn cultured cells cancerous. The retrovirus will give a living mouse leukemia, but even when Smotkin infected cells with the whole virus, he couldn't turn a plate of normal tissue into a heap of malignant immortals. Thus, Smotkin never expected to profoundly alter cultured cells by transfecting them with leukemia virus DNA; the best he could do was to seed a crop of viral particles.

Other mouse viruses, the scientists recalled, can transform cultured cells into cancer cells. For example, the Harvey sarcoma retrovirus, a puissant agent of leukemia and solid tumors in mice, has a similarly devastating effect on plated mouse tissue. Normally, cultured cells lie happily in a smooth layer one cell thick, a monolayer. When the cells are infected with the Harvey tumor virus, however, some members of the cell population begin proliferating wildly, cells heaping atop cells until the background monolayer is pocked with little pimples of overgrown cells called foci. For ten years, scientists had considered the appearance of such foci to be a reasonable laboratory facsimile of tumors: each focus is the replicative outcrop of one infected cell, just as an animal tumor is the malignant result of a single cell gone mad.

Weinberg and his student discussed whether they might be able to use the transfection technique to duplicate the petri dish equivalent of cancer. Would DNA from the Harvey sarcoma retrovirus form malignant foci in NIH 3T3 cells? They saw no reason that it shouldn't. "At that point, transfection seemed to have the same effect on cells as infection," said Smotkin, "and I went into the experiment expecting success."

In transfecting the DNA from the Harvey tumor virus, Smotkin followed the same steps he had before: isolating the DNA, snipping it into pieces, crystallizing it with calcium phosphate, and dropping the coated genes onto mouse skin cells. His cockiness proved justified. Two weeks later, when he fished his plates from the incubator and scanned the cells under a microscope, he found that several of the monolayers were dotted with dark clusters of cancer cells. Transfection was looking more and more powerful, its potential applications ever more diverse. Smotkin was one of the first people in the world to transform cultured cells into cancer cells, not by infecting the cells with tumor viruses, but by slipping in raw DNA.

Yet none of that mattered to David Smotkin. He had finished his Ph.D. and he was leaving the lab, this time on a permanent vacation. He was leaving science. For all his luminous successes as a molecular biologist, he had made up his mind two years earlier that he couldn't stay in basic research. He wanted to go to medical school. He applied to Yale Medical School and was accepted. "I was kissing science goodbye," he said.

"Bob was really mad about that. For those entire two years, he tried to talk me out of it, but it was no use.

"I really don't understand myself why I made that decision. I guess I thought that research was too strange and too cold. To be successful in science, you have to be completely obsessive about it. There's no limit to how much you put into it, how many hours you can spend doing science, or how much of your ego gets wrapped up in it. My entire sense of self was consumed by science, and that exhausted me. I didn't want to put my whole life and ego on the line for years and years, doing experiments I could never be sure would work. I wanted more immediate and assured gratification, and I wanted a normal life."

"To the extent that I had spent so much time training him in basic research, I felt that I had failed," said Bob Weinberg. "He had become a first-rate molecular biologist, and I was sad to see science lose such a good scientist."

As disappointed as he was, Weinberg accepted Smotkin's pending departure in a spirit appropriate to the ephemeral nature of an academic laboratory. Massimo Gianni had left to return to Italy. David Smotkin would be only Weinberg's second loss, yet Weinberg already had decided that the built-in obsolescence of the scientific system might not be such a bad thing. Regardless of their career plans, graduate students and postdoctoral fellows never stay in a lab more than a few years. For some lab directors, the guaranteed employee turnover is a source of constant anxiety. For Weinberg, change is the wellspring of a lab's vitality. "I went to visit a scientist in Berlin," he said, referring to a trip he made in 1985, "and he told me that most of the people in his lab had been there for ten or fifteen years. He sounded *proud* of the lab's stability, but it made me shudder. I thought, This man's lab is like a stagnant pond."

Toward the end of Smotkin's tenure, Weinberg gave up trying to persuade him to stay in science and concentrated instead on ensuring that Smotkin passed on his transfection expertise to somebody else. The honor fell to Mitchell Goldfarb, a new graduate student and another first-rate mind. Whenever Weinberg saw Smotkin in the hallway, he'd ask, How's Mitch Goldfarb coming along? To which Smotkin would mutter that he wasn't anybody's baby sitter.

Mitch Goldfarb, who's now an associate professor at Columbia University, is funny and friendly and buck-toothed, but most of all he is quick. Mitch Goldfarb is as quick as a Saturday night, and he learned how to do transfections in about that amount of time. Indeed, he learned so quickly that when I spoke with him, he couldn't quite remember whether Smotkin was still at the lab, let alone whether Smotkin had been his transfection tutor. Transfection is one of those things, said Mitch, that, as a molecular biologist, you can't imagine *not* knowing how to do. It's like learning to talk.

However his skills were acquired, Goldfarb soon became highly talented in the art of transfection. He began with the Harvey sarcoma virus that Smotkin had abandoned in his haste to leave science, but Goldfarb took the experiments one critical step farther. Gradually and imperceptibly, Mitch Goldfarb's project shifted the Weinberg lab from the study of viruses and petri dish cancer to the study of the cancer that concerns us most: our own.

David Smotkin had shown that purified viral DNA, when transfected into cultured cells, had the same rancorous effect on the tissue as did live tumor viruses, that DNA caused focal formation. Goldfarb addressed a slightly different question. Was cancer mobile? He wondered whether he could take cultured mouse cells that had bloomed into foci at the presence of tumor viruses, isolate the DNA from those cells, transfect it into a sample of healthy cells, and cause cancer once again. The query was not shallow. It was one thing to demonstrate that purified, powerful viral genes could transform healthy tissue into malignant tissue. It would be quite another to show that viral genes, when diluted by all the mouse genes that are in a cancerous focal cell, could rally for an encore performance and spawn another crop of focal cells. Such a demonstration would prove that transfection was a very sensitive assay indeed.

The results came with a speed and clarity rare in biological research. Goldfarb infected mouse cells with the Harvey sarcoma virus, fed the cells plenty of calf's serum and nutrients, and set the dishes aside. In a couple of days, the retroviruses had plundered and marauded as expected, and the monolayers of cells were overrun with foci. He then purified the DNA from the malignant focal cells, obtaining long taffy strands of nucleic acid. The isolated DNA was a mixture of a few viral genes and tens of thousands of host mouse genes, but Mitch didn't much care about the rodent DNA. He knew that if he transfected normal mouse genes into mouse cells, the extra mouse genes should have no effect, but he hoped that the scattering of retroviral genes would retain their cancerous potential.

Mitch followed the same Graham–van der Eb recipe that Smotkin had employed for pure viral transfection. Goldfarb carved up the mass of mouse and virus DNA into manageable fractions, pinched in calcium phosphate, and transfected the solution onto NIH 3T3 cells. Two weeks later, he passed the cells under a microscope and looked for foci.

And found them. There weren't many — the efficiency of transfection was still rotten — but they were there. Two or three of the two million cells that he had transfected with the blend of mouse and virus DNA had proliferated into cancerous clusters. Cancer could be passed not just from virus to host, but from host chromosome to host chromosome. The Harvey viral genes didn't care what sort of company they kept — whether they stood alone or were accompanied by a rash of mouse genes. The viral genes remained tumorigenic.

Of equal importance to the success of the experiment was Mitch Goldfarb's ability to distinguish the signature of tissue culture malignancy: an outbreak of foci. Mitch could look at a dish, see where the bumps were, and know which cells encased foreign, cancerous DNA. But it's a good thing that Goldfarb's eyes were as quick as his mind, because the focus test can fool you. Although NIH 3T3 cells are nominally normal and are supposed to lie in a single layer of cells when untouched, the skin cells occasionally misbehave and start mimicking cancer cells.

"Looking for foci in a monolayer is not a trivial exercise," said

Weinberg. "The monolayers of untransformed cells tend to accumulate spurious clusters, here and there, that are often indistinguishable from genuine foci."

I know what Weinberg means: I saw for myself. That is, I didn't see for myself. One afternoon an extremely patient scientist presented me with two dishes of NIH 3T3 cells, one containing real, transfection-induced foci and the other containing "spurious clusters," and asked me to guess which had the foci. I carefully studied the plates under a microscope. Dish A was mostly a soft filigree of healthy cells, speckled here and there by dark spots. Dish B was mostly a soft filigree of healthy cells, speckled there and here by dark spots. I put one sample under the microscope. I put the other under the microscope. I switched the plates again and again, grunting in frustration. Finally, convincing myself that its clusters were a smoke-ring darker, and knowing that I had a fifty-fifty chance of being right anyway, I chose Dish A. "This one has the foci," I mumbled, and beamed with pleasure when he complimented me on my acumen.

Mitch Goldfarb was rarely bamboozled by spurious clusters. When he identified a focus as a focus, he could screen the cells of the focus and recover viral particles, the surest sign of integrated viral genes. Transfected DNA, he told Weinberg, was as potently cancerous as infected DNA and could be detected in culture by the focus method.

Weinberg was deeply impressed. On the face of it, transfection seemed a barbaric procedure. His student was dumping foreign genes into cells virtually by the beakerful. Yet transfection said something very refined about cancer. It said that a tiny amount of tumor DNA had supremacy over a mass of normal DNA. It said that transfection was a magnificent assay for discriminating between the few cancer genes and the multitudinous healthy genes.

Weinberg wanted numbers. He wanted quantification. How much viral DNA, he asked Mitch, is needed to produce foci? That is, how many copies of the sarcoma virus must be present in the donor mouse cells to allow the virus to bellow above the background noise of the donor DNA? The overwhelming amount of the DNA that Mitch transfected into the NIH 3T3 cells was piggyback mouse DNA. What ratio of sarcoma DNA to mouse DNA was required to deface a monolayer of recipient cells with foci?

To satisfy Weinberg's curiosity, Mitch Goldfarb screened the infected donor cells with a radioactive probe fashioned from the genetic material of a tumor virus. He determined that every infected cell contained one Harvey sarcoma virus. The figure was boggling: the transfection protocol was so sensitive that when a solution of DNA containing one sarcoma virus per 40,000 mouse genes was splashed onto NIH 3T3 cells, the cells that consumed the stew of viral and rodent genes were as fully transformed as they would have been if infected with live virus. It's sort of like the story of the princess and the pea. The princess could feel the presence of a single pea that had been placed under twenty mattresses and twenty featherbeds; the recipient skin cell could feel the presence of one sarcoma virus amidst more than 40,000 donated mouse genes.

The fact settled into Weinberg's cerebrum: one Harvey virus per donor mouse cell was enough to form foci. The fact was both simple and profound; it was the sort of fact that kept Weinberg pacing around his living room after hours. For his part, Mitch fell under the thrall of the Harvey sarcoma virus, and he decided to do his Ph.D. thesis on the way in which the virus integrates itself into the host chromosomes and ignites cancer.

But in 1975, just as Weinberg and Goldfarb were mulling over DNA ratios and dissertation subjects, there came an announcement from a rival camp on the other side of the continent. Michael Bishop and Harold Varmus, of the University of California at San Francisco, reported a discovery that was so spectacular nobody could believe it at first. By the following year, however, few doubted that what Bishop and Varmus had found was real, and it changed the way Bob Weinberg and his lab thought about cancer.

Bishop and Varmus (who are, incidentally, often spoken of as a hyphenated construct, Bishop-Varmus, although they've had separate labs for years) studied the most famous retrovirus of all: the Rous sarcoma virus. The virus was first isolated from a chicken tumor back in 1911 by Peyton Rous, a young and somewhat dyspeptic researcher at the Rockefeller Institute in Manhattan. Rous's contemporaries scorned the notion that a virus could cause cancer, and Peyton Rous eventually abandoned the experiments, deeming cancer research "one of the last strongholds of metaphysics." His eponymous virus, however, lived on, in tissue culture and in legend. Rous was awarded the Nobel Prize in Medicine in 1966, and it was through the analysis of the descendants of the original chicken virus that scientists got their first glimmers of how tumor viruses, in the words of Michael Bishop, "wreak their mischief."

Extending from 1960 to the mid 1970s, and entailing the contributions of scores of scientists, the story of the Rous sarcoma virus is another shaggy dog epic with a slam-bang conclusion. In their first grope toward understanding, one team of researchers found that the Rous sarcoma virus came in two varieties: an innocuous, noncancerous virus that infected cells and replicated without doing harm, and a pernicious version that was incapable of replicating but instead sparked malignant growth in its host cell. Other biologists measured the viral cousins and determined that the cancerous pathogen was significantly longer than its mild-mannered relation. They also discovered the reason for the size difference. The tumor virus had a whole extra gene tacked onto its tail end. The addition of the gene had shorn off a small piece of the virus's replication gene — hence its reproductive defect — and substituted a malignant potential. Because the extra gene was the segment that caused sarcomas, or tumors of the connective tissue, in chickens, it was designated *src* (pronounced "sark"). Other sequences of the virus seemed to play no part in causing cancer.

The clincher came in 1975, when Bishop-Varmus, along with Bishop's postdoctoral fellow Dominique Stehelin, figured out precisely where the tumor virus had picked up its annexation. The *src* gene, they said, came from the chicken. Whey they screened healthy cells of hens, they found nesting in the DNA the same *src* gene they had spotted in the Rous virus. Thus, the gene that lent the tumor virus its malignant brawn wasn't a native component of the virus, the way the genes for reverse transcriptase and protein coat were. Rather, at some point during the Rous sarcoma virus's cellular peregrinations — either as the pathogen moved from cell to cell in one host animal, or as it passed from one chicken to another — a mistake was made in the production of infant viruses. The parent virus was so careless in dictating the packaging of its offspring that an RNA copy of a chicken gene was stuffed into the hindquarters of a viral particle. And not just any chicken gene, not a gene that designates the shape of the beak or the color of the feathers, but a gene that is somehow intimately involved in the growth of the bird.

The California scientists went into a frenzy, screening every species of cell they could find for evidence of the *src* gene. They sifted through the DNA of ducks, of turkeys, of geese, even of the flightless Australian emu, among the world's largest and most primitive birds. Everywhere they looked, they found the *src* gene. They checked the cells of mice, of cows, of rabbits, of fish; all animals of land, sea, and air had the *src* gene. Finally, they screened the DNA of humans, and there, too, was the *src* gene. "That was something many big-name biologists just couldn't accept at first," said Mike Bishop. "It was fine when we said the *src* gene was in birds and cows, but when we said we'd found it in man, they balked. You can't imagine the intellectual tussle we had."

The tussle ended when other researchers were able to repeat the Bishop-Varmus experiments; scientists then turned their minds to considering the implications of the *src* gene. The gene found in creatures as evolutinary ancient as fish, which meant that the *src* gene had persisted for half a billion years. "When something survives the process of natural selection for that long," said Bishop, "you know the gene has to be pretty important." Moreover, the gene was shown to be active in most tissues of the human body: it made protein, which meant that it was performing a fundamental duty. Noting that the gene, when pilfered by a retrovirus, turned malignant, scientists assumed that the normal role of *src* must be to help orchestrate some phase of cell division; later, they'd get a better idea of how the *src* protein operated in cells.

The discovery of the *src* gene on so many bands of the phylogenetic spectrum roiled the scientific community and warranted the introduction of new and provocative terminology. The version of the gene borne by retroviruses had been called an oncogene; it caused cancer. For the cellular copy of the gene, Bishop coined the term *proto-oncogene*; it had the potential of causing cancer. Many scientists grumbled about the word proto-oncogene, because it implied that the primary function of the indigenous *src* gene was to become tumorigenic. The native *src* gene has a real and presumably key role in cells, they said; evolution hadn't gone to so much trouble to preserve the gene just to set the stage for cancer. But the term allowed Bishop to make his point. "We showed what had long been suspected," he said, "that we cary the seeds of our cancer within us."

In the wake of the *src* finding, retrovirologists began testing their particular pathogens for the presence of pirated genes. A dozen or so species of viruses turned out to contain oncogenes that had proto-oncogenic counterparts in healthy cells. Mitch Goldfarb's Harvey sarcoma virus was one of them. Edward Scolnick of the National Cancer Institute, another scientist working with the Harvey pathogen, discovered that the virus toted around an oncogene called *ras* (pronounced "rass" or, rarely, "razz"). Like the Rous sarcoma virus, the Harvey virus had pinched its cancer gene from its host cell, in this case the cell of a rat. And the *ras* gene is as evolutionarily hoary as the *src* gene. Fish, flies, mice, monkeys, and humans possess the proto-oncogenic *ras* gene.

More than ever, Mitch Goldfarb was convinced that the Harvey sarcoma virus was worth his while. He continued to do transfections, but his interest veered toward recombination: the mechanism by which a noncancerous retrovirus moves into the host DNA and purloins the cellular *ras* gene. "Mitch did some beautiful work on the Harvey virus," said David Baltimore. "He was really the first person to show how recombination between a virus and an oncogene occurs."

Yet Weinberg's thoughts were drifting elsewhere. Although he too was dazzled by the Bishop-Varmus work, Weinberg had less interest in the Rous virus or the Harvey virus than he did in the cellular genes that the pathogens embezzled. Animal cells contain at least twelve genes, he thought, that are susceptible to the subversive activity of retroviruses. He considered the meaning of such proto-oncogenes as *src* and *ras*. The genes are vital to life; that's why evolution had so carefully conserved them. When attacked by viruses, however, the genes become agents of death. They become cancer genes. It struck him that a tumor virus was not so much a cancer-causing pathogen as it was a mutagenic event, like a blast of ultraviolet radiation or a dose of a chemical carcinogen. A retrovirus in itself is harmless. Only by mutating a cellular gene can a virus plant the seeds of a tumor.

In the story of cancer, thought Weinberg, retroviruses are merely a narrative device; the proto-oncogenes are the real protagonists. They are the heroes and the antiheroes.

Bob Weinberg decided that there should be a way to take viruses out of the picture. There *must* be a way. Despite the many insights into tumorigenesis that retroviruses afforded, most types of human cancer had nothing to do with recombining pathogens. And the problem of human cancer had increasingly insinuated itself into Weinberg's subconsciousness.

Besides, he was weary of retroviruses. Even before the Bishop-Varmus work, the field of virology had been waspishly competitive; with the detection of cellular *src*, the competition grew frenetic. Weinberg was tired of hustling. Between Smotkin and Goldfarb, his lab had marched forward on several technical fronts. It had joined the handful of groups able to perform transfections; it had discovered that NIH 3T3 cells were the best, though not the only, transfectable cell type; and it had applied the focus test to the art of transfection. Still the lab's efforts fell short of revolutionary. Others were doing very similar experiments. There were Bishop and Varmus, of course, and Ed Scolnick. Most irritatingly, Geoffrey Cooper of Harvard Medical School — one of MIT's traditional rivals — was transfecting the DNA from a chicken leukemia virus onto the Weinberg lab's favorite NIH 3T3 cells. "It seemed as if from the beginning," said Smotkin, "that guy was forever on our tail."

What was really needed was one good Boston blizzard to rattle Bob Weinberg's brain.

3

The Blackest Magic

SCIENTIFIC IDEAS, like mythical women, come in two varieties: virgins and whores. Virgin ideas are profound and require great streaks of insight on the part of their formulators. The structure of DNA was not at all obvious. There was no predicting that genes should be arranged along a double helix of matching bases. In fact, before James Watson and Francis Crick proposed their model, it had been suggested that DNA consisted of three nucleic strands rather than two. Only after the sublime Watson-Crick scheme was reported in *Nature* did biologists slap their foreheads and cry, "Of course! How could it be anything but?" The creation of monoclonal antibodies — the so-called magic bullets that target specific proteins in the body — was likewise profound. Perhaps nobody but Cesar Milstein of Argentina and Georges J. F. Koehler of West Germany, and their lab associates, would have thought to combine mouse spleen cells with mouse antibodies, although the impact that monoclonal antibodies have had on basic and medical research is now clear to all in the business. (And in 1984 the two men accordingly each won a Nobel Prize for their effort.)

Other ideas are there for the taking. They are apparent enough so that a number of researchers who have no contact with one another may hit on the same notion simultaneously or sequentially, albeit not with equal clarity. And if a good idea is really self-evident, a scientist may believe it so vigorously that he doesn't even bother gathering honest data to support his idea before rushing it into print.

The idea of transfectable oncogenes was a whore.

Even before the discovery of the cellular *src* gene, scientists knew that cancer began with the mutation of DNA. There was much evidence that human genes were riddled with fragile sites that could be susceptible to carcinogens and radiation. In the 1960s Bruce Ames, of the University of California at Berkeley, demonstrated that chemicals known to turn cultured cells malignant were also highly mutagenic to DNA. Whenever he spritzed carcinogenic compounds onto plates of the bacterium *Salmonella typhimurium*, he'd generate large populations of mutant microbes: not fast-growing, cancerlike bacteria, but bacteria able to synthesize amino acids they weren't able to make before. Any time an organism starts synthesizing a novel compound, it's certain that its genes have been tampered with. Carcinogens and mutagens were chemical cousins, said Ames, and their mutual target was DNA. He did not try to guess the number of genes that might be affected by a single poison; for all he knew, the figure could be as small as one or two or as large as a hundred.

A biochemist unfamiliar with the minutiae of DNA, Ames also resisted speculating about how mutagens and carcinogens work on genes to elicit cancer. To molecular biologists, though, his experiments suggested two possibilities. The compounds could act by changing genes into different chemical configurations: shuffling around base pairs, substituting *a*'s for *g*'s or *g*'s for *c*'s or whatever. The mutant genes would then encode different, and more dangerous, types of proteins. Or carcinogens could operate by killing genes: cutting out pieces of the DNA and destroying the genes' ability to express any proteins at all. If the task of the gene is to control the growth of a cell, a destroyed gene would leave nothing to stay the course of a tumor. The theoretical difference between the two outcomes was to prove of paramount importance as the field of oncogenes evolved.

Working on the larger scale of the chromosome, Jorge Yunis, of the University of Minnesota, identified a half-dozen or so points along the human chromosomes that consistently broke when isolated chromosomes were exposed to carcinogens. Seeing that certain chromosomal spots were prone to snap like an old woman's bones, Yunis suggested that perhaps individual genes at those spots were involved in cancer. Again, the question was how many genes. A single chromosome contains thousands of genes; the sites flanking one of its identifiable break points, hundreds of genes.

But even if it were arbitrarily assumed that a very limited number of genes participate in the genesis of a given tumor — say, six or fewer — how could scientists find those genes? What assay, what test, could detect the DNA fragments that had been puckered and wounded by a noxious chemical? Tumor cells frequently are an undifferentiated mess, a genetic Beirut: some chromosomes are missing, others are running wild, still others are doubled over or partly amputated. Much of the chaos, scientists assume, is the consequence, not the cause, of cancerous transformation; as malignancy progresses, the DNA in the nucleus of the tumor cells starts curling and duplicating out of control. How, in that rubble of chromosomes, can scientists find the commando genes that first set the cancer spinning?

Michael Wigler, who was then at Columbia University, had a great idea. Wigler was a master at transfection. Unlike the virologists who transfected viral DNA into animal cells, Wigler transfected animal genes into animal cells. In a landmark experiment, he'd shown that when he slid a chicken gene for the enzyme thymidine kinase into mouse cells that lacked the ability to produce the enzyme, the imbibing rodent cells would thenceforth be capable of making fowl thymidine kinase.

Perhaps, thought Wigler, it would be possible to perform a similar type of experiment with cancer genes. Perhaps he could transfect genes from mammalian tumor cells into a dish of normal recipient cells and turn the recipients malignant. Afterward, he could go back to the beneficiary cells and fish out the donated segment of DNA that had done the deed: the oncogene.

No, he decided. It wouldn't work. "I'd had a lot of medical training," he explained, "and that training taught me that cancer was likely to be a disease of missing genes, not of genes gone awry." In other words, Wigler believed that carcinogens or radiation acted by murdering genes — ripping out their protein-making ability — rather than by changing the genes into an alternate form.

There was good evidence to buttress Wigler's conviction. Biologists had found that when they fused normal animal cells with tumor cells, the resulting hybrid cell was normal. It seemed that a factor in the normal cells compensated for the deficiency in the tumor cells — that the healthy cell had the gene which the tumor cell had lost through mutagenesis. By the same logic, thought Wigler, if you transfect DNA from malignant cells onto a plate of normal cells, you'll never see any effect. In the original tumor cell, the destruction of the growth-control genes may have resulted in cancer, but you can't duplicate the disorder by putting worthless genes into healthy cells that are equipped with a full set of growth controllers.

Wigler didn't even bother getting out his petri dishes to give transfection a whirl.

Charles Heidelberger, a molecular biologist then at the University of Wisconsin, was a little more optimistic than Wigler. He not only thought about the possibility of transfecting genes from tumor cells into normal cells, he tried doing it. He tried, and nothing happened. The recipient cells stayed smooth and calm, as though they had absorbed no alien DNA at all. Heidelberger concluded that either the transfection assay wasn't sensitive enough to detect cancer genes or that he wasn't sensitive enough to do the transfections properly.

As early as 1975 or 1976, David Smotkin contemplated transfecting DNA from human tumor cells. "It was a natural progression in thinking, from transfecting leukemia virus DNA to transfecting the Harvey DNA and getting foci," he said. "Transfecting human tumor DNA was an obvious thing to want to do. But I couldn't see a way that I would be able to pull the gene back out from the foci once I put it in, so I never took the idea anywhere. It was just a thought game I was playing. Who knows? If I had pursued the whole thing, maybe I would have stayed in basic research."

Yes, the idea of transfectable oncogenes was obvious — as obvious as a crisp winter wind. Many people thought of it. But few were willing to waste their time shooting at the breeze.

Bob Weinberg is not an athletic man, and he's a tiny bit proud of his allergy to exercise. The only times he sweats are when he's hammering nails into his New Hampshire house or raking leaves around his Massachusetts house or shoveling snow from the driveway of either house. He doesn't run, except to catch up with one of his children, and he doesn't much like walking, because it hurts his feet.

One February morning in 1978, Bob Weinberg had no choice. He had to walk from his home, then on Beacon Hill, to his lab at the Cancer Center, a mile away. The night before, a record blizzard had muffled the entire Boston area — this stolid city of snow-quilted winters. There was no public transportation, the roads were unplowed, and Bob Weinberg wanted to go to work. Donning a down parka and rubber boots, he headed toward the Longfellow Bridge, which spans the Charles River over to Cambridge and which is known locally as the Salt and Pepper Bridge, because its stone piers look like giant salt and pepper shakers. As he walked, wiping swirls of snowflakes from his thick mustache, Weinberg thought about cancer. He thought about the oncogenes that had been found in the bosoms of viruses. He thought about the DNA of the Harvey sarcoma virus, which Mitch Goldfarb could transfect into mouse cells and later descry through the appearance of foci. Again, the idea of the virus vexed him: most animal and human cancers have nothing to do with retroviruses. Most tumors are caused by chemicals, and those chemicals must assault a limited number of genes. He wondered how he could identify, amidst the gaggle of cellular genes that have no relation to cancer, the few genetic targets of carcinogenic chemicals.

Perhaps subliminally aware, as he trudged across the Longfellow walkway, of the symbolic significance of bridges, Bob Weinberg had an idea.

"It was an idea of numerology," he said. "I thought about the sensitivity of the transfection technique and about the fact that we could detect the presence of one sarcoma virus amidst roughly a millionfold excess of unrelated DNA sequences. The DNA of a single sarcoma virus was all that we needed to produce a focus of transformed cells.

"It was only a small logical leap, then, to ask whether we'd get the same result through transfection if the cell had been transformed by a chemical carcinogen. If the carcinogen had affected one gene in a hundred thousand — that is, one of the two copies of the putative oncogene within a cell, a cell having all genes in duplicate — would transfection be sensitive enough to pick it up?"

There was no reason, he thought, that transfection shouldn't be able to do for animal genes what it had done for viral genes. There was no reason that his lab couldn't isolate the DNA from animal tumor cells and, through the aid of calcium phosphate, smuggle the genes into a collection of healthy cultured cells. The active cancer genes from the tumors would then integrate into the recipient chromosomes, switch on as though they felt right at home, and begin producing cancer proteins. The result: the healthy cells would become malignant and replicate into foci. Later, the Weinberg lab could examine the cells of the foci and find the foreign segment of DNA that had transformed normal cells into malignant cells. Through transfection, the Weinberg lab could pass cancer from the original animal tumors into a plate of flat, healthy cultured cells. They'd have their assay for detecting single cancer genes. Weinberg would be able to get viruses out of the picture. He wouldn't have to think anymore about the *ras* gene in the Harvey sarcoma virus or the *src* gene in the Rous sarcoma virus. He would be working with cellular genes that had been mutated by chemicals; he would find the genes that were the true protagonists in most animal tumors. He could even find human cancer genes.

"I was, as you might imagine, very excited by this notion," said Weinberg, "and I decided that it was worth the attempt."

The walk from the Cambridge end of the Longfellow Bridge to MIT's Cancer Center takes about seven minutes. Add another seven minutes for the difficulty of tramping through knee-high snow. For Bob Weinberg, fourteen minutes never flew so quickly. Mitch Goldfarb was among the first to hear the Longfellow idea. Bob Weinberg wanted him to try transfecting the DNA from tumor cells into noncancerous NIH 3T3 cells, and he wanted him to try it fairly soon. Say, within the next couple of days.

No, said Goldfarb. It wouldn't work.

He and Weinberg quibbled over the reasons that it wouldn't work. Goldfarb believed that there was "something fancy" about viral DNA that allowed it to transform cells. The fancy somethings were called long terminal repeats, or LTR's. Scientists had discovered that when viral DNA slithered into host chromosomes, it brought with it a series of about 600 base pairs that was identical at both the front and rear ends of the viral DNA. The mysterious repeated sequences appeared to be important in allowing viruses to blaze on as sets of active genes once they were comfortably ensconced in the cellular DNA. Goldfarb thought that the reason the viral genes could be so easily transferred from one cell to another and still retain their ability to switch on was that they had LTR's. Genes from tumor cells do not have long terminal repeats, said Goldfarb, and it seemed very possible that the mammalian genes, when passed into recipient cells, would not be expressed and therefore would not cause cancer in their new station. "I was pretty ignorant at the time," said Goldfarb. "If I'd known about Mike Wigler's work with transfection, I would have realized that animal genes could be transferred from cell to cell and expressed with ease. But, hell, I was only a graduate student."

Weinberg and Goldfarb were also aware that chemical carcinogens could operate by puncturing the double helix and killing genes, in which event transfection would be worthless; transfection could not detect the presence of a dead gene.

But maybe that isn't what carcinogens do, said Weinberg. Maybe they activate silent genes or mutate healthy genes into more potent genes, and maybe we can find those altered genes by transferring them into healthy mouse tissue.

What the hell, he said. Let's give it a try.

No, said Mitch. It won't work. Mitch had his Ph.D. project to finish, on the Harvey sarcoma virus; Harvey was what he was interested in. "I was reluctant to forge ahead," he told me, "and Bob was reluctant to push."

Several weeks later, push fell to shove, Weinberg secured a tumor cell line from his colleague Charles Heidelberger, and Mitch gave it a try.

The cells Weinberg handed to his student were connective tissue cells that had been severed from normal baby mice. Heidelberger had cultured the kidney cells in dishes and then transformed them into cancer cells by dousing them with 3-methylcholanthrene, a potent chemical carcinogen. Mitch isolated the DNA from the mouse cells, applied it to NIH 3T3 cells, and incubated the cells for two weeks. As a control, he mock-transfected a group of cells with calcium phosphate alone, an exercise not expected to yield foci. "Quite frankly, I didn't know what I was doing," said Mitch. "I knew the procedure, of course, but I wasn't sure what I should be looking for. When you're doing virus work, you know what to expect, but what could I expect from unknown genes?"

Goldfarb's befuddlement did not soon disappear. For months, he strove to transfer the cancer trait from the kidney cells to mouse skin cells. Sometimes he thought he saw foci on the transfected cells, and his spirit soared. Then he'd go back to his control dishes, see clusters there as well, and his spirit slumped. Other people in the growing Weinberg lab were at least as skeptical as Mitch. "I thought it was the craziest thing I'd ever heard," said David Steffen, Weinberg's third postdoc. "Transferring DNA from tumor cells? It made me wonder what sort of a crazy lab Bob was running."

After repeated failures, Mitch Goldfarb stopped paying much attention to transfection and focused his energies on his dissertation project, which never disappointed him. He went out of his way to avoid Weinberg, who'd invariably ask how the transfection experiments were going, to which Goldfarb could only reply, They're not. And then, out of the gray, transfection worked. "It was my twenty-fifth try," he said. "I remember that number because we were coding the experiments by the letters of the alphabet, and this was the next-to-the-last letter. The foci in this case looked exactly like the foci you got when you transfected the Harvey sarcoma virus, and the foci were unmistakably stronger than in the controls."

Transfection *worked.* Mitch Goldfarb had taken genes from a tumor cell and passed them into flat mouse cells. A segment of DNA from the original cancer tissue had preserved its malignant strength, expressed its aberrant protein in its new location, and fanned the growth of bulging foci. For what could be the first time in history, someone had bracketed a specific portion of mammalian DNA involved in chemical carcinogenesis, an achievement of boundless relevance to anybody who is concerned about human cancer.

But then something odd occurred. At least, I thought it was odd. Mitch Goldfarb decided to drop the project. Here was a man who'd spent hundreds of hours trying to induce foci with tumor DNA. He knew how important the results would be if the experiment worked and how gilded his name would be should the results come to publication. Yet when the experiment finally worked, he tipped his hat and said good night.

I was so puzzled by his decision that I asked him about it on two separate occasions.

"Yes, I can see it might strike an outsider as strange," he said, folding his skinny arms against his chest and crossing one skinny leg tightly over the other. "It would strike an *insider* as strange. But look at it from my point of view. What is one in twenty-five times? That's not exactly an encouraging number. It's more like an artifact than a finding — it could have been the result of viral contamination; it could have been anything. Besides, I was finishing up my Ph.D. and thinking about where I was going to do my postdoc. You can't stay in graduate school forever, even when you have a good project.

"But the real reason I didn't pursue the transfection stuff was Asilomar. Under the Asilomar regulations, I wouldn't have been able to go on and clone the oncogene. Without being able to clone the gene, there wasn't much I could do beyond growing a few acres of foci."

Today, when molecular biologists fairly freely manipulate and recombine the most repellent pathogens, from human leukemia viruses, the AIDS virus, and the smallpox virus to the bubonic plague bacterium and the malaria parasite, the Asilomar regulations sound almost quaint, like a passage from Burton's *Anatomy of Melancholy*. In the middle and late 1970s, however, the regulations forestalled, for better or worse, an astonishing amount of basic research. First conceived as an earnest effort at social responsibility, the Asilomar ordinances stemmed from a 1974 conference of the same name, held in Pacific Grove, California. More than 150 scientists from fifteen countries attended the convocation, and during four days of often acerbic exchange debated the merit of regulating the nascent field of genetic engineering.

By Asilomar's end, the conferees had drafted strict guidelines for recombinant DNA research; these were adopted by the National Institutes of Health and became United States law in 1976. Among the provisions: any experiments involving the use of viruses and other dangerous microbes must be performed in containment facilities, equipped with air-lock doors and coats, hats, gloves, and booties for anybody entering; scientists should work toward developing a "fail-safe" strain of *E. coli* that would die outside the laboratory; under no conditions should scientists attempt any cloning protocols that entailed transferring cancerous DNA from one organism to another.

Although the NIH guidelines had relaxed somewhat by 1978, when Mitch Goldfarb was thinking about isolating the oncogene he'd detected through the transfection assay, the procedure of recombinantly inserting a cancer gene into a cloning virus and then into *E. coli,* and amplifying that cancerous DNA several million times to get the quantities required for analyzing the gene, was still strictly forbidden. It was permissible to attempt transfecting cancer genes into cultured, immobile donor cells, but to clone the gene into a microbe like *E. coli,* which was capable of infecting human beings, was unquestionably against the law. "There really was nothing I could do," said Mitch. "The interesting experiments obviously had to wait until the law was changed."

Bob Weinberg thought that Mitch's objections were beside the point. He too believed that one success out of twenty-five attempts was not a number to hang a career on, and he wanted better evidence that transfection worked. To him, Asilomar had no immediate relevance; he saw no reason to put off, for God or Caesar, what were obviously the most important experiments he had ever formulated. "Mitch's argument didn't make sense," he said. "We weren't talking about cloning any oncogenes at that point. The only thing we had to do was to show that transfection was reproducible.

"That transfection could work was all that counted. The rest was just commentary."

Bob Weinberg fantasized about how astonished and impressed the scientific community would be if his lab succeeded in transferring cancer genes from tumor cells to healthy cells and reproducing the experiment. He was not, under any circumstances, going to drop transfection.

But before Weinberg had a chance to utter a word in public about his new experiments, another researcher convinced the scientific community that transfection was a scam.

There are more than four thousand scientific journals and newsletters that cater to the biomedical profession, but for molecular biologists the most prestigious publications by far are *Nature, Science,* and *Cell.* When DNA researchers have something important to say, they try first to say it in one of the three. *Nature* is published weekly by Macmillan Ltd. in London, *Science* is the weekly journal of the Washington-based American Association for the Advancement of Science, and *Cell,* which now comes out twice a month, was until recently published by MIT Press. Although one might think that any scientist at MIT would be entitled to a free subscription to a university-related journal, *Cell* is fat and glossy and too expensive to be distributed as a faculty perk. Unable to afford a personal subscription on his associate professor's salary, Bob Weinberg, in the late seventies, was obliged to read David Baltimore's leftover copies of *Cell* months after publication. Had he kept up with the literature, he might have seen a brilliant paper by Demetrios Spandidos, a postdoctoral fellow at the University of Toronto. And had he seen the paper, Bob never would have bothered brainstorming on the Longfellow Bridge. A month before Weinberg had his great idea about transfectable oncogenes, Demetrios Spandidos reported in *Cell* that he had successfully performed the same sorts of experiments that Weinberg was planning to do.

"Historians would say, Oh, come on, he *must* have known about that paper. He was at MIT, and *Cell* is published by MIT Press," said Weinberg. "But the fact is that I didn't get *Cell,* and I'd never heard of Demetrios Spandidos."

Weinberg did not remain ignorant forever. Soon after Mitch Goldfarb began achieving his ambiguous results, Demetrios Spandidos was invited to give a talk at Harvard. Bob Weinberg attended the seminar, and he couldn't believe what he heard. Spandidos was dynamic, polished, forceful — and he had done everything that Bob was then struggling to do. Not only had he done the experiments; he had done them much more cleanly and reproducibly than Weinberg's lab had done or could hope to do any time soon.

"It was a fabulous piece of work," said Weinberg. "As I sat there listening, I was both enormously impressed and enormously depressed. I was impressed because it was extraordinarily well done. He presented data that were elegant, comprehensive, encyclopedic in their addressing of the problem. I was convinced that the laboratory from which he'd come had launched a major series of experiments. Theirs was a major effort, an enormous amount of effort.

"At the same time, I was depressed. For obvious reasons."

Spandidos's approach to transferring cancer genes from one cell to another, as he described it during his seminar, was not identical with the transfection technique being used at the Weinberg lab. Spandidos, too, had purified DNA from tumor cells (in this case, malignant hamster ovary cells), precipitating the DNA with calcium phosphate and adding the DNA to normal cells. Instead of looking for foci, however, Spandidos relied on another criterion of cancerous transformation to separate those cells which had taken up oncogenes from those which had not. He tested the recipient cells to see whether they could grow in a soft, puddinglike agar base. Normal cultured cells require the feel of a firm petri dish beneath them before they'll divide; cancer cells will grow happily in suspension. Spandidos declared that a large percentage of his transfected cells thrived in soft agar, but his control cells did not. In his hands, the efficiency of transfection was truly remarkable — perhaps ten times better than what the Weinberg lab had managed. And Spandidos had done something else that Weinberg couldn't have dreamed of doing at that point. He had made a preliminary analysis of the donor DNA. By passing the nucleic acid over a separation column, he'd determined that only DNA fragments of a certain size would turn the receiving cells cancerous. Spandidos was no longer thrashing around in the dark with large hanks of DNA; he seemed to be well on the way to isolating individual oncogenes.

His listeners were dazzled. The seminar room purled with enthusiasm. Spandidos was a genius and his work was flawless. After the lecture, many prominent scientists approached Spandidos to congratulate him; there was talk that Harvard had offered him a job. Depressed though he was, Weinberg also offered Spandidos his somewhat inhibited praise. "I told him that I was interested in his work and that I was beginning to get similar results." Demetrios Spandidos, said Weinberg, had an unexpected reaction. He was overjoyed.

"Well, you might say, that's quite natural," Weinberg said to me. "But, indeed, it's *not* natural. When you go up to a scientist who's just announced a discovery, and you tell him that you've gotten similar results, that person usually has very mixed feelings. On the one hand, it's nice to know that your ideas have been confirmed. On the other, it's very unsettling to realize that you have a competitor.

"Spandidos had no such diluted joy. He was obviously delighted."

Weinberg was faintly taken aback by Spandidos's pleasure, but worse was yet to come. Soon Weinberg heard that Spandidos had been invited to the Cancer Center to give a talk there, too: a job seminar. Scientists usually give job seminars — descriptions of their work-in-progress — at universities where they're likely to get jobs. Like Harvard, the Cancer Center, Weinberg's employer at that time, was considering offering Demetrios Spandidos a position. "All this occurred without my consent or knowledge. They knew what I was working on, and they were thinking of hiring somebody else who was doing almost exactly the same thing," said Weinberg. "This was almost insulting to me."

Bob Weinberg was "almost insulted" by MIT the way Henry V was almost insulted when the Dauphin of France sent him a barrelful of tennis balls. Weinberg was livid. He didn't know whom to blame or whom to appeal to. For days, he could obsess about nothing but Demetrios Spandidos. He talked about how it wasn't fair or right that MIT contemplate hiring Spandidos. He stewed about how the Cancer Center supposedly was predicated on faculty cooperation, not competition. He considered leaving MIT. "He tried to hide his anxieties," said Amy Weinberg, "but Bob is always bad at pretending to be happy."

"It was the closest I ever came," said Weinberg, "to losing a night of sleep."

Bob Weinberg's angst would dissolve as quickly as it had formed. The next week, David Baltimore came into his office and closed the door behind him. He asked Bob whether he'd heard the news about Spandidos.

Bob said no, he hadn't heard.

You're not going to believe this, said Baltimore, but Spandidos has just been thrown out of the lab in Toronto. People are saying that the work he did was all made up.

Baltimore was right: Bob Weinberg couldn't believe it. Those beautiful experiments that he'd just heard about the other week were made up? Not just fudged here and there, but woven from whole cloth?

That's what I've heard. That's apparently why Lou Siminovitch has booted him out, said Baltimore, referring to the head of Spandidos's lab.

Now Bob Weinberg and everybody else in the field really had something to chatter about: a bona fide, delectable scandal, the likes of which they had not heard since William Summerlin of the Sloan-

Kettering Institute in New York painted the backs of mice and called the patches skin grafts.

Perhaps because reports of research fraud have received extensive coverage, the general public may assume that cheating in science is as rampant as it is in New York City government. Scientists, after all, are under great pressure to get results. Their grant money, job opportunities, and egos depend on facile and consistent successes. But the reality is that genuine, fill-in-the-blanks, color-the-DNA-x-rays fraud is rare. "Most of us have now and again convinced ourselves that we saw something that wasn't quite there," said Mike Wigler. "And we're constantly having to decide whether a questionable piece of information is relevant to the problem at hand or not. But if you're careful and conscientious, you'll repeat the experiment until you're sure. You don't rush to publish on the hope of the moment." Outrageous fraud is rare because most scientists are romantic. They go into science committed to excavating fundamental truths about the natural world; truth may wear many guises, but a fabricated experiment isn't one of them. Scientists are also pragmatic. They know that if they're doing an experiment that's worth cheating on, and if they fall to the temptation to cheat, they're likely to get caught. Should they produce "revolutionary" results in a field of intense activity, other scientists will seize on those results to launch experiments of their own, quickly revealing the initial findings as wrong. Cheaters can always hope that a scientist who uncovers their fakery will assume that an honest mistake was made and not start slinging around defamatory accusations. Alternatively, cheaters, when confronted, can try the familiar line "I'm sorry the technique didn't work for *you*. In my very experienced and talented hands, it works every time."

Serious chicanery, however, is rarely worth the effort. Sooner or later, cheaters get caught. And when they're caught, they're tarred, feathered, ass-paddled, and river-dunked. Mark Spector, a brilliant graduate student who claimed to have discovered the biochemical pathway of cancer, got caught because his protein gels had been doctored; where there was supposed to be radioactive phosphorus-32, there was radioactive iodine-125. Spector was forced to resign from Cornell University just before receiving his Ph.D. Summerlin confessed to his paint job and was fired from Sloan-Kettering. In 1980, Boston University ousted two scientists accused of falsifying some data in the course of a three-year, $1 million cancer research project. Demetrios Spandidos got caught — if cheat he did — and expelled from his lab. Whether and how much Spandidos cheated may never be known. There was no official investigation into his case, as there was with the Spector debacle. Spandidos's experiments have not been shown to be unequivocally false, and no cooked data were found. To this day, Spandidos insists on his innocence, and he is now a practicing, hard-working scientist, at the new Pasteur Institute in Athens, who publishes regularly in *Nature* and elsewhere. The majority of scientists, however, suspect that Spandidos did something unkosher. They're not willing to speculate on the extent of his infractions, because none is about to take the time to reproduce his experiments and see what's what. Spandidos's former boss, Louis Siminovitch, has accused Spandidos of being an unregenerate charlatan. "He's a dangerous character," said Siminovitch. "I believe that he's been cheating from the beginning of his scientific career, and I believe that he'll cheat again."

Regardless of the truth or falsity of Spandidos's data, his experience and his dramatic expulsion from Siminovitch's lab deeply affected the oncogene field. After the Spandidos affair, biologists were leery of any result that mentioned the word *transfection*. Scientists who wanted to discuss transferring cancer genes from one cell to another knew they had better produce some convincing data — but not too convincing, please. "I was skeptical about transfecting oncogenes before Spandidos," said Wigler. "He helped fan my doubts." Weinberg had no doubts about transfection, of course, but Spandidos didn't make any easier his task of disseminating his data. "If you're going to talk about Bob Weinberg," said Phil Sharp of the Cancer Center, "you can't leave out the story of Spandidos."

Demetrios Spandidos came to Louis Siminovitch from Angus Graham's lab at McGill University in Montreal with the highest recommendations. As a graduate student, Spandidos had published eight papers, two or three times the average number for a Ph.D. candidate. Siminovitch instantly liked and admired his new postdoctoral fellow. Spandidos was impishly handsome and had a jolly, if somewhat intense, disposition. "I found him to be very interesting," said Siminovitch. "He was very intelligent, brilliant really, and very personable. He read the literature avidly, and he knew everything in the literature." Other people in the Toronto group were taken with him as well. "He was very friendly, yet he was independent," said Pamela Stanley, a Siminovitch postdoc at the time who now works at Albert Einstein College of Medicine in the Bronx. "He had the sort of character that fitted into the lab."

Through his perusal of the literature and his private theorizing, Spandidos conceptualized the experiments that he wanted to work on, and Siminovitch gave him free rein. "I've always run a pretty permissive lab," he said. "People come in, I tell them roughly what's available, and then I let them do what they want. I've been in business for thirty years, and I think I've done very well at that." With his decision to attempt something as bold as transferring the cancerous phenotype from tumor cells to normal cells, Spandidos labored heroically. "He was like a house on fire," said Siminovitch. "He worked from nine A.M. to eleven P.M., seven days a week, never stopping. He was always there at the bench."

In less than a year, Spandidos had accrued very elegant data on transfection, and he started publishing papers. Although Siminovitch was a little surprised by the amount of work that Spandidos managed to accomplish, he saw that the postdoc was always in the lab and assumed that Spandidos's technical abilities were as superlative as his mind. Nancy Stokoe, a technician, mentioned to Siminovitch that, on the contrary, Spandidos was a careless benchworker, but still Siminovitch said nothing, nor did he ask to look at the raw data. "It's not my policy to look at raw data in any detail unless they don't make sense," said Siminovitch. "And his data made a lot of sense. His figures were always phenomenal."

Then, said Siminovitch, he began to notice a disquieting change. Spandidos was spending less time at the bench and more time in his office. Still Siminovitch said nothing. "I thought to myself, I'm not in on weekends or at night. Maybe that's when he was doing all this work. He'd have experiments set up and running, and they always looked very convincing. The discussions that we had together were very realistic as well. He would say he should do thus and such, I would suggest a change, and he would say, Oh, that's very reasonable; what sort of volume should I use? He'd express concern that if he was out of the lab somebody should check a column for him. Doesn't that sound like somebody who's doing his experiments?" Siminovitch casually asked a couple of people in his lab to try repeating Spandidos's efforts. They were unable to, but Siminovitch knew that the protocols Spandidos used were very complicated and that it was quite possible the others had done the experiments incorrectly.

Spandidos's fame mounted swiftly. Scientists everywhere requested reprints of his *Cell* and *Nature* papers, and young researchers wrote grant applications to attempt projects that would extend Spandidos's findings. Although he'd been a postdoc for only two years, Spandidos confidently took to the job circuit. He applied to five places for a faculty position: MIT, Harvard, Cold Spring Harbor, the University of Toronto, and a new biomedical facility in his native Greece. Siminovitch wrote glittering letters of recommendation, and all five institutions responded with interest. Demetrios Spandidos looked like a scientific supernova. But there's a problem with supernovas: the brighter they are, the more violently they collapse.

In the spring of 1978, another postdoc in the Toronto lab came to talk to Siminovitch. The postdoc was very upset. He declared that Spandidos couldn't possibly be doing everything he claimed to have done. "I work as hard as he does," said the postdoc, "and I don't get that much done." It was the first complaint that Siminovitch had heard from one of Spandidos's peers. Until then, the scientists in the lab had expressed nothing but praise and respect for their successful colleague. Siminovitch could not ignore the postdoc's insinuations. When a new postdoc, William Lewis, joined the group, Siminovitch set him working side by side with Spandidos. Lewis attempted to duplicate every experiment. He watched Spandidos's techniques. He repeated them carefully, and he obtained no results. "He told me that things weren't working the way Spandidos said they should, that something funny was going on," said Siminovitch. "Eventually, Lewis realized that when he used Spandidos's technique for preparing chromosomes, the chromosomes would be completely inactive." If chromosomes are inactive, they can't very well be used to transfer anchorage independence or any other genetic trait of cancer.

Around the same time, Richard Axel of Columbia contacted Siminovitch about a paper that Spandidos had just submitted to *Cell.* Axel was a referee for the report — a peer who reviews a paper and recommends to the editor whether it should be published — and he had studied the data closely. "Conceptually, I couldn't find anything wrong with the paper," Axel told me. "But the techniques, I calculated, would require for one experiment the number of tissue culture dishes that I might use in ten years." Ten years' worth of culture dishes amounts to several thousand dishes, and the cost of the liters and liters of medium required to feed the cells in those dishes far exceeds the annual revenues of an average laboratory. "I asked Siminovitch how that could be," said Axel, "and my point made Siminovitch review the data more carefully."

Siminovitch reviewed the data in the paper and then checked the laboratory supply sheets for the past year. He agreed with Axel's calculations but saw that no untoward number of petri dishes had been ordered. Siminovitch came to a horrible conclusion. Demetrios Spandidos had been fabricating his experiments. Reeling with the potential gravity of the affair, Siminovitch frantically called Spandidos into his office. "I said, 'I don't know how much cheating you've done, but I don't want you in the lab. I know you have a wife and child, and I don't want to put you in the street, so I'll let you finish out your grant here. But I don't want you doing any more experiments.'"

Spandidos vehemently denied the accusations. He reminded Siminovitch how hard he worked, and he maintained that he had done everything he claimed to have done — not just some of it, but all of it. The other people in the lab, he said, were simply jealous of him. They were playing tricks on him. Spandidos was so agitated by the encounter that he had to stop several times to go out to the hall for a drink of water.

If you're innocent, said Siminovitch, let me see your raw data. Spandidos fetched his notebooks. "They were practically empty," said Siminovitch. He demanded to see more data. Spandidos said that he'd thrown most of the data away.

Well, then, said Siminovitch, what about the tissue culture plates? You used so many, you must have poured them somewhere. Spandidos said that he had poured them elsewhere. Going over to the incubator, Siminovitch looked inside and saw very few plates. I want to know why there aren't more dishes around, he said. Spandidos explained that he had done most of his experiments not with standard plates, but with maxi-size petri dishes and bottles.

"I told him that as far as I was concerned," said Siminovitch, "he hadn't done the experiments."

The next day, Spandidos came back for a second meeting with Siminovitch. He begged to be allowed to remain in the lab long enough to prove that he'd done the work. Siminovitch refused, saying that it would take too many years and that he didn't want a cheat around, anyway. Their relationship rapidly disintegrated. Siminovitch telephoned all the scientists in his broad acquaintance and told them what had happened. In self-defense, Spandidos circulated what Siminovitch calls a poison pen letter to everybody he could think of. Spandidos wrote that Siminovitch was a "big fraud" who was out to get him. He said that his career had been ruined and that Bill Lewis could not be trusted. He followed up his first letter with additional letters. "I try to describe as precisely as I can," he wrote, "a personal experience which I believe uncovers a most cruel barbarity that occurred in the scientific community . . . I am still confused and perplexed about the motivations of Dr. Siminovitch. It is a fact that he attacked me in a cruel and ruthless manner at the time he realized I was interested to leave his lab for institutions of higher achievement. It is however possible that Dr. Siminovitch was misled by certain envious, jealous and incompetent people." Those who had met Spandidos were shocked by the near hysterical tone of his letters. Could the author of such ravings be the same genial, intelligent scientist who gave magnificent seminars and was familiar with the details of a vast body of literature?

Every story has as many sides as there are protagonists, and, not surprisingly, Demetrios Spandidos has a very different view of the affair from that of his former supervisor. For the past decade, he has sought to vindicate himself, and he has succeeded in salvaging a workable career from potential disaster. "It was one of the dirtiest stories in science," he said. "So as painful as it is for me to talk about it, I am driven to get my side out there." Spandidos claims not to know why Siminovitch turned against him, but he attributes his troubles to three factors: the jealousy of others in the lab; Siminovitch's resentment that Spandidos was planning to leave so quickly for an independent position after Siminovitch had been his biggest supporter and, in essence, his PR agent to the larger scientific community; and Siminovitch's frequent absences from the lab.

Spandidos described to me how he conceived the idea of transfecting cellular oncogenes, a narrative that convinced me of the importance of bad weather to scientific inspiration. "It was one of those cold winter mornings in 1977," he said. "I had to walk for ten minutes in the snow to cross the Queen's Park to get to the lab. Then it came to my mind that cancer genes should exist in tumor cells and that I could get those genes by using gene transfer techniques." He said that once he had the idea, he worked harder than any other person in the lab. "I never had coffee breaks, and I ate lunch at the bench in five minutes," he said. "I had an operation to have a large kidney stone removed in December 1977, and I amazed everyone by returning immediately to the lab." Although there were others who were equally productive, he said, their work was "mundane" by comparison.

Spandidos insists that he had no knowledge of Siminovitch's suspicions when Siminovitch asked him to demonstrate to Bill Lewis exactly how he did his experiments. "I told Siminovitch that I would be very happy to show anyone everything that I knew," he said. "Everything went nicely for the first couple of days, but the experiment, I knew, would take up to three weeks before we got results.

"But only four days had passed when Siminovitch called me into his office and attacked me. He said that the experiment had not worked and that he believed I had cut corners in my previous experiments. I was shocked. I told him that I was certain I was right and that time would tell. Because I thought that some people in his lab might be jealous, I told him it would be better to investigate thoroughly before making any decisions. He refused. I offered to show him my notebooks, but he didn't want to look at the data. Later I wrote a letter to the Medical Research Council [the governing body of the University of Toronto], asking for an investigation, but Siminovitch told me, 'Don't waste your time. That letter will never reach the committee.' For the first time in my life, I felt a tremendous pain in my stomach, and I could not eat anything.

"He did not want an open investigation. I don't understand why. I am still confused and perplexed. I think he was thirsty for power. Or maybe he made a mistake in his judgment and then he found it impossible to investigate and reconsider. After all, it was hard for him to know what was going on in his lab. He was there one or at most two hours a day. There were postdocs in his lab for more than a year whom he had never talked to or asked what they were doing. Another possibility was that he was angry at me. He told me, 'And why did I write that strong letter of recommendation to get you your three-year fellowship if you weren't planning to stay on in my lab?' He was a very good salesman, and he probably felt that if I left, he would have nothing to sell."

Spandidos admits that some of his initial reactions to the scandal were "rash," and he regrets the tone of some of his letters. Nevertheless, he believes that his transfection experiments were important. "As far as I know," he said, "my work was the first to show that cancer genes exist in tumor cells, as can be shown by gene transfer. All the work that has happened since then has only proved me right."

As poorly as Spandidos handled his disgrace, Louis Siminovitch did not do much better. "I made two or three errors," he said, "which, if given another chance, I would turn around." He admits that he should have started paying closer attention to the data the moment his suspicions were aroused. Siminovitch never published retractions to any of Spandidos's printed reports, an omission that many people, including Benjamin Lewin, the editor of *Cell,* have criticized harshly. "I didn't know which experiments he'd done and which he hadn't," said Siminovitch. "Not knowing what was right or real, I didn't know exactly what to retract." Perhaps most damaging in the long term, Siminovitch's refusal to allow the Medical Research Council to conduct an independent investigation of the allegations has ensured that the truth will never be known. Siminovitch explains that he did not want to put his lab through the prolonged upheaval of an inquiry that might have taken up to two years.

Among the most embittered critics of Siminovitch was Bill Lewis, the postdoc who was unable to reproduce Spandidos's work. Lewis, who has since died, was convinced that Spandidos did cheat, yet he believed that Spandidos should have been given a fair hearing nonetheless. Lewis also believed that he himself paid the price of Siminovitch's secrecy. "When I went in to tell Siminovitch that I couldn't reproduce Spandidos's results, what I said to do was to write to the journals that had published the articles and *say* that these results are not reproducible, period," Lewis said to me in 1986. "I felt as though I wanted to be defended, and the only way I could see that it could be done was through a retraction. Siminovitch said he didn't want to do it because he would open himself to libel. I think that's always what he falls back on, and I don't think it's right.

"I also felt it was important that there be an inquiry. I was sort of sitting back and waiting for it to happen, and it never did. So even though I believed the results had been fabricated, I was in some ways sympathetic to Spandidos, because I thought he was ill treated. He was victimized. You have to ask yourself, Why was there no inquiry? And you don't have to be very bright to realize that for a man in a position of power, like Siminovitch, there was a risk to his grants, his reputation, and so forth.

"I think it's a wonderful illustration of science gone wrong and the complexities of the issue. What does a supervisor do when something has gone wrong in the lab, and what's his responsibility? My feeling is that if I were in the same position — and I hope I never will be — it would only be in my best interests to be very public about it, that I could only benefit. These things don't go away. They only fester."

In the wake of the scandal, pending job offers to Spandidos from MIT, Harvard, and elsewhere evaporated. Yet Spandidos was not entirely abandoned. His adviser from McGill, Angus Graham, rallied to his aid and helped secure him a position at Glasgow University, in Scotland. "My feeling is that he didn't get a fair shake," said Graham. "The whole thing should have come to a university court, and the whole thing should have been thrashed out. He was one of the best students I ever had, and I thought he deserved another chance. You bet your boots I was going to help him." Working with John Paul, a biochemist at Glasgow, Spandidos kept a low profile for several years but then began publishing again and holding seminars at which he emphasized his innocence. The new Spandidos impressed people just as the old Spandidos had. Robert Hoffman, of the University of California at San Diego, invited Spandidos to spend a summer sabbatical in his lab, and told him afterward that he would always be welcome. "He's bright, enthusiastic, and has incredible energy," said Hoffman. "At first, a few of the people in my lab were nervous about having him there, but when they saw how he worked, and how excited he was by the experiments they were doing, they changed their minds.

"Since there was no investigation into the case, I don't see how anybody has the right to fault the guy," said Hoffman. "He's been treated worse than a criminal. In the U.S., a person accused of a crime is supposed to be innocent until proven guilty. Spandidos has been presumed guilty until proven innocent."

For weeks after hearing the news from Toronto, Weinberg talked about Spandidos as much as he had when he thought he'd be saddled with a next-door competitor. Scientists like gossip, they admit to liking gossip, and this was the best piece of gossip they'd heard in years. Baltimore received copies of Spandidos's panicky letters, which spurred a new round of "Can you believe its?" Weinberg, at least, had few doubts about Spandidos's guilt, and he later called Toronto to commiserate with Siminovitch. "He was very sympathetic to me," said Siminovitch. "I was going through a terrible time, and Bob told me that he thought Spandidos was a pain in the neck." At home, in the lab, and with friends, Weinberg pondered the strange case of Demetrios Spandidos. "The ghost of Spandidos," said the postdoc David Steffen, "was at our group meetings."

"Bob was relieved that Spandidos wasn't going to be coming to MIT," said Weinberg's friend Tony Komaroff, "but at the same time he worried about the effect the whole business would have on the field. He couldn't understand why anybody would be driven to cheat like that."

Bob Weinberg was driven, but never to cheat. He was driven to prove that he was right, and when the Spandidos affair had been talked to tedium, he wanted to get back to the job of oncogene transfection: genuine, reproducible, scientifically sound transfection.

But with Mitch Goldfarb resolutely opposed to the project, who would assume transfection's mantle? Bob Weinberg was too busy administering the lab to bother with benchwork, and the other people in his group, who were studying various details in the retroviral replication cycle, flatly refused to abandon their projects for the chanciness of transfection, particularly in the aftermath of the great scandal.

"He was constantly pestering me to get involved," said David Steffen. "But I'm as stubborn as he is, and I resisted. Number one, I didn't see how I could do anything that Mitch hadn't already done. I thought that Mitch's work was good, and I didn't see what I could add to it. Number two, I told Bob that I believed what I was doing was worth doing, and I didn't feel like just walking away from it. I said, 'Bob, why are you always willing to drop what you're doing the minute you find something new? You dropped SV-40 for retroviruses, and now you want to drop retroviruses for transfection.' He told me, 'There are two kinds of scientists: the kind who dot their *i*'s and cross their *t*'s, and the kind who skim the cream off the top and move on. I'm the kind who likes to skim the cream and move on.'

"I still tried to argue. I said, 'Why can't you just add to what you're doing?' I said, 'Look at David Baltimore. He's still working on his thesis project. He has a whole group working on the polio virus. He never drops anything.' Bob said, 'That's not my *style*. I'm not David Baltimore.'"

Bob Weinberg was not and never wished to be David Baltimore, but he took one piece of Steffen's advice. He wouldn't force anybody to change direction; Bob never likes to strong-arm people. Instead, he'd find new blood. "When Chiaho Shih came to the lab," said Bob, "he was at my mercy."

On several occasions, Weinberg described this or that new person in the lab as being at his "mercy," and the oscillating pattern of his successes does seem to parallel the arrival of an unencumbered and impressionable graduate student or postdoc willing to toy with his ideas. Yet it's hard to imagine Chiaho Shih being at anybody's mercy. Before I met him, I'd consistently heard the opposite characterization: that Shih was severe, possessive, and often cantankerous. Graduate students normally complete their doctoral training under the aegis of a single senior scientist, but Chiaho Shih had transferred to Weinberg's lab after three years with Howard Green, a biologist then at MIT, because Shih and Green did not get along. Shih did not fare much better socially in the Weinberg lab. "He was one of the best people to come out of Bob's lab from a technical standpoint," said David Steffen. "But he fought all the time, with Bob and everyone else. He was very paranoid about his experiments. None of his collaborations ever seemed to work out." Shih had emigrated to the United States from Taiwan in the mid 1970s, and he experienced difficulty in mastering English. "That was very painful for everyone," said Cliff Tabin, a graduate student at the time. "At group meetings, Bob would talk to him in a very slow, precise way, as though Chiaho were retarded. He'd say, 'Now-Chiaho-do-you-understand-what-you-must-do-next?' Chiaho would grin and shake his head, but the rest of us knew that he was just playing dumb."

When I went to visit Chiaho Shih at the University of Pennsylvania, where he's now an associate professor, I found him to be more good-natured than his reputation had led me to expect; he seemed almost shy. Yet I could see that he had a quick temper and stores of resentment. At one point, recalling his months of difficulties at MIT, which he believed were the direct consequence of a postdoc's interference in his work, Chiaho turned hard and brittle, his whole body twisting in his chair, his limbs angling like a trapezoid, as though he were physically restraining his fury. When I asked Chiaho how he had found the environment in the Weinberg lab, he stared at my tape recorder and said, "Are you going to turn that thing off?" and brayed with laughter.

Chiaho Shih was far from retarded, and when he arrived at the Weinberg lab, he was not as immediately malleable as Bob had anticipated. Chiaho didn't like the looks of transfection any more than the rest of the Cancer Center group did. "At that time the whole lab was working on viruses," said Shih. "And so I really wanted to join the mainstream, because that transfection project seemed like a dead project. I was starting my fourth year of graduate study at the time, so I didn't want to waste time. I was really under pressure. I told Bob, Let me work on viruses, too, but he said, Well, give transfection a try. Even though it's a deserted project, maybe they missed something. He told me that if it didn't work, he would give me a fast project to get my thesis done. Finally, he convinced me. I just really had one choice."

Having no alternative, Chiaho approached his project in a manner that suited Bob well. He worked demonically hard, and his moroseness proved useful. "One of the things that I was very concerned about, during the early stages of the transfection experiments, was that word of it might leak out," said Weinberg. "With Chiaho, I didn't have to worry about telling him to keep mum. He is, by his nature, secretive and withdrawn."

Admittedly, Chiaho Shih didn't have much to leak for a while. His first few months on the experiment recapitulated Goldfarb's woes. Shih spun the eggdrop strands of DNA from a batch of chemically transformed mouse cells similar to the donor cells Mitch had used, although the cell line was created independently of Mitch's in case the first cells had been contaminated with viruses. He applied 75 micrograms of DNA (lubricated with calcium phosphate) to every one million NIH 3T3 cells, and he incubated the cells for two weeks for each transfection. He spent many hours peering through the microscope at control cells that matched the transfectants, focus for focus. "It wasn't working for me, and I was nervous," he said. "I asked Bob again for a different project, but he said, Let's try a little longer." Weinberg suggested that Chiaho obtain a new line of NIH 3T3 cells, on the off chance that the lab's cell supply had gone bad.

Chiaho ordered more mouse cells. The experiments continued to balk. But just as the clouds suddenly had broken for Mitch, Chiaho one day found a bristle of firm foci straining up from a dish of transfected cells. "The cells in the foci looked really cancerous," said Chiaho. "Now that it had happened with two independent cell lines, I began to believe it was true."

Chiaho began to believe it was true. The other postdocs and graduate students in the lab began to believe it was true. Bob Weinberg didn't believe it was true — he *knew* it was true. Transfection was real, and it was the most sensational thing that had ever happened to him. "Any doubts that people may have had, any complaints they may have made about the riskiness of transfection didn't matter two bits to me," said Weinberg. "I thought I was right." The lab needed more cells, more proof. The time had come to take transfection seriously. A new postdoctoral fellow, Ben-Zion Shilo, had joined the lab, and he too wanted to work on the transfection project. Orders were placed for a medley of tumors. Charles Heidelberger sent more cells treated with 3-methylcholanthrene; the National Institutes of Health shipped a line of mouse cells that had been transformed by the carcinogen DMBA; x-irradiated and ultraviolet-blasted cells arrived from the University of California; scientists at the University of Chicago freight-expressed live mice with throbbing, chemically induced fibrosarcomas, tumors of the muscle tissue. All told, the lab assembled fifteen kinds of cancer cells. Bob Weinberg was ready to start playing the petri dish equivalent of three-card monte. He'd be the dealer, and he'd play his hand with the lights out.

From Weinberg's perspective, three challenges had to be addressed in demonstrating the authenticity of transfection: that neither the recipient nor donor cells were corrupted by viral particles; that transfection worked for the DNA of more than one type of tumor cell; and that the transfection-induced foci were distinct from spontaneous clusters. The first issue could be resolved, at least roughly, by screening transfected cells for the presence of viruses, keeping a particular watch for stray retroviruses from the labs down the corridor. The second problem was simply a matter of mincing the tumor cells Weinberg had amassed.

To show that there was a real difference between cancerous foci and spontaneous clusters was harder. Scientists would demand rigid assurance that the Weinberg lab wasn't seeing what it wanted to see. Bob Weinberg decided that the surest way to discriminate between legitimate and felonious foci was to perform the experiments double-blind. Medical researchers and social scientists often do double-blind studies when they're testing new drugs or questionable psychological theories, but the method is almost unheard of in molecular biology.

The two blind men were Bob Weinberg and Chiaho Shih. Chiaho prepared DNA from the fifteen cell types, scissoring the material into segments and transfecting it onto skin cells. As controls, some NIH 3T3 cells were left untreated; others received doses of calcium phosphate but no foreign DNA. Chiaho marked each lid of the dishes with a color and number and placed the cells in an incubator. Deliberately unaware of Shih's coding system, Weinberg strolled into the lab midway through incubation and switched all the lids around, scrawling in a ledger what lid had gone where — blue 10 to yellow 14, for example. Bob was exacting about the ritual. He'd get furious if the wrong kind of felt-tip pen was used or if the slashes of color were drawn too thinly. "I'd never seen Bob get so wrapped up in something," said Ann Dannenberg, a technician at the time, "as he did with colored Magic Markers."

After the two weeks necessary for the cells to absorb foreign DNA, Chiaho hunched over a microscope and counted foci. He had no preconceptions, no reason to convince himself that he beheld heaps of cells when none was there. Comparing his tally with Weinberg's notations, Chiaho determined that transfection, finally, was reproducible and unassailable. The cell type that worked most reliably was still the Heidelberger 3-methylcholanthrene line. The DNA of the majority of the fifteen tumor varieties mulishly resisted transfer, but two types of the DMBA-transformed mouse cells engendered foci as well. Chiaho counted a total of sixty-one individual foci on the NIH 3T3 cells that had received tumor DNA. Most significant, he scored virtually no foci on the control cells.

Bob Weinberg was skimming somewhere above the cloud cover. Most studies of the molecular basis of cancer heretofore had centered on tumor retroviruses, flukes of nature that had little to do with commonplace cancer. The genes that Chiaho was now able to transfect into healthy cells were the real article: they were the target genes of chemical carcinogens and ultraviolet radiation, known cancer dangers in the environment. They were the genes that had suffered a mutagenic blow in the original cancer cells, and they had been so potently transformed that they could provoke cancer a second time around. The skeptics were wrong. Cancer was not, or at least was not always, the result merely of the death of genes or of the loss of crucial growth-controlling genes. In the cells that Chiaho Shih transfected, certain genes obviously had been mutated to a more aggressive form. Weinberg's bulldog persistence had paid off, and now he was absolutely convinced of the value of transfection.

"The whole field had been stigmatized by Spandidos. It was all viewed as being a lot of nonsense, and we had to make sure we were absolutely right before we talked about it," said Weinberg. "When I did the double-blind studies and Chiaho Shih came up with the answers, then I was sure, and I was ready to talk."

4 Pyrrhic Victories

WEINBERG announced his results at a Gordon Conference in New Hampshire on June 20, 1979. He may have expected clarion peals of exuberance, but what he heard was closer to a chorus of kazoos. "They were highly skeptical, highly dubious," said Weinberg. Oh, it was worse than skepticism. "People made fun of Bob," recalled Mariano Barbacid, who was then at the National Cancer Institute. "To most of them, it seemed like a very stupid experiment." Questions tumbled on the heels of criticisms. Some scientists didn't believe that the donor DNA was responsible for the transformation of the NIH 3T3 cells; they still thought the foci could have arisen spontaneously. Others wanted to know about all those tumors which hadn't yielded foci. Really, the only tumor types that worked were cells that had been treated with carcinogens in tissue culture. Was transfection limited to a tiny group of test tube cancers? And what about viral contamination? The Weinberg group had eliminated the possibility of interference by several types of retroviruses, but what about DNA viruses? What about the Epstein-Barr virus, which has been implicated in a human cancer of the nose and throat? What about herpes virus or the related cytomegalovirus? What about a species of virus that had never been described before? Any DNA pathogen, blown by the wind, lurking on a beaker, or resident in the donor cells could explain focus formation. Everybody in the audience was playing David Hume.

There was scant applause when Bob Weinberg had finished answering questions. So he returned to MIT to gather more sensational proof.

The most pressing item on the agenda, he told his researchers, was cloning the oncogene. The lab had to take cells from the foci and isolate the specific gene that had turned them cancerous. They must enfold the entire malignant segment in a cloning vessel such as a bacteriophage (a virus that infects a bacterium like *E. coli*) or a plasmid (a little circle of DNA and protein). They needed to show that the isolated segment, but no other piece of DNA from the focal cells, could transform a fresh plate of healthy NIH 3T3 cells. No scientist would argue with the evidence of a cloned gene.

Weinberg asked Benny Shilo to try isolating the oncogene of the 3-methylcholanthrene cells that Chiaho had so consistently used for his transfection experiments.

When Chiaho Shih heard that Benny was assigned to the cloning project, he was outraged. Chiaho insisted that he be allowed to clone the first oncogene. After all, he'd been the one to get transfection to work. Why shouldn't he be allowed to finish what he had begun?

No, said Weinberg. Let Benny clone the mouse gene while you continue working on transfections. Weinberg had something big and sexy in mind. He wanted Chiaho to start transfecting DNA from human tumor cells. Admit it, he said. That's the *really* important next step.

Weinberg promised Chiaho that if he was able to transfect human tumor DNA, he would be given first shot at cloning the human oncogene. Reluctantly, Chiaho Shih agreed.

"That was about the time," said Weinberg, "when my hair started turning gray." Two years later, Weinberg would be surprised to find a brown strand left on his head.

Benny Shilo knew from the outset that the task of cloning the mouse oncogene would be enormous. The donor cancer cells were mouse cells; the recipient normal cells were mouse cells. Benny Shilo was surrounded by a monochromatic sea of mouse DNA. What signpost would there be, while screening the DNA of the transfected NIH 3T3 cells, to alert him to the presence of one oncogenic mouse fragment out of thousands upon thousands of normal mouse genes?

Bob Weinberg suggested that Shilo attempt to tag the donor DNA with little snips of bacterial DNA. Before Shilo transfected the malignant material into the recipient cells, he should mix in *E. coli* DNA. Any mouse cells that had imported an oncogene should also have taken in a few pieces of bacterial DNA, which looks quite different from rodent genes. After transfection, Shilo could riffle through the DNA of tumorigenic foci for signs of *E. coli,* with the hope that the bacterial gene would be sitting right beside the foreign oncogene. Benny liked the idea. Such a technique could be used not only to clone this particular oncogene; Shilo and the Weinberg lab could flaunt it as a paradigmatic protocol for cloning any gene from any cell.

As Benny Shilo contended with cloning, Chiaho Shih continued his transfections. The lab could not immediately procure human tumor cell lines, so Shih passed the time by transfecting the DNA from a widening miscellany of other animal tumors. A proud moment came when he finally managed to break the species barrier. Slipping DNA from a rat neuroblastoma (brain tumor) into the mouse receptacles, he witnessed his most perfect foci ever. "They were tiny foci, but they were really, really good," he said. "They were very firm and rounded. For me, that was the turning point. Now we had a rat gene expressed in mouse cells, and best of all, those rat donor cells didn't come from Charlie Heidelberger. I was beginning to think there was something funny about Heidelberger's cells, since those were always the ones that worked best in transfection."

Chiaho was happier still when he successfully transfected DNA from mouse lung cancer cells. The other types of cancer cells that had produced foci were sarcoma cells. Sarcomas, malignancies of connective tissue, are the rarest of human cancers and thus of the least interest to general biology. Lung cancer is a carcinoma, a neoplasm of the epithelial tissue that covers the skin, makes up such glandular organs as the breast, and lines the lung and gastrointestinal tract. More than 85 percent of all human cancers are carcinomas.

With the species barrier demolished and the first carcinoma DNA transfected, Chiaho was confident that transfecting genes from important human tumor cells like lung and colon could be mastered. When, at last, Weinberg received several human malignancies from pathologists at nearby hospitals, Chiaho set to work trying to transfer genes from human colon, lung, and breast carcinomas to mouse cells.

But this time the species barrier appeared to be insurmountable. The human DNA would not yield foci.

Moreover, after months of scuffling with his mouse DNA and bacterial tags, Benny Shilo still had not cloned the rodent oncogene. In screening the entire DNA of the transfected cells, Shilo was able to detect ribbons of *E. coli* genes; the tagging trick seemed to work. Yet every time he tried inserting separate fragments of the mouse-plus-bacteria DNA into cloning viruses, the oncogene would somehow get lost in the process. He couldn't isolate the gene. None of the cloning viruses was able to transform cells, as they should have done if they'd contained a whole mouse oncogene. Benny didn't know why he couldn't clone the gene, and Weinberg didn't know why. Much later they would learn that the mouse oncogene was too big for the cloning virus that Benny relied on, but at the time all Benny knew was misery. "It was a terrible period for Benny," said Weinberg. "He deserved so much better."

At joint weekly group meetings of the Weinberg, Sharp, and Baltimore labs, people would recommend all sorts of things. Why aren't you doing *sib* selection? they asked. Why aren't you doing this or that? Bob would reply, We're trying to develop paradigmatic methods for getting the gene. We don't just want to clone the gene; we want to do it in an elegant way.

"The bottom line was that we had somehow convinced ourselves that this was the best way to do it, but it kept proving not to be," said Luis Parada, then a graduate student. "We all felt the frustration. It was infectious. There was an enormous amount of depression in the lab, and in my view that's because we were so unproductive."

Having few new data to report to the scientific community, Bob Weinberg talked repeatedly about transfections. He showed Chiaho Shih's results from the mouse lung tumors and the rat brain tumors. His audiences were not impressed. They didn't want to hear any more about clusters in dishes. They wanted to see an oncogene. Bob Weinberg had been working on transfections for a year or more. Where was the cloned gene?

"People were beginning to believe all over again that there was something wrong with transfection, since the genes could not be cloned," said Luis. "They were saying, you know, shit or get off the pot."

Bob Weinberg's hairs were graying quickly. He had, however, a trustworthy way of coping with stress. Whenever experiments in his lab stalled, he would retire to New Hampshire to work on his house. "There's nothing like sawing wood to put the world back in perspective," he said. His friends thought he was insane. Any other scientist caught in the swirl of the biggest project of his career would haul a cot into the lab and stay there until the work got done.

"I told him, Bob, I don't know why you're spending this time up in Rindge, building the house, wasting a lot of time away," said Tony Komaroff. "You should be spending every Saturday and most of Sunday in the lab. He said, 'I wouldn't be any more effective. I can only be there so much and then I start to blank out. I think I honestly have some of my best ideas when I'm up hammering nails.'"

While sawing and hammering and putting the world in perspective, Weinberg decided that Benny Shilo should drop the mouse gene–cloning project. A technician in the lab, who, along with Chiaho Shih, had been transfecting DNA from human colon cancer cells, thought she was beginning to see an occasional focus. The results weren't solid yet, but Weinberg hoped that transfection might yet succeed for human oncogenes. He told Benny to wait until the lab had a gene worth cloning.

Eight or ten months earlier, Bob had said the same thing to Chiaho Shih. "Chiaho hit the roof when he heard about what Bob told Benny," said Cliff Tabin. "Bob had forgotten that he'd promised Chiaho he could clone the first human gene, but Chiaho didn't know what a rotten memory Bob sometimes has. Chiaho's upbringing had taught him to revere authority and to trust people in authority. Now he felt that he'd been deceived."

Chiaho responded by marching over to Harvard University to visit an old friend of his, a fellow scientist from Taiwan. The friend gave Shih some human cancer cells that Chiaho could call his own. They had been taken long before from the bladder tumor of a man named Earl Jensen and were, accordingly, called EJ cells. Chiaho was determined to clone a human oncogene. He would have nothing more to do with the colon carcinoma cells. He'd transfect the Jensen tumor DNA, and if he managed to produce foci from the maneuver, he'd fish out the gene himself.

Chiaho Shih transferred the DNA from the human bladder cancer cells to the chromosomes of the mouse cells. In ten days' time, he detected foci. Black foci, fat foci, foci that weren't on his control dishes, and foci that were far superior to the occasional foci from the colon cancer DNA. No longer was Chiaho manipulating rodent genes from cells that had been artificially transformed in tissue culture. He was clipping and framing real human cancer genes, the sorry bequeathal of real cancer victims. Transfection had been real before, but now transfection talked. Chiaho proudly displayed his latest data to his boss. Weinberg returned to the scientific stage with his lab's most spectacular results yet.

The reaction was predictable: the audience asked to see the cloned gene. They demanded proof that the human EJ gene was indeed responsible for the remarkable appearance of foci.

Bob Weinberg wanted that gene. He was tired of making a fool of himself in public. He discussed with Chiaho the best way to clone the human bladder oncogene. There was no reason to think, said Weinberg, that isolating a human gene would be any easier than luring forth a mouse fragment. As far as Weinberg knew, Earl Jensen's bladder genes looked exactly like mouse skin genes. Chiaho had watched Benny Shilo flail around with the bacterial tagging procedure, and Chiaho didn't relish the thought of repeating his tribulations.

All was not lost. Bob Weinberg had neighbors. Next door at MIT's Cancer Center, David Housman and his postdoc, James Gusella, were searching for subtle variations in human DNA, or genetic polymorphisms, that might allow them to predict a person's susceptibility to inherited diseases. Their experiments entailed making somatic hybrids — crosses between human and rodent cells. Housman and Gusella had found that they could detect human genes in a background of rodent DNA by screening for the nucleic hallmark of human DNA: *Alu* sequences. Housman suggested to Weinberg that *Alu* sequences might be one way of wresting human cancer genes from a tangled skein of rodent chromosomes.

The klieg lights lit up in the Weinberg lab. The scientists reviewed a paper about *Alu* sequences. The peculiar sequences are repeated patterns of about 300 nucleotides found scattered throughout human DNA. The sequences don't inscribe protein; they're located in the introns, the junk parts of genes that are shorn away when a message of RNA is cut and pasted before its exodus from the nucleus. Researchers speculate that *Alu* sequences are the archaic remnants of viruses that jumped into the chromosomes of our protohuman ancestors, became permanently deactivated, and have been passed along from generation to generation ever since. The Weinberg scientists paid no heed to the origin of *Alu* sequences. Only one fact about the repeated bases caught their interest. There are so many thousands of *Alu* sequences distributed among the human chromosomes that the chances of a sequence lying within or adjacent to any single gene are excellent.

Weinberg's lab could exploit *Alu* sequences as a marker. In his aborted attempt to clone the mouse oncogene, Benny Shilo had tried to add artificial markers to his DNA. But *Alu* sequences aren't fake tags, imposed from the outside; they're an integral part of most human genes. The lab could divvy up the DNA from transfected cells into cloning viruses, or bacteriophages, and then pick through the individual viruses for signs of *Alu* sequences. Those cloning bacteriophages which held segments of uninteresting mouse DNA would lack *Alu* sequences. If Chiaho Shih found a phage that enclosed an *Alu* sequence, *and* if that phage was able, on its own, to turn healthy cells cancerous, Shih would know he had cloned the human oncogene. David Housman already had a radioactive probe designed to ferret out *Alu* sequences, and the Weinberg lab could begin cloning the human gene immediately.

To Weinberg, immediately meant yesterday. Transfection was not Bob Weinberg's private garden anymore. The Asilomar regulations had relaxed considerably, and Mitch Goldfarb, now a postdoctoral fellow with Mike Wigler at Cold Spring Harbor, was pursuing transfection as ardently as he'd once resisted the idea. Bob Weinberg felt betrayed by his old graduate student. "I was for a while upset that he'd go to a lab that was in direct competition with mine and transfer everything we learned over to them," said Weinberg. "It was obvious that Wigler was very interested in what we were doing and would start working on oncogenes. But I soon realized that I was being petty and small-minded." Besides, Weinberg couldn't afford the luxury of stewing. In the spring of 1981, Wigler's lab announced the successful transfection of transforming DNA from human bladder cancer cells — not Jensen's tumor cells, but another class of bladder cells stodgily dubbed T24. Through protocols known only to them, the Cold Spring Harbor scientists were scrambling to clone the oncogene they had detected through DNA transfection.

Not long after Chiaho Shih began succeeding with his transfections, Geoffrey Cooper of Harvard Medical School also managed to propagate foci reproducibly. Shih's initial transfection paper appeared in the November 1979 issue of the *Proceedings of the National Academy of Science*; five months later, Cooper's first oncogene transfection paper was published in *Nature*. Because *Nature* has a far larger circulation (and reputation) than *PNAS*, Cooper's work received broad attention. People now spoke of Cooper and Weinberg in the same breath as the twin fathers of oncogene transfection, and the linkage set Weinberg's teeth on edge. Cooper had transfected DNA from chicken leukemia cells, and Weinberg knew that Cooper was trying to clone the leukemia oncogene buried in the rodent foci. Later Weinberg heard that Cooper also had transfected DNA from human bladder carcinoma cells. Everybody in the business, it seemed, was studying bladder cancer.

So when Chiaho Shih said that he wanted to clone the bladder oncogene himself, Weinberg was dismayed. Chiaho argued that he had gotten the EJ cell line from his Taiwanese friend for just that reason. Weinberg pointed out that Benny Shilo had been working on cloning for a year and a half and that his skills would be of great benefit to the bladder oncogene project; Chiaho, after all, had no experience in molecular biology. Weinberg also wanted Mark Murray, another postdoc, to contribute to the cloning effort. We're in a race now, Weinberg emphasized, and we must do whatever it takes to win. Chiaho, Benny, and Mark must collaborate. Again, Chiaho Shih had no choice but to agree.

The collaboration lasted three weeks.

The trouble started when Benny Shilo suggested that they keep open the authorship of any future papers on the cloning of the oncogene, deferring decisions on the order of the three authors until they saw who had done what. Scientists are as touchy about the order of authors' names on a paper as actors are about the sequence of names in movie credits. As a general rule, the first author is the scientist who performed the bulk of the experiments, and the last author is the director of the lab in which the work was done. The authors in between — and there can be as many as fifteen or twenty on major scientific reports — may have contributed anything from hands-on experimental help to a few words of advice over the telephone. Postdocs and graduate students need first-authorship citations for their résumés, and nothing provokes a skirmish as surely as disputes about accreditation. Chiaho Shih didn't like the idea of keeping the authorship order open. As the first one to transfect DNA from the human bladder cells, he felt very proprietary about the EJ gene. "Mark and Benny were postdocs, and they were already close friends," said Chiaho. "I worked in a different bay, and I was at a disadvantage. I was a graduate student. I knew I didn't have a chance to be first author, but I didn't want to have done all that transfection work just to hand it over to them and act like a technician."

Instead of demanding a showdown, however, Chiaho simply felt resentful, and the sniping between bays quickly grew nasty. Benny Shilo had the benefit of his greater knowledge of molecular biology, and he was determined that the EJ project move as quickly as possible. When Chiaho Shih told him that he would not be in the lab for one weekend, Benny asked for his DNA samples so that Mark and he could continue working. Chiaho refused to give up his reagents, declaring that he had been planning to do that part of the experiment himself. Benny argued that it was ridiculous for Chiaho to slow the cloning because of his paranoia and possessiveness. Chiaho Shih hid the samples in his desk. At other times, when Mark or Benny asked Chiaho for something, he'd give them a biological compound similar to what was requested rather than the proper reagent; if confronted, Chiaho would plead lack of knowledge of English, and nobody knew whether to believe him or not. "You didn't want to be around those guys," said Cliff Tabin. "It was a totally unpleasant display."

After several acrimonious exchanges, Chiaho went into Bob's office to ask for his own project. "I told him that it wouldn't work out to have the collaboration," said Chiaho. "I said that it would make life in the lab very unhappy for as long as it took to clone the gene, maybe a year, who knows? I thought in the end we'd be slower this way. I wanted to incubate my own project. That's what I told him."

"What did Bob say?" I asked.

"I'm not sure he believed it. What does it matter to him who the first author of a paper is? The important thing for him is that the work gets done. The important thing is that his lab gets the paper. That's true of every lab in the world — in Taiwan, everywhere.

"He told me he saw my point of view," Chiaho went on, "but that he thought things would move quicker if I cooperated."

Bob Weinberg would not have to make a decision about what to do with Chiaho. Several weeks into the problematic collaboration, a technician found that the DNA from a human colon cancer line reproducibly produced foci in NIH 3T3 cells. Benny Shilo and Mark Murray decided to chuck the EJ project in favor of cloning the gene from the colon cells. Colon cancer is a far more prevalent human malignancy than bladder cancer, and the cloning of a colon cancer gene would have a greater scientific impact. Benny and Mark stopped coming around to Chiaho's workbench to ask about EJ DNA.

Thus was launched an intralaboratory competition to be the first to clone a human oncogene: Mark Murray and Benny Shilo on one side; Chiaho Shih, in his preferred solitude, on the other. Bob Weinberg was not happy with the arrangement, but he saw that he had no choice. He couldn't force people to collaborate against their will, and he couldn't tell one researcher that he should bow to the prerogatives of another. "I have always tried to prevent people in my lab from competing with one another," said Bob Weinberg. "But good intentions are not always sufficient."

• • •

With all the glib talk about the revolution in biotechnology, it would be reasonable to assume that cloning a gene is as difficult as finding new subatomic particles or mapping distant galaxies. In essence, though, gene cloning is simple; undergraduates today learn how to isolate genes in second-year biology courses. The contending teams of Shilo-Murray and Chiaho believed that cloning the first human oncogene would be easy as well. After all, they had everything they needed. They had a population of cells that they knew contained the oncogene. They had a probe to tweeze out the appropriate gene — a labeled *Alu* sequence probe. They had a monomaniacal drive to beat Cooper, Wigler, and each other. What they didn't yet have was the Weinbergian magic. They didn't have any luck. Everything that could go wrong went worse.

The Weinberg scientists knew that the key to cloning a gene was the comprehensiveness of their bacteriophage libraries. A phage library is a staple of molecular biology, containing every tome, or gene, from the cell that is being screened; it's a genetic Library of Congress. If Chiaho Shih wanted to find the EJ gene, and Benny and Mark wanted to isolate the colon gene amidst the mouse genes of the NIH 3T3 cells, they needed to make a library of all the genes in the DNA of the transfected cells. Only then would they have the genes arrayed neatly enough, and in sufficient quantity, to have something worth reading.

To build their expansive libraries of genes, the sparring researchers followed just such a strategy. They isolated DNA from the malignant mouse cells on their dishes and shredded the DNA with restriction enzymes — all-purpose enzymes that slice the DNA at defined points along the double helix. Using a different set of enzymes that weld pieces of DNA together, they snapped the tens of thousands of individual segments into the bellies of phages, the viruses that infect bacteria. They then allowed the machinated phages to infiltrate dishes of *E. coli,* where the viruses dependably multiplied in their bacterial hosts. Each time the viruses multiplied, they replicated the enclosed pieces of mouse, EJ, or colon DNA as well as their own genes. Eventually, the swarms of new phages got so big that they burst through the bacterial cells, creating a series of clear spots, or plaques, on the plates. Every one of those plaques, the Weinberg scientists knew, was a unique family, populated by the offspring of the original viruses that infected the bacteria, and every offspring was chubby with one or another fragment of the mouse or human DNA. The original fragments of DNA, in other words, had been copied many millions of times over, which made the scientists' analyses of the individual genes that much easier. The researchers only hoped that the entire collection of plaques on their dishes of bacteria represented the complete set of genes from the NIH 3T3 cells — including the human oncogenes.

Their hopes were in vain. The phage libraries they created were not Libraries of Congress. They were branches of the New York Public Library, and the most important volumes were missing.

By this time, Benny Shilo was so accustomed to disappointment that he merely sighed and tried again. He knew all the things that could have gone wrong. He and Mark Murray may have used the wrong restriction enzyme; they may have cut the oncogene in half, rather than snipping it on either side. They built another library of focal DNA, this time throwing in a different cocktail of restriction enzymes at the beginning.

Still nothing. The collected phages didn't have a complete oncogene buried in their stacks. The viruses seemed to have pieces of the colon gene: by using their *Alu* probe, Benny and Murray were able to detect the signature of human DNA. But pieces of a gene were little better than no gene at all.

Well, then, perhaps the experimental conditions had been wrong. If the enzyme reactions didn't all work in perfect harmony, the oncogenic segment of DNA may never have found its way into the torso of a phage. A little too much salt or a few too many free-floating hydrogen ions in the solution, and the oncogene could have been left behind in the test tube. Benny and Mark tinkered with all the variables they could think of. They even tried making libraries by throwing in identifiable pieces of bacterial DNA, just in case the human colon oncogene didn't have any *Alu* sequences that would make it stand out from the background of mouse DNA. The Shilo-Murray phage libraries were so good that other biologists borrowed their plaques to seek mouse genes unrelated to tumorigenesis, but for Benny and Mark the pages were blank.

Two or three times a month they wandered into Weinberg's office and groaned that there was nothing in the libraries. Bob would assure them that it wasn't their fault. Something was going on that none of them understood, he said, but he was confident that they would work things out. He was lying. "I had never been so consistently frustrated in my career," he said. "I was certain that our lab would end up as a footnote in the history of science."

Across the corridor, Chiaho Shih was also pounding out asterisks. Chiaho had never done any molecular biology, and he hadn't even reached the point in the construction of his phage library where he could twiddle with variables, but he was determined to figure out everything for himself. His own cloning project skidding, Benny began discussing Shih's experiments with Weinberg. Together Bob and Benny came up with an idea for a way that Chiaho might clone the EJ gene. Bob told Chiaho what he had to do to design the best library. You don't know how big the bladder oncogene is, he said, so you have to follow a procedure that will lessen your chances of cutting the gene in half. Bob explained to Chiaho which restriction enzyme he should use: MBO-1; and he warned him to use the enzyme sparingly enough to cut the DNA only every so often rather than into tiny pieces.

Chiaho thought that Benny Shilo was interfering again. Chiaho told Bob that *he* had thought of the best way to make a library, and it wasn't the way Bob and Benny recommended. He wanted to use a different restriction enzyme and different cleaving conditions. "The thing I want to say," Chiaho told me, his body twisting with anger, "is that I was very upset. I was forced to do something that I thought was the wrong way. Bob wouldn't listen to me. He only listened to Benny." Chiaho drew a diagram to show me the difference between his library idea and that of Weinberg. After he'd finished his sketch, he continued to tap the paper with his pencil and finally made a slash through the entire drawing.

"I want you to know that I tried those libraries for six months, and that I was depressed the whole time. It was too hard, and I thought it was too stupid.

"So finally I bluffed Bob. I told him that I heard Cooper and Wigler almost had cloned the gene. He said, Okay, do whatever you want, just get the oncogene. Then I used my approach, and I got the gene" — he snapped his fingers — "just like that. It took me less than three months." In December of 1981, Chiaho Shih found a virus in his library of NIH 3T3 DNA phages that held the entire EJ gene. When Chiaho transfected the isolated phage back into normal mouse cells, he seeded foci, demonstrating that the complete cancer gene was indeed nested within the span of the virus. The human oncogene had been cloned.

Bob apologized to Chiaho. "He admitted that he had been wrong," said Chiaho. "But I didn't accept his apology."

"Didn't accept it?" I asked.

"No, I was still angry. I didn't say anything. I just stood there, and then I walked away. I was very immature and foolish about it. He didn't have to apologize. It's very rare for a senior scientist to apologize to a graduate student."

"By all rights, Benny Shilo should have gotten the first clone," said Luis Parada. "He's the person who began to work on it, and he worked very hard. But in many ways it was poetic justice that Chiaho got there first. He was the one who helped the field break loose." Benny Shilo never did clone the colon carcinoma oncogene. As it turned out, the colon carcinoma oncogene is huge, extending for over 40,000 bases of DNA, and a gene that size could never have been fitted into a bacteriophage. Benny Shilo returned to his native Israel vaguely dissatisfied with his postdoctoral fellowship at MIT.

Yet Chiaho Shih's great victory was a parochial triumph. He beat two of his lab mates, but Benny and Mark were the only people he trounced. "The whole cloning issue was academic," said Weinberg. "That was one horse race we lost."

To this day, Chiaho Shih refuses to believe it, but he was *not* the first scientist to isolate a human oncogene. Six weeks to two months before he had the EJ gene safe within the boundaries of a bacteriophage, Mike Wigler, Mitch Goldfarb, and their colleague Kenji Shimizu cloned a human oncogene from their T24 bladder tumor cells. Rather than screen for *Alu* sequences, they had tagged the human DNA with bacterial sequences, but their bladder cancer gene was far more tractable than Benny Shilo's long-sought colon cancer gene: a bacterial fragment clung close to the oncogene right through to the packaging of the DNA into individual viruses, permitting the Cold Spring Harbor scientists to pull out the gene from a phage library with a bacterial probe. "Mike was very lucky to have Kenji around," said Carmen Birchmeier, a scientist in Wigler's lab. "Kenji was probably one of the only scientists alive who could have cloned the gene using that bacterial tag technique. Anybody else would have been crazy to try it."

"It *can't* be that they had the gene first," said Chiaho. "Bob announced the cloning at a Gordon Conference, and Manuel Perucho from Wigler's lab was there. If Wigler's lab had the gene, too, Perucho could have said that at the conference. People do that all the time. But he didn't. He made no comment at all."

"If you're talking about who physically had the gene in their hands first," said Weinberg, "Wigler's lab was more than a month ahead of us."

I asked Mike Wigler whether it was true that his lab was the first to clone a human oncogene.

"Were we?" he replied. "Who said that?"

"Bob told me," I said.

"How does he know?" asked Mike.

"Maybe one of our postdocs told one of his postdocs," said Daniel Broek, a Wigler postdoc.

Regardless of who physically had the gene in hand, Mike Wigler cloned the gene first according to the record that counts. His paper was published first. He submitted it to *Nature* on December 31, 1981, and it appeared in the April Fool's Day issue of the following year.

Chiaho was not even the *second* scientist to clone a human oncogene. Three or four weeks after Wigler had the bladder fragment, Mariano Barbacid reported the isolation of the same T24 oncogene.

Mariano Barbacid? said everyone in the field. Who's he? Weinberg and Wigler had barely heard of the man, and they certainly didn't know he was in the running to clone an oncogene. "It's as if he'd crawled out of the woodwork," said Luis Parada.

Mariano Barbacid, a Madrid-born scientist, may have crawled out of the woodwork, but like a termite, he had not been idle in his anonymous cranny. At the National Cancer Institute, he'd been working on transfections as long as Weinberg had, and he'd started out with human tumor DNA, not with mouse genes. Barbacid, however, deliberately kept quiet about his experiments. "When I started getting foci, I was skeptical of my own results," said Barbacid. "That's why I didn't believe Bob's first paper — because I didn't believe what I was seeing myself.

"I didn't have the nerve to tell anybody what I was trying to do. They were laughing at Bob, and I thought they would laugh at me, too. Bob had enough of a reputation that he could afford to take some heat, but who was I? I was a nobody. I was an outsider, a guy from Madrid."

Barbacid eventually came to believe in his results, and, on hearing reports of *Alu* sequences, decided to try cloning the human oncogene by the grace of an *Alu* probe. Neither he nor the sole postdoc working with him had ever cloned a gene, yet they managed to pull out the T24 bladder oncogene within a matter of eight months. He announced his results at a conference by tacking his data on to a makeshift billboard during a poster session. "People would come over and read the poster," he said, "and they'd ask me if I worked with Weinberg. I said no, I worked with myself. Then they would say, 'How could you do that?' Even now, my friends who know me don't give me credit for cloning the gene two weeks before Bob did.

Only Weinberg and Wigler are always careful to reference me."

Within two months of one another, three labs had cloned two distinct and completely new human oncogenes. They had planted little flags on unclaimed wilderness. Or so they believed. A month after Chiaho Shih secured his EJ phage, Luis Parada would disclose to the Weinberg lab what neither Bob nor anybody else in the field wanted to hear. All the cloning heartaches had been for nothing. The human cancer genes were not new genes, but the same old genes they'd been working with for years. They were the genes that Weinberg had tried so desperately to put behind him.

Luis Parada is very athletic. That was the first thing I noticed about him, and he was one of the first people in the lab I noticed. He barreled down the corridors wearing white shorts and a tight T-shirt, and I saw that he had muscular legs and a broad, muscular chest. He played squash a lot and rode his bicycle a lot and sometimes made up for his sporting afternoons by working late into the night. Luis is from Colombia, and his eyes, hair, and trim mustache are all the color of Colombian espresso beans. Before he went into molecular biology, Luis was a stage actor in a small Bogotá troupe. Perhaps it was his training as an actor that taught him to speak, even in casual conversation, as though to an audience much larger than one person. And perhaps it was his father, a general in the Colombian army, who instilled in Luis an extraordinary degree of self-confidence. Luis's self-confidence is as broad as his chest, but he has a right to it: Luis Parada was one of the most successful graduate students to emerge from Bob Weinberg's lab.

"In a short amount of time, my projects have worked," he told me. "I don't know what to say. I think it's luck. People tell me it wasn't. They say you make your luck, and you wouldn't have so many successes if it were luck. But I think I was in the right place at the right time."

Luis came to the right place when he entered the Weinberg lab in 1980. He was put on a project that nobody thought was of pressing importance at the time. Luis's job was to provide the negative control that Weinberg, and most of the people working for him, believed would only confirm the obvious. It was the sort of project that postdocs shunned because they didn't think they could skim publication-worthy data from it. It was a project that made Luis Parada, a first-year graduate student, a very famous young scientist.

According to the thinking of the time, there were two kinds of oncogenes. First, there were the oncogenes that retroviruses had filched from their hosts. By 1980 scientists had identified some fourteen retroviral oncogenes and their cellular counterparts. About half a dozen of those genes had been cloned and analyzed in detail.

Then there were the oncogenes that had been detected in human and other animal tumors through transfection, among them Earl Jensen's gene, the colon carcinoma gene, a mouse lung tumor gene, a rat brain tumor gene, a chicken lymphoma gene, and the genes from the cells that had been chemically transformed in petri dishes. Luis arrived at MIT in the middle of the race to clone one of those transfected oncogenes. As a former retrovirologist, Weinberg knew that many of his colleagues were wondering whether the transfected oncogenes weren't the same genes that others had found in the sticky fingers of retroviruses. It was certainly possible that the gene which had been mutated badly enough to help ignite Jensen's bladder cancer was a human version of the gene that had been accidentally lifted by Peyton Rous's chicken virus or any of the other known retroviruses. No one could offer a logical reason to explain why the oncogene that had been passed into NIH 3T3 cells couldn't be the human *src* gene or the human copy of the *ras* gene that was found in the Harvey sarcoma retrovirus.

No logical reason, yet Weinberg refused to take the notion seriously. "I knew it was a control we had to do," he said, "but I didn't give the control my highest priority."

Month after month, Weinberg postponed comparing retroviral DNA with transfected DNA, although the experiment would have been far easier to manage than the effort to clone the transfected oncogene. He came up with all kinds of excuses for his scientists not to bother. For one thing, he reasoned, there were only those fourteen genes that had been found in retroviruses. Human DNA has at least 50,000 genes. Surely the chances were tiny that the gene which went amok in Earl Jensen's bladder cells was one of those few genes known to turn cancerous when plundered by viruses. "You'd think that with so many genes available," said Howard Temin, "there would be more than fourteen genes that had the potential to contribute to cancer."

For another thing, in 1980 no human retroviruses had yet been found. Our cells have indigenous versions of the same genes embezzled by animal tumor viruses, but viruses didn't seem able to pick up the genes from one human host, infect another human host, and give the second person cancer. If retroviruses were incapable of sparking human cancer, that could mean the genes regulating human cell growth are somehow different from, or at least more complicated than, the growth genes in other animals that are subject to virally induced tumors. Perhaps only in chickens and related species, speculated Weinberg, does the *src* gene modulate cell division. Perhaps in humans other genes control cell growth — genes such as the ones his lab had disclosed through transfection.

The biggest, if not the most persuasive, reason that Weinberg didn't believe his transfected oncogenes were the same as retroviral oncogenes is that he didn't *want* them to be the same. He didn't want to find any so-called homology between the two classes of cancer genes. "Bob had left the field of retroviruses because he was sick of the competition," said Luis. "He wanted to make a career for himself."

"The last thing I wanted to think about once I'd finally gotten out of retroviruses," said Weinberg, "was retroviruses."

"That's why the project was given to me," said Luis. "The people in the lab didn't think there would be any homology, but they knew the control had to be done. So it seemed like a good way for a new graduate student to get his feet wet. I thought it was great. I kept saying, 'Why not? Why shouldn't they be the same?'"

Luis got more than his feet wet; he was in over his head. If the genes in the transfected cells were related to viral oncogenes, Luis had little idea of how to prove it. For an experienced postdoc like Benny Shilo or David Steffen, the project would have been almost trivial. Luis Parada, however, was an experimental neophyte. Chiaho Shih hadn't known any molecular biology when he began cloning the EJ gene, but at least he'd worked as a cell biologist for three years, studying different components of the cytoplasm. Luis debuted in the Weinberg lab with a baccalaureate in theater and basic biology. And what with the pressure to clone the human oncogene before another lab did, and the tensions between Shih and Shilo, there was nobody around willing to take Luis under his or her wing. "I was left on my own, and I wasn't very directed," said Luis. "I was easily distracted by other things that were going on. My main project was checking for homology, but if I got bored or frustrated, I'd try to clone other oncogenes or just mess around."

Luis didn't need an isolated copy of the transfected oncogene to test for gene similarities. He had at his disposal cloned versions of several human cellular analogues of retroviral oncogenes, including the *src* gene, the *ras* gene, and a gene called *myc* (rhymes with "sick"), an oncogene originally sighted in a chicken leukemia virus. All he had to do was to perform a Southern blot, one of the Romper Room exercises in molecular biology. The technique is among the quicker and cruder ways to screen for the presence of genes in a sample of DNA. Luis purified the complete repertoire of genes from mouse cells that had been transfected with human tumor DNA. After slivering the DNA with restriction enzymes, he dribbled the gene solution onto a slab of agarose gel; this separated the pieces according to size. Luis then applied radioactive copies of the retroviral oncogenes to the gel and blotted the gel with a piece of specially treated nitrocellulose paper. If any of the viral probes had the same structure as the rough-cut genes from the transfected cells, the nucleotides of the probes would stick to the nucleotides of the mouse DNA during the transfer of the genetic material from gel to paper. Luis could detect the nucleic marriage by taking an autoradiograph of the completed Southern blot. If the radioactive probe had clung to a gene in the sample of DNA, it would have left behind a black smear of gamma rays.

With an almost methodical determination, Luis Parada fumbled every step of the blotting. He had difficulty purifying DNA from the mouse cells. When he ran a gel — pouring liquid agarose into a mold before adding the fractionated DNA — the gel often ran right out the bottom of the plastic frame. He couldn't get the hybridization conditions right: the reaction that would boost the likelihood of the probe DNA sticking to its homologue in the transfection DNA. He didn't know how much radioactive compound to add to his probe DNA. He was flummoxed.

As the months passed without any firm data from Luis, Bob Weinberg became increasingly nervous. He needed the control results to include in several papers being written at the lab. He checked in with Luis almost daily. If he was out of town, he'd call and ask for Luis. Luis would sigh and say that he was doing the best he could.

"The poor guy had so many technical problems, but no more problems than you'd expect from any first-year graduate student," said Cliff Tabin, one of Luis's closest friends. "Ordinarily a new person at the lab would never have been given such an important project, and when Luis first came nobody thought it *was* important. But after a while, Bob started thinking of it as more and more of a big deal. He wanted that negative control, and Luis couldn't give it to him."

"No doubt about it," said Luis. "I was screwing up. And the more I screwed up, the more anxious I became, and the more I screwed up."

Eventually, Luis managed to tame his Southern blots, and he began to see that there was no apparent analogy between the transfected genes and the retroviral oncogenes. Only once did he find a wisp of a potential mating on an autoradiograph. He thought there could be a kinship between a segment of DNA in the transfected mouse cells and the *ras* gene from the Harvey sarcoma virus (Mitch Goldfarb's old specialty), but he didn't tell anyone of his tentative result other than Cliff Tabin.

"At that point he was so unsure of himself," said Cliff, "that he didn't believe his own work."

Because Luis's results were overwhelmingly negative, Weinberg began telling his colleagues that the two classes of oncogenes did not seem to be related. It was the result that Weinberg had wanted, and he was glad that the preliminary data were in his favor. "Bob sounded fairly confident that there was no homology," said Mike Wigler, "so we didn't even bother looking ourselves." Bob Weinberg, however, had spoken too soon.

"I have many preconceptions, and my preconceptions almost always make sense," said Bob Weinberg. "But the fact is that most of them are wrong."

Chiaho Shih cloned the EJ oncogene in January. With an isolated gene now available, Luis Parada decided to repeat the homology searches. This time, the experiment was easy enough for the greenest of students. Cloning had eliminated the interference of background mouse DNA, and all Luis had to do was compare a single pristine strip of human DNA with his three or four retroviral probes.

By February, he had an unequivocal result. In the middle of his autoradiograph glowed an ink-black smear, the sign that one of his radioactive probes had met its match. Luis's lonely, tentative result of six months earlier had been correct. Jensen's bladder oncogene was none other than the *ras* gene, the tumorigenic portion of the Harvey sarcoma virus. Like the Rous sarcoma and other cancer viruses, the Harvey pathogen had become malignant by plundering a copy of a cellular gene — the *ras* gene. The Harvey virus had picked up its *ras* gene from rats, but our cells, too, contain copies of the normal, proto-oncogenic *ras* gene; the healthy *ras* gene is indispensable to every organ of the body. And, as Luis discovered, *ras* is also a hot spot for cancer. Of the 50,000 genes in Jensen's bladder cells, it was the *ras* gene that had been the target of an unknown mutagen, just as it was *ras* that had been pocketed (and mutated) by a rodent retrovirus.

All those months of blindly swapping and coloring petri dishes, of confronting doubtful colleagues, of constructing vast gene libraries, of watching brown hair turn silver or disappear altogether, had culminated in the rediscovery of a viral cancer gene.

"My first reaction," said Weinberg, "was disappointment."

"Bob was upset. Everyone in the lab was upset — except me," said Luis, grinning so broadly that his mustache looked as though it were about to take off. "I was overjoyed at having found the homology. It was my first big discovery. But I had to keep my gloating to myself."

The Weinberg scientists were justified in their despair. They had recently convinced themselves that they were breaking new ground in cancer biology, and they thought they had to share their field with only a scattering of investigators. In light of Luis's discovery, however, the Weinberg lab knew that every retrovirologist in the United States and abroad would converge on human oncogenes. Whenever scientists sniff the possibility that their work relates to human cancer rather than just to rat or chicken cancer, they pounce. Human cancer is where the grants are, and cancer research gets noticed. Earl Jensen's bladder carcinoma gene — that is, the *ras* gene — was about to become the most fashionable gene in molecular biology. A scientist wouldn't even have to be a passable transfector to study the oncogene; he or she could approach it through the experimental graces of the Harvey sarcoma virus.

Most upset of all was Chiaho Shih. He'd sweated and sulked and squabbled, only to clone a gene that Ed Scolnick of the National Cancer Institute had cloned long before. Chiaho grumbled that if Luis hadn't been so slow, he would have been saved months of grief.

More sulking and surprises were to follow. Luis compared the Harvey *ras* gene with the bladder oncogene that Mike Wigler and Mariano Barbacid had cloned from the T24 cells, and found that those two genes were also analogous. Stated somewhat differently, the T24 gene was the same as the EJ gene — and each was the *ras* gene. Wigler's triumph in the cloning marathon wasn't much of a victory after all. "I called Mitch Goldfarb and said, 'Mitch, you know that T24 gene you guys cloned?'" Weinberg said. "'Well, it's the same Harvey virus gene you worked on for your Ph.D. thesis. There's nothing new under the sun.'" Nor was there anything proprietary under the sun. At the same time that Luis discovered the similarity between *ras* and the bladder carcinoma segment, other scientists reported detecting the genetic filiation. Luis's paper was published back to back with a similar article from Douglas Lowy's lab at the National Cancer Institute in the June 10, 1982, issue of *Nature,* and Geoffrey Cooper's group at Harvard Medical School published a more comprehensive report on the homology between *ras* and human oncogenes in the June issue of the *Proceedings of the National Academy of Sciences.*

Once the transfection artisans had accepted the threat of stepped-up competition, they realized that the homology between their genes and viral genes was, in fact, a scientific bonanza. Now they didn't have to bother with messy biochemical means to find the protein encoded by the oncogene: Ed Scolnick had already done that for them during the ten years that he'd worked on the *ras* gene. Scolnick knew the size of the *ras* protein: on a molecular scale, it weighs 21,000 daltons, the equivalent of 21,000 hydrogen atoms. He and others knew the protein's idiosyncracies: it has a little chain of fat on one end. *Ras* researchers even knew where in the cell the protein lingers: the fat cap of the protein is buried in the soft inner layer of the cell membrane, and the body of the protein dangles into the cytoplasm. "Really, the connection between EJ and *ras* was of benefit to us all," said Weinberg. "It pushed the field ahead by several years."

There were many things, however, that Scolnick didn't know about *ras*. He didn't know what was amiss with the oncogene. He hadn't determined the difference between the normal, cellular *ras* gene and the mutant, cancerous *ras* gene. The most salient question about the *ras* oncogene was how the protein that it produced caused cancer. Before scientists could address that mammoth problem, they needed to pinpoint the defects in the gene.

The opening events were barely over, and already the jockeys were headed for the Kentucky Derby. "Finding the mutation in the *ras* gene," said Weinberg. "Now *there* was a horse race."

5 No Reason to Stay in Science

THE RACE to identify the lesion in the *ras* gene became a grand event quite rare in basic research; MGM or Columbia Pictures might have turned it into a film, if Hollywood wasn't so thoroughly convinced that scientists are a group of unphotogenic maladapts. The race had a defined goal: to find, for the first time in history, the molecular origin of human cancer. The rules for the race were clear. On the one hand, there was the oncogenic *ras* gene, which initiated tumors; on the other, the healthy cellular *ras* gene, which didn't. Somewhere along the DNA of the malignant *ras* gene lurked a mutation that had turned it cancerous. All the scientists had to do was compare parts of the healthy gene with parts of the oncogene until they pinpointed the error. Three contestants participated in the race — enough for complexity and intrigue, but not enough to confuse the audience. There were unexpected twists, blighted hopes, and tragedy. And, in the best tradition of Hollywood, the most thrilling element of the race was the astonishing prize at the end.

But let's start a few steps back. When Bob Weinberg first thought about finding the mutation in the human bladder oncogene, he wasn't sure just how close his competitors were to being in a position to look for the same thing. At the time nobody knew that Weinberg, Wigler, and Barbacid were all working on identical oncogenes. Nobody knew that the oncogene from Earl Jensen's bladder carcinoma and the oncogene from the T24 bladder cancer cells were the same genes, and that the gene had been identified years earlier in the claws of the Harvey sarcoma rat tumor virus. In that brief, benighted era before the discovery of the consanguinity between human and retroviral oncogenes, Bob Weinberg felt confident that *his* lab was in control of the EJ gene.

The group already had one important piece of information about the bladder oncogene mutation: it was fairly small. Right after Chiaho Shih cloned the EJ gene, he looked for gross discrepancies between his oncogene and the normal proto-oncogene. He clipped his cloned cancer gene with restriction enzymes and then clipped the normal gene with the same set of enzymes. Restriction enzymes do their shearing by clipping at defined landmarks along the gene; each enzyme recognizes a different pattern of nucleotides. If the oncogene was completely different from the healthy cellular gene — if large pieces of the gene had been juggled around during mutagenesis, or if hundreds of nucleotides had been lost or changed — then the restriction enzymes that snipped the proto-oncogene would no longer slice up the cancer gene into the same number of pieces. But Chiaho Shih found that the enzymes cleaved the DNA of the two genes in a similar manner, which meant that the majority of the nucleotides had remained the same, from placidity to malignancy. The EJ bladder oncogene had not been totally debauched. Some subtler DNA mutation lay behind its tumorigenicity. Chiaho Shih received his Ph.D. and left the Weinberg lab soon afterward, disappointed that he hadn't been the one to find the lesion in his gene.

In early 1982, Randy Chipperfield, a first-year graduate student, arrived at the lab. When Bob asked him what he wanted to do, Randy said that he'd like to take over Chiaho Shih's project and look for the oncogene mutation. Chipperfield was obviously bright and ambitious, so Weinberg said, sure, he could do it. Randy was, of course, new to molecular biology, but Bob was not yet in a terrible rush to get the project done.

Then came the desolate day in February when Luis Parada discovered the connection between the EJ oncogene and the retroviral *ras* gene. After Weinberg had emerged from his gloom and recognized the homology as the scientific boon it was, he decided to collaborate with Ed Scolnick, the king of *ras* and an old friend of Weinberg's. Scolnick had some biological tools that Weinberg wanted: a working copy of the normal human *ras* gene and antibodies that could discriminate between the *ras* protein and the tens of thousands of other proteins in human cells. Weinberg called Scolnick, who was moving from the National Cancer Institute in

Bethesda, Maryland, to Merck Laboratories in West Point, Pennsylvania, and asked for a copy of the normal *ras* gene and *ras* antibodies.

Scolnick replied that he'd be happy to provide Weinberg with the antibodies, but as for the normal clone, Esther Chang, who'd isolated it, had left the lab and taken the clone with her.

Weinberg called Esther Chang at her new home in another corridor of the National Cancer Institute. You give us the clone, Weinberg said to Chang, and I'll agree to make you last author on the paper reporting the discovery of the mutation. The offer was too good to pass up: Chang sent the clone. Weinberg and Chang further agreed that, as independent confirmation, she would look for the *ras* mutation in parallel with the Weinberg lab.

Luis Parada had not yet compared the T24 bladder oncogene studied by Wigler and Barbacid with the EJ oncogene of the Weinberg lab, so Weinberg wasn't aware that he was in a race. The project to find the *ras* mutation remained Randy Chipperfield's, and now Randy had almost everything he needed to begin the experiments. The only thing he didn't have was experience. Weinberg turned to Cliff Tabin, who had the reputation of being one of the lab's most talented DNA manipulators and, more important, was a nice guy. Weinberg asked Cliff to take Randy by the hand.

"He knew I was the sort of person who wouldn't mind spending forty percent of my time helping the guy out," said Cliff. "He said, Don't worry, don't worry, I guarantee you you'll be an author on the paper. I told him, Don't be ridiculous. I don't care about authorship. If I help Randy do some cloning, that's not exactly a big intellectual contribution to the project. Forget about authorship."

Cliff Tabin really didn't mind spending most of his time assisting Randy. His own projects weren't going anywhere, and when he'd been a new graduate student, the postdoc David Steffen had mollycoddled him. Helping Randy was to be one of the luckiest things that happened to Cliff's career, but one that would later bring him deep sorrow.

Clifford Tabin is delightfully handsome in a husky, masculine, linebacker sort of way. Like his friend Luis, Cliff is a generic athlete. He plays rugby, soccer, squash, baseball, and football; he used to bring a football into the lab and, between experiments, would throw it around the bays. "People would yell at me to stop," he said, "but I never broke any equipment." Cliff can eat more food than the rest of the Weinberg team combined. When I took him out for lunch, he ordered a salad — and returned to the salad bar twice to pile up a dinner plate with potato salad, egg salad, Jell-O, vegetables, thick slices of bread smeared with butter. He ordered a large bowl of soup, and then he ordered a triple-decker sandwich of various meats and cheeses. Robert Finkelstein, another Weinberg graduate student, told me that he's seen Cliff eat six hamburgers and three hot dogs at a single laboratory barbecue.

Cliff has boundless stores of nervous energy. He'll talk and talk, about evolution, about politics, about grand scientific models, about Weinberg: "I'm a Bob Weinberg type of scientist," he said. Cliff is as famous for talking as he is for eating. "Once you get him going, you realize he doesn't have an 'off' button," said a roommate of his. All I had to do to set him talking was ask him a single question, perhaps periodically throwing in a grunt to let him know I followed him. I was surprised by his talkativeness, because David Steffen had told me how much Cliff disliked reporters. When journalists began crowding into the Weinberg lab and disrupting everybody's work to report on those wondrous new things called oncogenes, Cliff would do anything to sabotage their stories. If he knew television crews would be filming the lab in action, he'd show up in what he thought were his most outrageous clothes: jeans that were patched from crotch to cuff, fluorescent-green shoes, punk wraparound sunglasses. Reporters would ask him questions, and he would fabricate answers. "They'd want to know things like what does all this mean in terms of a cure for cancer?" said Cliff. "I'd start babbling about different types of cancer and when different ones would be cured. I said, Well, we cloned a gene from a bladder tumor" — he stopped to gesture at the area of his midriff just above the organ in question — "so it was likely that cancers of the intestinal area, the midsection, would be cured first.

"I kind of felt sorry for them. They were just trying to do their job, and they wasted hours of footage on me."

Cliff Tabin loves sports and conversation but has no taste for doing experiments, so he wanted to turn Randy into a self-sufficient researcher as soon as possible. He and Randy discussed what they had to do to find the mutation in the *ras* gene. There were two parts to the experiment. One part had a pleasant, interior decorator ring to it: mix and match. Cliff and Randy had to take pieces of the normal *ras* gene, link them together with pieces of the transforming gene, and search for hybrid clones that would, on transfection, produce foci in cultured cells. By matching ever larger segments of the proto-oncogene to ever tinier sections of the oncogene, they could narrow down the region of the oncogene that contained the guilty mutation. The *ras* gene is 6600 nucleotides long — a lot of matches to mix.

The second thing that had to be done, said Cliff, was sequencing, or spelling out, small regions of the *ras* gene. Once they knew approximately which region of the oncogene housed the lesion, they had to look at every nucleotide of the segment to find the bases that had been altered during tumorigenesis. That part of the project I can't help you with, said Cliff, because I've never sequenced anything in my life. To learn DNA sequencing, Randy approached Michael Paskind, the Cancer Center's resident sequencing ace ("and a damned good shortstop," said Cliff).

All at once, Randy Chipperfield had a slurry of new techniques to learn. He had to learn transfection, he had to learn cloning, and he had to learn sequencing. Cliff volunteered to start the mixing and matching while Randy set up the sequencing system; once Randy had the sequencing under control, Cliff would hand the cloning over to him. Cliff also agreed to take care of Bob Weinberg, who was beginning to be somewhat of an irritation.

Not a single mutation experiment had been done yet, but Bob Weinberg *knew* what was wrong with the *ras* gene. He had a model in mind, and his model was all he wanted to discuss. At group meetings, he'd leap over to the blackboard and draw messy diagrams predicting the results of Randy Chipperfield's project. He'd draw a line representing the *ras* gene and arrows pointing to the location of the lesion. The arrows were clustered around a single section of the gene: the promoter.

There was really only one logical explanation for the *ras* mutation, said Weinberg. Chiaho had shown that there was no major disruption of the *ras* DNA to explain the gene's tumorigenicity; hence, the mutation must have damaged a critical node of the gene. And what's the most critical node of any gene? he asked rhetorically. The promoter, the sequence of DNA that controls the gene's expression. A mutagen had damaged the switch of the gene, and as a result the *ras* gene never turned off. RNA transcripts kept rolling off the mutant *ras* template. *Ras* protein continued peeling off the messenger RNA. Earl Jensen's bladder cell bulged and squirmed with so much *ras* protein, and so many growth signals from the excess of the protein, that it had no choice but to divide uncontrollably.

No doubt about it, said Weinberg, a promoter mutation lay behind the power of the *ras* oncogene.

Cliff Tabin and other scientists in the lab protested vigorously.

There was another type of mutation, they said, that could explain *ras*'s malfeasance: a structural mutation. There could be a mutation in the protein-coding sequence that follows the promoter of the gene. A handful of chemical bases that determine the amino acid subunits of the *ras* protein could have been changed, resulting in the wrong amino acids being stacked together. If a few inappropriate amino acids were inserted into key positions, they pointed out, the protein would end up with a crooked spine or with knobs, kinks, and side chains that it's not supposed to have. That mutant protein would then behave in an unacceptable manner, either stimulating cellular enzymes that it should have nothing to do with, or mistreating the enzymes that *ras* is designated to tweak.

So there were two possibilities, said Weinberg's subordinates: a promoter mutation that spawned the right *ras* protein at the wrong time, or a structural mutation that synthesized the wrong *ras* protein at the right time. And both possibilities, they declared, were equally likely.

Nonsense, said Weinberg. A promoter mutation was far more likely. Proteins usually don't mind if a couple of amino acids are substituted for a couple of others, but they *do* mind if their on-off button doesn't work.

"He had an absolute prejudice about it because of his background in retroviruses," said Cliff Tabin. "If you put a viral promoter in front of an oncogene, you get tumors. So by analogy, the cellular promoter of a human oncogene must be behaving like a retroviral promoter. And Bob wasn't alone in his conviction about the *ras* gene. If you'd taken a survey of everyone in the field, all would have said yes, the mutation must be regulatory. The only people who would have said 'structural' would have been the people who knew about Bob's belief and wanted to go against it.

"But Bob would be so dogmatic and condescending about it that you couldn't help kidding him."

"It was the same tune at every group meeting," said Luis. "Bob would talk about how we have to understand the promoter mutation of the *ras* gene. And we'd all chime in, 'Or the *structural* mutation.' It was an ongoing joke."

Weary of the dogma, Cliff Tabin thought of a little scheme to prove Weinberg right or wrong even before Randy Chipperfield got his project under control. If the mutation was in the promoter of the gene, then cells transfected with the oncogenic *ras* gene should produce much more *ras* protein than cells treated with the normal *ras* gene. Weinberg told Cliff Tabin that Scott Bradley, a medical student, was coming to work in the lab for the spring and summer, and Weinberg wanted Cliff to come up with something to keep him occupied. Cliff suggested that Scott could compare the amount of protein synthesized by the cancerous *ras* gene with the amount spewed out by the normal gene. That experiment, said Cliff, would demonstrate immediately whether the promoter theory was correct.

"Bob said yeah, yeah, that would show it," said Cliff. "But he wasn't interested. He thought the answer would be trivially obvious. So, on the sly, I had Scott go ahead and try it anyway."

Within a couple of weeks of his arrival, Scott Bradley reported the results of his experiment to Cliff. Bob Weinberg had predicted that the cancerous *ras* gene would pump out many times more protein than the normal gene, but Scott had found otherwise. The protein levels, he said, were the same. Well, that settles it, Cliff replied. If the mutation was in the promoter of the *ras* gene, he thought, it wore an awfully convincing fake nose and glasses. "God, that was fun," he said. "It's always great when you can prove somebody in authority to be wrong."

Cliff and Scott displayed the protein data to Weinberg. "He still didn't believe it," said Cliff. "He wanted us to repeat the experiments, to do all sorts of controls. Oh, his arguments went on and on, and his points were all legitimate. But I knew how good Scott was. He was brilliant, and very gifted technically. Once I saw his protein data, I was sure the mutation was structural."

Another graduate student, Cori Bargmann, repeated the protein measurements, using a different procedure, and got the same results. Bob Weinberg was finally convinced. The *ras* mutation was not in the promoter of the gene.

"I'm entitled to my preconceptions," said Weinberg. "But the important thing about preconceptions is not to let them blind you. When data prove me wrong, I shut up." Weinberg shut up about promoter mutations. Cliff and Randy Chipperfield would mix and match, swap and stitch, to localize a structural mutation.

Yet now that he had a clue to the sort of lesion he should be seeking, Randy couldn't find it. He was having difficulty learning how to sequence. He couldn't purify single strands of DNA; his sequencing gels weren't coming out right — nothing was coming out right. As Cliff began the easy parts of the mix-and-match experiments, he hoped that Randy would look over his shoulder to see what was involved. Randy didn't look over his shoulder, because he was spending more and more time agonizing about the sequencing. Randy felt completely incompetent and unhappy. He fell in love with a woman who worked at the lab, but she rejected his advances. Randy's depression crippled him.

Bob Weinberg learned that Mike Wigler and Mariano Barbacid were racing to find the mutation in the T24 bladder oncogene. By that time, everybody knew that the T24 gene was the same as the EJ gene and that both were the human *ras* gene. The pistol had been fired, the Weinberg lab was in the middle of a horse race, and Randy Chipperfield was close to collapse.

"He was too depressed to work," said Cliff. "He wanted to do the mix-and-matches, but he knew that the whole thing had shifted into high gear. He said, 'Forget it; you go ahead and do it all.' I argued with him for a while, but I could see that he wasn't very happy.

"He felt that he had two choices, to stay around and feel altogether useless or to take a couple of months off. He decided that taking a vacation would be less hard on his ego. Right before he left, he told me, 'If we lose this race, it will have been my fault.'

"One morning I came to lab and heard that Randy was gone."

Randy Chipperfield's abrupt departure took Weinberg by surprise. He hadn't known just how upset Randy was, and he hadn't known when or even if Randy was planning to take his extended leave of absence. Bob called an emergency meeting of the people working on the mutation project. Cliff was to assume the entire mix-and-match effort, he said, and the sequencing would have to be farmed out. Weinberg asked Ravi Dhar, in Doug Lowy's lab at NCI (where Esther Chang supposedly was duplicating the mixing and matching), to take on the sequencing. "Bob decided that sequencing was mysterious and impossible and that it could never be done by our lab," said Cliff. "So now we were involved in this whole Boston-to-Bethesda political circus. Boy, was I sorry about that. I had named the normal *ras* clone EC, after Esther Chang, but if I'd known what a hassle it would be to work with her, I'd have named it after my grandmother."

It was early summer of 1982 when Cliff began constructing clones that were a hybrid between the normal and cancerous *ras* genes. "I worked my ass off, fourteen hours a day, seven days a week," he said, "and I hated every minute of it." Using restriction enzymes to cut the DNA, and ligation enzymes to sew it back up again, he mated the right half of the oncogene with the left half of the proto-oncogene, and the left half of the oncogene with the right half of the proto-oncogene. Those unions revealed that the mutation lay in the left side of the oncogene. The clone that had the left half of the oncogenic *ras* gene was cancerous. Cliff had narrowed his search to something around 3000 nucleotides.

He tacked the first 1500 bases of the oncogene onto the final 4100 nucleotides of the normal gene. That diptych clone transformed cultured cells into malignant cells.

He reduced his search to 1000 bases. Esther Chang began reporting results that varied from Cliff's. *Her* hybrid clones that were the equivalent of *his* clones did not transform cells, as he claimed they did. Cliff politely informed her that she didn't know what she was doing, and he told Weinberg to ignore her data.

Now the cutting and pasting of normal gene to transforming gene was getting slippery. As he scaled back the size of the oncogenic portion of the clone, he had less room on the transforming DNA to carve into with enzymes. He had to order rare and expensive restriction enzymes, and the laboratory supply bill spiraled skyward. Sometimes Cliff had to cut two pieces of DNA, join them, and slit them open again to slip in a third fragment. There was no guarantee that any of the pieces would match one another in the proper order or in the proper direction, or even that they would find one another at all. During the rejoining reactions, a stretch of normal DNA could as easily get hooked up with another piece of normal DNA, even though it was supposed to connect to the cancerous counterpart. Each time Cliff thought he had a working clone, he had to stop and test it by dropping it into mouse cells and searching for foci. If no foci materialized, he threw the clone out.

He continued to slim down his search. He tailored a clone composed of 6000 bases of normal *ras* and only 600 of transforming *ras*. He transfected the hybrid gene into cells; cancer clusters bloomed.

Threads of normal DNA and wisps of cancerous *ras* were flying in test tubes. The kinetics of the cloning reactions approached the absurd. There was no chance that the pieces would all fall into place.

In Bethesda at NCI, Ravi Dhar was sequencing segments of those genes which Cliff suggested were most likely to be the region of the mutation, concentrating on the 600-base area that Cliff had determined was actively oncogenic when stitched into a complete *ras* clone. Of those 600 bases, a little fewer than half lay in the promoter region of the gene. Cliff already knew that the mutation was not a promoter error, so he told Dhar to examine the sequence of the 350 nucleotides lying within the protein-coding region of the gene. In early August, Dhar flew up to Boston, his sequencing data in hand.

I've got a very weird result, he told the Weinberg scientists. In this 350-base segment, there's only one difference between the normal and the oncogenic *ras* genes. I'm finding it hard to believe the data, he said, but only a single nucleotide appears to have changed. He pulled out his sequencing films: each *g*, *c*, *t*, and *a* of the gene had left behind a little black slash of radioactivity on the x-ray, creating a ladderlike design. He pointed to three of the fuzzy rungs of the ladder. This is the twelfth codon of the normal *ras*, he said, referring to the triplet of nucleotides that dictates the twelfth amino acid brick of the *ras* protein. You see how this reads *g*, *g*, *c*?

Weinberg and Cliff nodded. They had to take Dhar's word for it; they didn't know the first thing about reading sequencing films.

Dhar pulled out another autoradiograph. Here's the sequence of the transforming *ras*, he said. Now take a look at the twelfth codon of this gene. It reads *g*, *t*, *c*. The guanine of the normal *ras* has been replaced by a thymine. As a result, said Dhar, the twelfth codon of the *ras* gene now inscribes the production of a valine amino acid instead of a glycine. All the other 349 nucleotides, said Dhar, seem to be identical in both genes.

Weinberg had expected that the lesion would encompass a much greater area of the oncogene: ten, twenty, fifty bases. Ravi Dhar's data suggested that the chemical difference between a healthy, compliant gene and a marauding cancer gene was as small as a difference could be: one nucleotide out of 6600. It was a point mutation.

Cliff Tabin and Scott Bradley eagerly watched Weinberg's face to see his reaction to this incredible bit of news. Bob Weinberg has often said that the hallmark of his work is "simplicity," but the simplicity of the *ras* lesion was too much even for him. Weinberg didn't believe it. He refused to believe it. To learn that the mutation was structural rather than regulatory had defied logic. But to say that the structural flaw was a point mutation, a mutation of a single nucleotide, defied fantasy. Something was obviously wrong, he declared, either with Ravi's sequences or with our clones. If a point mutation was the basis of *ras* activation, he told Cliff, there was no reason to stay in science.

"He was being dogmatic again," said Cliff. "So I had to take the other side. Why *shouldn't* it be a point mutation?" Cliff knew that there was a biological precedent for point mutations in genes exerting grave effects. Sickle cell anemia, for example, is the result of a single nucleotide change in the gene for hemoglobin. Cliff believed that point mutations could explain the genesis of cancer as well.

Cliff returned to his mixing and matching to whittle down the oncogenic segment further. He was determined to show that a clone of 6250 bases of normal *ras* and 350 bases of cancerous *ras* — the 350 bases where Ravi had spotted a change in the twelfth amino acid —was fully tumorigenic.

Ravi Dhar called back to say that the point mutation wasn't there after all. He was resequencing the entire territory.

Cliff paid no attention to Ravi's latest result. His final clone was proving to be the most difficult of all. He was trying to weld four separate fragments of DNA, all in the right order and orientation. No gene cloner had ever been able to link that number of scraps together into an operative whole. For twenty days, none of the enzyme reactions worked. Weinberg retreated to his house in Rindge with his friend Tony Komaroff. He told Tony that he was afraid he would lose the race; this time, his relegation to history's annotations was assured.

In mid August, Cliff thought he had produced an active clone that was one part oncogene, nineteen parts proto-oncogene. Ten days later, the magic foci appeared on his plates of mouse cells. "One of the reasons why I have more of a reputation than I deserve," said Cliff, "is that people look at me and say, He's the guy who got four-part ligations to work."

The same night that Cliff was sure his unwieldy construct was cancerous, Ravi Dhar called Weinberg at home. He'd resequenced the whole 350-base region, using two separate sequencing techniques. On every film, there was a point mutation at the twelfth codon: glycine had become valine. Dhar was certain that a single nucleotide change was behind the power of the *ras* oncogene.

Bob Weinberg admitted that his preconceptions about point mutations were wrong. His lab had just made one of the most astounding discoveries in the history of cancer research. Not only had the researchers found the first example of a molecular mutation that led to cancer; they had discovered that the lesion was smaller than anybody would have dared imagine. A point mutation was indeed the basis of *ras* activation. Far from leaving science, as he'd threatened to do, however, Weinberg wanted to celebrate. He stayed up late. He bought champagne. He tried calling Cliff Tabin at the lab. Cliff wasn't there. Assuming Cliff was himself out celebrating with friends, Bob called Legal Seafoods, a restaurant near MIT. "It was a typically bizarre Bob Weinberg–type of thing to do," said Cliff. "For hours he was having me paged at Legal's.

"He finally reached me at home around midnight. I was just walking in the door, and the phone was ringing. It was Bob, saying, 'I'm glad I finally got you. Get Scott Bradley and come over. We'll have champagne.' My initial reaction was: Very nice, very cute, enjoy your champagne, I'm going to bed. In my book, the only times you celebrate with champagne are when you have a conceptual insight that turns out to be right or for the World Series. The mix-and-match stuff had been very straightforward and very boring to do. An exceptionally good technician could have done the same thing. I was sort of pleased that I'd overcome some technical difficulties, but I didn't feel any champagne-style triumph. But then I talked to Scott and he said, 'We really shouldn't pass this one up.' So we went over to Bob's. Some of Bob's friends were there, and they were just getting ready to break out the champagne.

"For Bob, it was a supreme moment of success, a culmination of five years of work. He believed in cellular oncogenes, he transfected them, he cloned them, and now he had found the mutation. We sat around, and Bob reminisced about how he had had some of his initial ideas. He talked about his walk across the Longfellow Bridge. He talked about his life, his philosophies. I was really glad to see him so happy. It was Bob's champagne moment."

After the private celebration came public exultation. The news was announced first at an American Cancer Society meeting in Buffalo, New York, and the scientific community was stunned. The conference buzzed like a symphony of chain saws with talk of the point mutation. One amino acid change in a protein of 190 amino acids? How could such a meager structural alteration have such a profound consequence? In human beings, most cancers take twenty, thirty years to develop. If it was so easy to activate an oncogene, the scientists mused, we should all have died of tumors in early childhood. The editors of *Nature* declared that the point mutation result was "one of the most startling discoveries so far in the long and frustrating search for an understanding of cancer." Other molecular biologists agreed. "We didn't yet know what the relevance of the mutation was to real human oncogenesis," said William Hayward of Memorial Sloan-Kettering Research Center in New York. "But Bob's work proved the power of a molecular biological approach to cancer research."

"What do you say when you first hear something like that?" said

Michael Bishop. "Your first impulse is caution, but you can't help being a little dazed by it."

"I think for many of us who had been working in oncogenes since the *src* days, the point mutation work was a turning point," said Arthur Levinson, a molecular biologist at Genentech in South San Francisco. "Now we could start saying, What's going on with these proteins that presumably contribute to human tumors? How does the mutation at the twelfth codon affect the ability of the protein to get to where it's supposed to go and do what it's supposed to do once it gets there? Those were the questions we wanted to answer, and the point mutation work gave us our first handle. It was a beautiful piece of work."

The tide of excitement swept up all in its path. Several days after the Buffalo conference, Ravi Dhar presented the sequence data on the *ras* gene to scientists at NIH. Who should be there but Mariano Barbacid, the termite from the National Cancer Institute? Barbacid stood up during the talk and said that the researchers in his lab were pretty sure they'd found the same point mutation at the twelfth codon position but that they were still cleaning up their results. Barbacid had lost the race. Three weeks later, however, he secured the proof and dashed off his paper on the mutation so quickly that, eventually, *Nature* published the Weinberg and Barbacid reports back to back. Nevertheless, the scientific community was alerted to who had truly won the race by the dates printed at the end of each article: Weinberg's paper was submitted on September 13, 1982, said *Nature;* Barbacid's on September 28. "We found the mutation first," said Weinberg, "and I'm very, very proud to be able to say that."

Not every member of the Weinberg platoon could share in the festive chauvinism, though. Randy Chipperfield returned from his two-month respite to find the lab in an uproar, the point mutation paper already in press, and Cliff Tabin the newest celebrity on the floor. Cliff tried not to play the peacock. Although he and Randy weren't close friends, Cliff could tell that Randy felt miserable. "All the fantasies he'd had about success when he'd started the sequencing experiments had come true for somebody else," said Cliff. "He was a graduate student. I was a graduate student. He knew that he could have been the star if he hadn't been depressed. So now he blamed himself not only for the troubles he'd had with sequencing, but for the way he'd reacted to those troubles." And when he first came back to the lab, Randy blamed Cliff because his name wasn't on the paper. Randy thought that Cliff, the first author on the paper, held a grudge against him for leaving at a critical moment and hadn't wanted Randy to be a co-author. Cliff explained to him what had happened.

"I didn't have any say in the matter," said Cliff. "We knew that Bob was drawing up a list of which authors were going to be where. We debated among ourselves for days about who was going to make it. Would Scott make it? He was only a summer student. Would Cori make it? She'd done one little protein measurement. Would Randy make it? He'd left the project early on, but I still felt that he'd contributed a lot to my early thinking about which sequences to look at, which restriction enzymes to use.

"Then I got a second draft of the paper back from Bob, and it had the list of authors. Scott was second, Cori was third. Esther Chang was the last author, just as Weinberg had promised, but that wasn't going to fool anybody out there. Bob played all these political games with NIH. Alex Papageorge was listed because he took some *ras* antibody out of Ed Scolnick's freezer and sent it to us. Scolnick was listed because he was Scolnick. Doug Lowy was listed because Ravi and Esther were in his lab."

But Randy's name was nowhere to be seen. Cliff went into Bob's office and demanded to know why Chipperfield had been excluded.

Did he contribute any data to the paper? asked Weinberg.

No, he didn't, Cliff admitted. But he did contribute ideas to the mix-and-matches.

Ideas are cheap, said Weinberg. If you didn't use any of his data, I don't think it would be appropriate to list him.

"I think he'd really been stung when Randy took off so suddenly," said Cliff. "And I had to agree that people do have ideas all the time. What counts is the data. But I'd have felt much less uneasy if he'd just put Randy's name on.

"At least Randy stopped holding it against me when I told him I'd lobbied for him all the way."

The energy in Weinberg's lab was headily distracting, but Bob wondered what had happened to the third jockey in the race. Where was Mike Wigler? Why hadn't Weinberg heard from him?

Mike Wigler hadn't disappeared — not by any means. The people in his lab were being very, very careful. They too had found the point mutation in position twelve, but they hadn't been satisfied that the valine change was the only mutation. They'd had trouble reading some of their sequencing data, not because they lacked the expertise

— they didn't — but because parts of the DNA molecule of the *ras* gene were so tightly coiled that the individual nucleotides smeared together on the x-ray films. Wigler, who prides himself on his scientific caution, was not satisfied with his lab's sequencing results. He had his researchers check, double-check, sequence, and resequence. When he was still not happy with the data, he asked his researchers to perform additional, highly detailed mix-and-matches. That operation generated a disturbing result: there turned out to be a second mutation in the promoter region of the *ras* oncogene.

The Wigler lab decided to figure out exactly what was going on. "In being careful, we fell behind," said Wigler. "That's always the tradeoff you have to make in science: caution versus being first." The Cold Spring Harbor scientists finally decided that the promoter mutation had no significance. By itself, the lesion was not cancerous, whereas the twelfth codon mutation was. Nevertheless, the promoter mutation was there. And when Wigler read the Weinberg paper in *Nature,* he saw that the MIT scientists had missed the second mutation. Wigler called Weinberg and told him that he thought his lab had made an error, and he spelled out the reasons.

Wigler could practically hear Weinberg blanching over the phone. Oh, my God, Weinberg replied. Nothing like this has ever happened to me before. He assured his colleague that he'd report the mistake to *Nature.*

Oh, forget it, Wigler said. This field has already suffered enough blows to its credibility. It's not an important mutation anyway. Just drop it.

After he hung up the phone, Weinberg huddled with Cliff Tabin to decide whether they should air their dirty laundry. They called Ravi Dhar and asked him to take another look at his sequencing gels. Dhar confessed that the promoter region was a little smeared and that it was difficult to discriminate unequivocally between *g*'s and *c*'s, *t*'s and *a*'s. Cliff and Scott decided to do one more restriction enzyme dissection of the DNA to see whether they could sort out the difficulties in the promoter region. To their mortification, they found a site in the front area of the oncogene that could be sliced by an enzyme — a site they hadn't noticed earlier. Wigler was right, they told Weinberg. Our published sequence on the cancerous *ras* gene was off by one nucleotide.

To confuse matters further, however, the Weinberg scientists found that their normal clone differed from Wigler's normal clone by one DNA base. Human genes sometimes have slight differences from one person to another, a phenomenon known as polymorphism. Tabin and Bradley ascribed the discrepancy between their normal gene and Wigler's to a polymorphism. "When all was said and done, we'd screwed up on the sequence of the transforming gene, but we got the normal gene right because of that polymorphism," Cliff explained. "I guess we were lucky that we weren't bogged down by the confusion in the promoter sequence during our initial experiments. It might have slowed us, just as it did Wigler." In any case, the Weinberg staff had found early on that a promoter mutation could not explain the power of the *ras* oncogene. They knew that the cancerous *ras* gene did not propagate any more protein than did the normal copy of *ras,* which had told them that the lesion had to be in the structure of the protein.

Because the promoter mutation was insignificant and had no effect on the tumorigenicity of the gene, the matter was dropped.

Wigler, for all his generosity in telling Weinberg not to bother with a public confession, still faults the Weinberg team for its precipitancy and carelessness. Wigler had lost the race because he was cautious; Weinberg's researchers had won because they'd been lucky enough not to notice an insignificant — but real — promoter problem that might have slowed them down. "They made a mistake," said Wigler. "They were rushing to get their data out first, and they didn't take the time that we did. The fact is that if Weinberg's lab had done the sequencing themselves, it would have taken them five years to do it. So of course they didn't know enough to be critical of the results of their collaborators.

"Bob will be pissed off at me for saying this, but I'm still somewhat bitter about the whole thing. This business about racing to beat the competition really annoys me sometimes. I don't see why he couldn't have consulted with me before rushing off to publish. We could have pooled our efforts and published jointly."

I expressed skepticism that such an idealistic proposal would win many followers. "Does that happen?" I asked. "I mean, do scientists who have been competing with each other change their minds and decide to cooperate?"

"Sure it happens," he said. "It happens all the time."

I asked him if he could give me an example.

"Well, it's never happened to *me,*" he said. "But I'm sure if I think about it I could come up with cases where it has."

Whether through sheer luck or an innate knack for ignoring irrelevant complications, the MIT group won the race, and that victory mattered profoundly to Weinberg. "In my mind," he said, "since we developed the system over the years, from 'seventy-eight to 'eighty-one, we deserved, in 'eighty-two, to bring it to its natural fruition, which was to find the activating lesion." The victory mattered to him because his lab had identified the first specific DNA mutation linked to human cancer. It mattered because the discovery of the point mutation was a clear and distinct apogee, on a par with the detection of the cellular *src* gene.

That's not to say the discovery of the mutation was Weinberg's greatest achievement. By the general assessment of the scientific community, Weinberg's most important overall contribution to cancer research was oncogene transfection. The ability to transfer cancer genes from tumor cells to normal cells essentially inaugurated the study of human oncogenes. It enabled scientists to begin specifying the molecular targets of chemical carcinogens. Without the proof that human cancer genes could be transferred from one tissue type to another, where they initiate a fresh outbreak of cell blight, conservative cancer researchers might have doubted that oncogenes found in animal tumor viruses had much relevance to human malignancies. By introducing oncogene transfection, Weinberg helped bridge the gulf between the abstractions of the viral oncogenes and the stark immediacy of a lethal disease. "Before transfection, Bob was a competent but unremarkable scientist," said Phil Sharp. "Oncogene transfer was *the* experiment that made him a superb scientist."

"Transfection was a very good idea," said Salvador Luria. "And it was to Bob's credit that he got such a good idea to work. The most outstanding thing about Bob is that he can make things *work*." Yet the beauty of transfection took years to prove. Only after the bladder oncogene had been cloned were scientists satisfied that transfection was more than hocus-pocus or an experimental byproduct of little biological import.

By comparison, the Weinberg lab slapped together the point mutation result in one summer, and the significance of the discovery was immediately vivid. The Weinberg scientists had discovered the molecular difference between a benign gene and a cancer gene. No longer were they saying, We have a gene that seems to be involved in carcinogenesis. They were saying, Here's the lesion. This is the origin of the problem. This is where the story of a cancer gene begins.

Scientists were now in a position to understand how specific DNA mutations introduce cancer. They could begin mapping the few fatal passages in the labyrinth of a tumor cell. In the fall of 1982, John Cairns and Jonothan Logan, of the Harvard School of Public Health, and among the sagest observers of cancer research, wrote in *Nature* of the discovery: "This year may one day be seen as the year when a strong sense of order came to cancer research and the long drive to understand the cancer cell really got under way . . . much that was impossibly obscure suddenly seems plain." Cairns and Logan predicted that "the technology for studying genes and their patterns of expression are now advancing so fast that the molecular biologist will soon be telling the [cancer] epidemiologist what to look for."

With the detection of the point mutation, the science of oncogenes became the most exalted specialty of cancer research. In June 1982, Frank Rauscher, senior vice president of research at the American Cancer Society, told the *New York Times* that he couldn't "think of anything more important right now" than oncogenes. "People have thought that with 100 different cancers we may have to deal with 1000 different things that can trigger them," he said. "Now that we are getting at the molecular basis of the turn-on [such as the *ras* mutation], maybe we can talk about far fewer than that." Money followed on the heels of praise. Between 1982 and 1983, the American Cancer Society doubled its research grants in oncogene work to $16 million. By 1983, the National Cancer Institute was disbursing $46 million to human oncogene research projects around the nation; in 1980, it gave virtually nothing.

As scientists converged on *ras,* they naturally sought to understand what the oncogene and its point mutation were all about. Many groups recapitulated the Weinberg-type experiments from beginning to end, transfecting tumor DNA into mouse cells and screening for cancer genes. The cumulative efforts revealed a striking pattern. Almost every time scientists scratched through the foci on their plates of mouse cells and pulled out human cancer genes, the genes turned out to be *ras* genes. Whether they began with the DNA of breast tumors, colon tumors, lung tumors, pancreatic tumors, ovarian tumors, or skin tumors, the one oncogene they could transfer from the malignancies into receptacle cells was *ras*. They didn't find the human version of the *src* gene through transfection. They didn't find the *fos* gene or the *abl* gene or any of the other cancer genes that had originally been descried in the embrace of tumor viruses. Again and again, the transfection assay served up versions of the human *ras* gene. For unknown reasons, no other gene in the tumor cells seemed capable of fomenting a fresh round of cancerous growth when removed from the original neoplasm. Every tale of pathological intrigue came down to *ras*.

More startling still, when scientists sequenced the *ras* genes isolated from tumor DNA, they repeatedly found the same mutation that Weinberg, Barbacid, and Wigler had first reported for Jensen's bladder oncogene. In cancers of the colon, pancreas, breast, and other organs, the twelfth codon of the *ras* gene had been changed from the incantation for glycine to the nucleic curse for valine. There were exceptions; sometimes the thirteenth or the sixty-first codon of the gene had been mutated rather than the twelfth. But often, the one-base switch at position twelve explained the tumorigenicity of *ras*.

The story was by no means a seamless narrative, however. The *ras* oncogene was baffling in its inconsistency. Although scientists kept coming up with *ras* oncogenes after they'd transferred tumor DNA into normal cells, the great majority of malignancies did not yield a *ras* oncogene through the transfection test. Instead, the neoplastic DNA yielded no oncogenes at all; the tumor genes simply disappeared into the background of mouse DNA, failing to foster the growth of foci and thus leaving scientists with no method for pinching out individual cancer genes. Nor could scientists attribute any particular kind of cancer to a mutation in the *ras* gene. Breast tumor cells from one patient might harbor a rampaging *ras* oncogene that showed up in transfection, but the same kind of breast tumor from another patient appeared to have a normal, proto-oncogenic version of *ras*.

Pooled together, the early data indicated that in 15 to 20 percent of all human tumors, *ras* is mutated, often at the twelfth codon of the gene. In other words, one out of five or six colon cancer patients had a mutant *ras* gene in his tumor cells; one out of five victims of lung cancer had a mutant *ras* in his lung tumor cells; and so forth.

Scientists had learned enough to realize that they had a lot to learn. They had to account for those 80 percent of tumors in which they could not find oncogenes through transfection. Obviously the DNA of the cells harbored mutant genes; otherwise, the cells wouldn't have turned malignant in the first place. But scientists knew that they needed better screening protocols to ferret out the cancer genes from those tumors that resisted analysis through transfection.

Scientists also had to determine when in the evolution of cancer the *ras* gene degenerated from a healthy, vital proto-oncogene into a twisted, dastardly oncogene. The only thing transfection told them about *ras* was that in some cancers the gene had been reconfigured into a more sinister format. It didn't say whether the mutation had occurred at the beginning, middle, or toward the end of the gestation of the tumor. In light of the relatively high percentage and diversity of tumors that tested positive for the *ras* oncogene, many biologists believed that the mutation must be an early and central event in carcinogenesis. Mariano Barbacid performed a series of elegant and oft-cited experiments demonstrating that, under carefully controlled conditions, 85 percent of lab rats injected with a mutagenic chemical contracted mammary cancer; and in every one of the rat breast tumors Barbacid examined, the *ras* gene had been mutated at codon twelve. "The fact that you always see the same thing," said Barbacid, "is evidence that a mutation of *ras* is responsible for the tumor." But scientists wanted *human* proof. They needed to know the moment of the devastating transition from glycine to valine in human cancer. Only then could they say at which point in cancer treatment doctors might be able to thwart a malignancy by attacking the mutant *ras* protein.

For five years, the human *ras* oncogene remained a giant Sphinx of Giza, a mocking riddle carved of stone. Scientists transfected human tumor DNA into normal cells, found the oncogene 15 percent of the time, found the mutation, repeated and refined and polished the results. Weinberg and other mavens of *ras* continued to insist on the importance of their oncogene to human cancer, but critics retorted that the statistics were so piffling and inconsistent as to be almost meaningless.

The mavens were gloriously vindicated in mid 1987. Working independently, Johannes Bos, of the State University of Leiden in the Netherlands, in collaboration with Bert Vogelstein's lab at the Johns Hopkins School of Medicine in Baltimore and Manuel Perucho's lab at the State University of New York, Stonybrook, discovered that *ras* is not only far more frequently mutated in human tumors than transfection had suggested, but that it's active at the earliest stages of cancer. The finding may have a profound impact on the treatment of colon carcinoma; at the very least, the breakthrough underscores the centrality of *ras* and the point mutation to human cancer.

Perucho (who earlier had been with Mike Wigler at Cold Spring Harbor) and Bos began by reasoning that while DNA is being handled before transfection, many oncogenes probably degrade and thus escape detection. To circumvent the brutishness of transfection, the biologists fashioned ultrasensitive probes representing tiny sections of the *ras* gene. Then, with their abbreviated nucleic segments, they picked through the raw genetic material of human colon tumors. If there were mutations in either the DNA or the RNA messages of the *ras* gene, the Bos-Perucho probes would detect the errors.

To their astonishment, the scientists found that *ras* was mutated in almost *half* of all human colon cancers, not in a mere 15 percent. Bos's group discovered mutant *ras* genes in eleven of the twenty-seven tumor samples probed, and Perucho's lab reported that twenty-eight of the sixty-six tumor specimens analyzed contained erroneous *ras* RNA messages. "I have calculated what that means in clinical terms," Perucho said. "Epidemiologists say that there are a hundred and twenty thousand people in the U.S. with colon cancer. Well, our studies indicate that fifty thousand of those people have an activated *ras* oncogene in their colon carcinomas. It's very remarkable when you think about such numbers."

Perucho and Bos took their probings further. Several of the cancer patients examined also had villous adenomas, the benign polyps that are believed to precede full-blown colon carcinoma. The scientists found that in more than 80 percent of the patients who possessed both malignant and premalignant colon growths, the *ras* oncogene already was active in the polyps. In other words, the mutation of *ras* is not a late-stage event that pops up along with the hundreds of other chromosomal aberrations thought to accompany the progression of cancer. It is an early lesion that sets the ghoulish game in motion.

The scientists' next step will be to determine whether there is any correlation between the mutation in *ras* and the aggressiveness of colon cancer. Many adenomas never evolve into carcinomas. Perucho and Bos plan to survey large numbers of polyp and colo-rectal tumor tissue to see whether a lesion in *ras* is a prognosticator of a more malevolent adenoma. If the answer is yes, then a simple screening test for the presence of *ras* could guide physicians in treating patients who are prone to the development of polyps. (Ronald Reagan may be one.) Adenoma victims who test positive for an active *ras* oncogene may be assigned an aggressive therapy to prevent the growth of unmanageable tumors. Already Perucho is trying to simplify his probe technique to allow hospital technicians to screen for mutant *ras* genes. Perucho is also training his RNA probe technique on other types of cancers in quest of *ras* errors that transfection has failed to detect. He's found that in cancer of the pancreas — among the deadliest of all malignancies — *ras* may be mutated in as many as three quarters of the cases. "What our new techniques are telling us," said Perucho, "is that *ras* activation is one of the most frequent events in tumorigenesis."

With the evidence amassing that the *ras* oncogene is some sort of a common denominator for cancer, scientists crave to know how a single amino acid switch can so radically alter the behavior of the *ras* protein. Weinberg and Barbacid could address the question only superficially in their initial reports on the twelfth-codon mutation. After consulting with protein chemists, the Weinberg scientists noted in their paper that glycine is the only amino acid that lacks a chemical side chain, and they proposed that the substitution of the normal glycine with a less flexible valine subunit might prevent the *ras* protein from folding at a critical juncture. Mariano Barbacid prepared computer graphics depicting how the insertion of a valine brick at position twelve could cause the *ras* protein to "stiffen," presumably into a permanently active shape.

Yet neither of the analyses tackled the fundamental issue of how a stiffer or improperly folded *ras* protein can hurl a cell toward cancer. Nor did either one clarify the role of the normal *ras* protein in the body. Scientists still have to learn which enzymes *ras* needles, and how its signal is sent from the inner face of the cell membrane — where the *ras* protein is — to the chromosomes in the nucleus. The search for the function of the *ras* gene, in both its normal and tumorigenic manifestations, has become the most pressing order of the day. Scientists must understand the biochemistry of *ras*. "All these pharmaceutical companies are desperate to know how *ras* works," said Manuel Perucho in the summer of 1987. "They have many drugs that they would like to try against the mutant *ras* gene, for colon cancer and other cancers. If we could get a drug that affected only the *ras* protein with the mutation, that would be a kind of magic bullet. But first we have to understand the signaling pathway of the gene. That is a hard biochemical problem of interest to everybody in oncogenes."

Nonetheless, in the years since the discovery of the *ras* point mutation, the importance of the result has only blazed brighter. When Weinberg told me about the Bos-Perucho data on the mutation of *ras* in adenomas and colon carcinomas, which he'd heard at a conference in West Germany, he sounded pleased and proud. "You know that I've long taken the position that *ras* is at the heart of the matter," he said. "Well, these new data are very encouraging indeed. We are moving rapidly toward a real understanding of the biochemical basis of cancer. Again, I can't say when or from whom the final answer will come. But it is safe to say that *ras* will be a big, big part of that understanding."

After the discovery of the point mutation, Bob Weinberg became a very famous man, and his friends and associates began placing bets on when Weinberg would win a Nobel Prize. Yet glad as he was to see the *ras* work extended and amplified by his colleagues, Weinberg was hardly prepared to leave the choice pickings for others. Transfecting the first oncogenes wasn't enough; numbering among the first to clone a human oncogene wasn't enough; linking the bladder oncogene to a retroviral gene was not enough. Not even the discovery of the point mutation sated him for long. Weinberg wanted to find the other DNA errors that may participate in human malignancies. More desperately still, he wanted to unravel the function of *ras*. He'd found the gene in human tumors; now he wanted to know what the gene did. He was growing accustomed to success, and his scientific style was settled. He liked big, paradigmatic problems, and as long as such problems remained, he would not remove his spurs.

Cliff Tabin, who'd given Weinberg his most precious result to date, tried his luck at mining for the mother lode. He spent over a year trying to find the function of *ras*, but without success. When he finally earned his doctorate, three years after he'd helped discover the point mutation, he was still most frequently cited in the field for being the guy who'd gotten four-part ligations to work. That distinction was one thing that helped earn him an excellent postdoctoral position at Harvard, but Cliff left the Weinberg lab with tainted memories of his celebrated experiment. He deeply regretted not having fought harder to have Randy Chipperfield's name on the *Nature* paper. Between 1982 and 1985, Randy had a series of personal and professional setbacks. Among the most distressing, Randy was scooped by Genentech, the biotechnology behemoth, after he'd spent eighteen difficult months on a project involving the *ras* gene. Randy was so upset about his failure that Weinberg asked Art Levinson, the director of the competing group at Genentech, to talk with Randy and attempt to console him.

Randy, however, would not be consoled, and his friends said that he was often extremely depressed. During my stay at the Whitehead

Institute, I heard quite a bit about Randy Chipperfield: fragmentary stories, rumors, scraps of sorrow. In the warm room were test tube racks that still had his name on them. Nobody could bear to move Randy's equipment; nobody wanted him to be forgotten. Late one night in March of 1985, Randy Chipperfield attempted to walk across the partly frozen Charles River, slipped through the ice, and drowned. According to the coroner's report, the level of alcohol in Randy's blood was enough to make many people his size comatose. "Randy died the way he lived," said Rudolf Grosschedl, a good friend of Randy's. "He was always walking on thin ice."

6

Cooperation and Collapse

CANCER is an ugly, painful disease, a disease of physical anomie and cell suicide, but it is also a disease of luxury. It is the consequence of the peculiarly luxuriant pleasure of pulling smoke into the lungs and keeping it there long enough for traces of tar, nicotine, carbon monoxide, hydrochloric acid, and 1079 other toxic chemicals to be absorbed by the blood cells and dispatched throughout the body. A shocking 30 to 40 percent of all malignancies is thought to be attributed to cigarettes, and nothing maddens cancer researchers more than the persistence of smoking. ("You want a cure for cancer?" said Richard Rifkind, of Memorial Sloan-Kettering. "Tell the bastards to quit smoking.") It is a disease of fatty food, pickled food, and too much food: obese people have a 20 percent greater risk of malignancy than do their ectomorphic counterparts. It is a disease of liquor and of promiscuity. Women who drink even moderately — three cocktails or glasses of wine a week — could double their chances of contracting breast cancer. Women with many sexual partners court exposure not only to the AIDS virus, but also to papilloma viruses — pathogens found in the tiny warts that sometimes appear on men's penises. There is a strong correlation between cervical cancer and papilloma infection; traces of the wart virus have been detected in the chromosomes of nearly every cervical tumor cell analyzed. And celibate nuns, by the way, have a zero incidence of carcinoma of the cervix.

But by far the surest way to get cancer is to live a luxuriously long time. Cancer is a disease of aging: the incidence curve plotting age against death rate from the illness is not, in fact, a curve but is a straight, steep diagonal that soars steadily upward from birth to ninety years. Only one in a million people die of cancer at the age of twenty-eight; for sixty-year-olds, that mortality figure leaps to a hundred in a million. Linking cancer to seniority is the average twenty- to forty-year latency period between initial exposure to a carcinogen and the onset of a tumor. Rubber workers inhale the fumes of vinyl chloride for two to three decades before tumors sear their livers and brains. Before they contract esophageal cancer, Bantu tribesmen in Southern Africa spend twenty-five years drinking a beer that, as a result of local brewing techniques, is spiked with nitrosamine. The U.S. Department of Energy predicts that twenty-one thousand more Europeans may die of cancer as a result of fallout from Chernobyl than would have otherwise, but many of the mortalities will not be seen for nearly half a century. Cancer moves with the speed of the Mendenhall Glacier.

Scientists have long theorized that the reason cancer correlates with age is that a cell must accumulate multiple mutations before it turns malignant. A single blow to the DNA, no matter how bruising, cannot undermine the cell's intricate system of growth control; only when a cell has stored up a string of irreparable genetic fissures and chromosomal rearrangements will it begin to reel. In 1947, Philippe Shubik and Isaac Berenblum of Oxford University demonstrated that cancer was a multistep disease by painting the shaved backs of mice with two classes of noxious compounds, called initiators and promoters. For their initiator they chose DMBA (dimethylbenzanthracene), a chemical that was known to be a carcinogenic component of coal tar. As their promoter they picked croton oil, a grease that can blister the skin but that is not, of itself, carcinogenic. The scientists found that if they slathered the mice with either DMBA or croton oil alone, no tumors grew. But if they stroked on a layer of the initiating compound and followed up the treatment with repeated daubs of the promoting oil, the mice developed cancer after sixteen weeks. The experiments suggested that cells had to undergo at least two changes before they surrendered to malignancy, and that there was some difference between initiating mutations and those changes induced by promoters. The Shubik-Berenblum theory of multistage carcinogenesis made sense, particularly when considered together with the extensive epidemiological evidence. Still, the argument was empirical: scientists could not *see* the individual changes.

When Weinberg, Barbacid, and Wigler identified the twelfth-codon defect in the *ras* gene, they reported the first individual DNA mutation, and the scientific community was astonished by the modesty of the lesion. The oncogenic *ras* gene was extraordinarily powerful. When cloned and transferred into the DNA of normal cells, the *ras* gene — and its single base-pair change — seemed to cause cancer all by itself. It didn't need the help of other mutant genes to turn flat plains of cultured cells into black hillocks of cancer cells. The story of *ras* defied the accepted model of multistep tumorigenesis.

Still, nobody really believed that the *ras* mutation was the alpha and omega of cancer. Even while swilling champagne with Cliff Tabin and Scott Bradley on V-night, Weinberg knew that his work wasn't through; there were other human cancer genes that had yet to be assayed and analyzed. He realized that if a point lesion in *ras* was sufficient to cause human cancer, there probably wouldn't be a Bob Weinberg or Amy Weinberg or Cliff Tabin around to celebrate his lab's discovery. But Weinberg and others in the forefront of the oncogene field were swept up in the competitive pace of their research. They were so busy with what they had — the first human oncogene — that they didn't stop to worry about what they lacked. The meaning of *ras* was challenge enough.

For many observers of the sundry races, however, the shortcomings of the *ras* work were as significant as the revelations. And as Weinberg and his colleagues came increasingly into the public eye, the complaints about their experiments grew ever more strident. Among Weinberg's critics was Ruth Sager, a respected biologist at Harvard Medical School. Sager argued that Weinberg's whole approach to oncogenesis was flawed. She criticized the method by which he had detected the great *ras* gene in the first place: focal formation in cultured mouse tissue. The NIH 3T3 cells that he'd used as recipients for cancerous DNA, she said, were not normal: before Weinberg fed the cells a single *ras* oncogene, they were already three quarters of the way toward cancer. That's why NIH 3T3 cells grow so beautifully on plastic dishes. They're considered "healthy" cells, said Sager, but after hundreds of generations of dividing in tissue culture, they've become more like cancer cells. Their DNA probably harbors multiple mutations right from the start. So what does Weinberg prove with his *ras* clone? That the oncogene provides the cells with a little extra oomph to achieve cancer?

Whenever Weinberg discussed the *ras* results at a seminar, critics like Ruth Sager would stand up and harangue him about NIH 3T3 cells. If the *ras* gene had any genuine relevance to human cancer, they said, Weinberg's experiments didn't show it.

"Bob wanted to shut people up once and for all," said Luis Parada. "That's why he asked us to try something different."

The first person Weinberg turned to for peace of mind was Cliff Tabin. During his mix-and-match experiments, Cliff had been transfecting the *ras* gene into NIH 3T3 cells on a daily basis, and Weinberg assumed it would be easy for him to perform a transfection control. Weinberg wanted Cliff to throw the cancer gene into a type of tissue other than the mouse skin cells that the Sager camp so despised. Weinberg knew that most cell types are transfectophobic: they refuse to accept foreign DNA into their chromosomes and thus could not be called on to demonstrate the potency of *ras*. But he knew as well that there was one cell type that does absorb alien genes and that would satisfy Sager's demands for better proof — rat embryo fibroblasts. Rat embryo fibroblasts (REF's) are cells scavenged from the connective tissue of prenatal rats. Unlike the NIH cells, rat fibroblasts haven't been bred in culture dishes over generations but must be prepared anew for each experiment. They're so-called primary cells, and they're reasonably equivalent to the fresh, healthy tissue of fresh, healthy animals.

If the *ras* oncogene managed to transform the normal REF cells into tumor cells, thought Weinberg, the case for our cancer gene would be solid. We still wouldn't know the other mutational events that contribute to tumor growth in humans, but at least we'd be secure in our assumption that damage to *ras* is one of the biggies. Should Sager and the other critics complain that the *ras* gene was meaningless to real cancer, Weinberg would be able to shoot back, Ah, but look. The *ras* gene causes cancer even in pristine tissue newly culled from fetuses. The *ras* gene is not a minor aspect of tumorigenesis, a little extra push that sends protocancerous cells over the precipice. A mutation in *ras* lies at the heart of mammalian malignancy.

The next time you do a transfection, said Weinberg to Cliff, why not pop the *ras* clone into rat embryo fibroblasts? It's no big deal. Just a minor control. We know what the answer will be. Of course the *ras* gene will have the same effect on rat fibroblasts that it does on mouse cells. *Ras* may not be the entire story of cancer, but it's potent enough to initiate malignancy in primary cells. We'll get clusters of cancer cells in our dishes of REF's, and the naysayers will shut up.

Cliff agreed that the oncogene would work on fresh tissue. He agreed that the outcome of the control was obvious. But this was one control that Cliff would not agree to do.

"I hate killing animals," said Cliff, "and the experiment would have involved killing pregnant rats and chopping up fetuses. The standard method for killing rats is to swing them by the tail and crack their heads against the edge of the table. I just wasn't interested in that." Lab rats are large and energetic — closer in size to cats than to mice — and pregnant rats are especially large and especially grisly to slaughter.

No problem, said Weinberg, I'll kill the rats for you. It's a control I want done.

Still Cliff refused. Bob might kill the first animals, but Cliff guessed that the experiment would require several rounds to work, and he didn't believe that Weinberg would agree to play executioner every week. Besides, years earlier Cliff had watched Weinberg try to kill a pregnant rat by dislocating its neck. The rodent seemed to be dead, but after the abortion, while Weinberg was in the midst of dissecting the embryos, the mother began to wake up, and Cliff had to complete the job of killing it. Said Cliff, "I was so disgusted by the process that I just didn't want to have anything to do with it."

Cliff refused to touch anything but the antiseptic NIH 3T3 cells, so Weinberg gave the control experiment to somebody else: Hartmut Land, a postdoc from West Germany. A large, jolly, ambitious man with the jolly nickname of Hucky, Land was willing to handle pregnant rats if it meant that he would learn some hard-core biology. He'd had to plead with Weinberg to win a position at the lab in the first place, and since his arrival Hucky had been distressed by the sluggish pace of his projects. The rat embryo fibroblast experiment seemed a good opportunity to score a quick success.

Hucky's success would not be quick, and the "minor" control that Weinberg had wanted to demonstrate the tumorigenicity of *ras* would be neither minor nor a control. The results of Hucky's experiments would signal the start of something that Weinberg didn't really need at all: another scientific breakthrough.

David Baltimore once complained to me that not enough of the scientists he worked with were appropriately eccentric and "off the wall," but the observation was made after the departure of Hucky Land. People remember Hucky as a rotund aggregate of idiosyncrasies. For one thing, he always wore the same outfit to the lab: green overalls and a too-short T-shirt that left a peekaboo glimpse of his pudgy midriff. When Hucky held his wedding reception in the Whitehead cafeteria, the bridegroom figure on the cake was decked out in green overalls and a T-shirt. He had a long beard but no mustache, and he might have looked Lincolnesque and wise had he bothered to comb his unruly mane. His English was uneven, and he often used words inappropriately, but he didn't care. If he decided that a slang word like *bummer* should mean something great as well as something terrible, a wonderful experimental result would thenceforth be referred to as his "good bummer." He jested with everybody and tried to keep lab spirits buoyant, but he had little reverence for authority. At a scientific conference attended by many members of the MIT nobility, Hucky caused a scandal when he publicly dismissed a question put to him by David Baltimore as silly and told Weinberg to stop annoying him. Was Weinberg such a lackadaisical manager, fumed the chairman of the MIT biology department, that one of his underlings would dare to pooh-pooh David Baltimore? The younger scientists, however, were delighted by Hucky's audacity. "You could be sulking and bitching and ready to quit," said Jay Morgenstern, a technician who eventually did his graduate studies in Land's lab in London, "and then Hucky would tease you till you started to laugh."

Hucky worked as hard as he teased, and he soon had a system established for transfecting DNA into fresh slices of embryonic tissue. He didn't like cracking the skulls of pregnant rats any more than Cliff did, so he jury-rigged a method for killing the animals humanely, with carbon dioxide gas. Laying little crepes of the fetal fibroblasts on culture dishes, he pipetted in pure clones of the *ras* oncogene, and then waited for the cells to pullulate into foci, as Weinberg predicted would happen.

After the two weeks required for transfection to work, Hucky screened his plates for the characteristic bumps of cancer cells. But the monolayers of cells were unnervingly serene, unbroken by the dense foci that the Weinberg scientists were accustomed to abetting with the *ras* gene. Hucky tried staining the plates and holding them up to the light. That was better; now he could make out some patches that were darker than the background cells. Better yet, the transfected rat cells could grow in soft agar, a standard test for cancer; truly normal cells stop dividing when suspended in the clear pudding. Better, however, did not mean ideal. The fibroblasts had been changed, but they didn't look quite cancerous. And *scheiss,* Hucky cursed to Weinberg, the cells weren't forming foci.

Hucky fiddled with the conditions of the experiments, transfecting in greater quantities of DNA and allowing the cells to incubate longer. Still, he could not make the fibroblasts look as satisfyingly tumorigenic as the transfected NIH cells had been. He always had to stain the cells to spy any wimpish clusters at all. More disturbing, the treated fetal cells didn't act cancerous. They didn't keep replicating on their plates, as cancer cells should; instead, they died out after about forty-five cell divisions. And when Hucky injected the cells into mice, he didn't give the animals cancer; mice inoculated with NIH cells that contained the *ras* oncogene quickly died of massive tumors.

Hucky didn't know what to make of his results. He and Weinberg frequently repaired to the cafeteria to munch on candy bars and discuss the possibilities. Something was going on, they agreed, but not everything was going on. Obviously, the *ras* oncogene had an effect on the fibroblasts. The cells were taking a few halting steps toward cancer: turning darker, thriving in soft agar. But then their flirtation with malignancy ended. They were Ur-cancer cells: they couldn't continue dividing into genuine foci. The *ras* gene didn't satisfy them, and they wanted something more.

Weinberg also wanted something more. Before Hucky went any further with his experiments, Weinberg wanted to build a model. He hated to think that his critics had a point and that his lab suddenly was in a position of defending the legitimacy of *ras*. Weinberg knew that *ras* was real and that the gene was vital to the understanding of cancer. Now he had to determine exactly what the gene had to say about malignancy in primary tissue. Panhandling for ideas, he frisked through the literature. He picked up the telephone. He attended a cancer forum at M. D. Anderson Hospital, in Houston, whence he emerged with fresh material. Robert Kamen, a biologist then with the Imperial Cancer Research Fund Laboratories in London, told the conference participants about his collaboration with François Cuzin on polyoma virus, a lethal pathogen that causes tumors in many mammals other than humans. As Kamen spoke, Weinberg realized that the polyoma virus could illuminate the evolution of human cancer. In the course of hearing one seminar, Weinberg moved beyond his simple obsession with the *ras* mutation to consider multistep carcinogenesis. He began to believe in immortality.

At his next lab group meeting, Weinberg discussed Kamen's polyoma results. He drew diagrams, and he introduced to his lab the Yin and Yang concepts of transformation and immortalization. The polyoma virus, he explained, synthesizes two proteins that operate synergistically to foster tumors in its host. One of those proteins, the so-called middle-T antigen, positions itself in the cytoplasm of the host cell, right beneath the cell membrane. This protein seems to transform tissue morphology — the shape and structure of the invaded cell. Under the sway of middle-T, the cell begins to secrete novel hormones usually confined to the effluences of cancer cells. And listen to this, said Weinberg: the antigen allows the cells to grow in soft agar. Sound familiar? *Ras* is a cytoplasmic protein, said Weinberg, and, as Hucky has shown, *ras* allows primary rat cells to survive in agar.

But, Weinberg continued, the polyoma virus needs its second protein, the large-T antigen, to propagate tumors in its unfortunate host. And here's what large-T does. The protein loiters in the nucleus of the host cell, and it somehow causes the cell to become immortal, to keep replicating. That's exactly the function the *ras* gene seems to lack. The *ras* clone cannot immortalize Hucky's embryo cells. Without immortalization, the cells can grope only halfway toward malignancy.

Weinberg now had his model for multistep tumorigenesis. A single cell, he said, must undergo at least two kinds of genetic errors to become fully neoplastic. It requires a mutation that will transform cell morphology. We don't yet understand exactly how morphology is changed, but we do know that a partly cancerous cell will grow in places where it shouldn't — a bowl of jelly, for instance — and that it will secrete hormones it shouldn't. The same cell needs another mutation to allow it to replicate to theoretical infinity. So let's apply Kamen's work in polyoma to our own system. We've got NIH 3T3 cells that will turn fully malignant at the nudge of *ras* alone. But cultured NIH cells already are immortal; they must already have been endowed with some mammalian equivalent of polyoma large-T antigen. Hucky's primary rat fibroblasts do *not* have that second mutational "hit," and that's the genetic event we've got to discover. We've got to figure out what sort of genetic mishap will work in concert with the transforming *ras* gene to turn a primary cell into a tumor cell. Then we'll have a working model for human cancer.

We know that human cancer is a multistep disease, he said. We know that when Earl Jensen contracted bladder cancer, a mutation in his *ras* gene was one of those steps. Now we must learn what the other step or steps may be. We have to find a gene that can *immortalize* primary tissue.

"I've got to give Bob credit," said Alan Schechter, a postdoc. "He was really one of the first people to talk about immortalization." And once he started talking, Alan added, "he beat us over the head with it."

Among those most receptive to Weinberg's exhortations was Luis Parada. A year earlier, Luis had made a name for himself by discovering that the EJ bladder oncogene was *ras,* but he still felt a little embarrassed that the experiment had taken him twice as long as it should have. He had not yet developed much confidence in his talents as a scientist, and sometimes he worried that he wasn't obsessive enough to be a great basic researcher. Nevertheless, in a lab focused almost exclusively on the *ras* oncogene, Luis had begun following the story of the *myc* gene, which rivaled the *ras* gene for the title of emperor of the oncogenes. Although the *myc* gene was first detected in the bowels of a chicken leukemia retrovirus, scientists subsequently isolated the human version of the gene. And once they found it in normal cells, they began spotting mutant forms of the gene in many human malignancies, particularly leukemias and lymphomas. Philip Leder of Harvard Medical School and Carlo Croce of the Wistar Institute in Philadelphia discovered that a remarkable genetic event took place in the tumor cells of patients with a cancer called Burkitt's lymphoma. The part of the chromosome that held the *myc* gene snapped off and reattached itself to the bottom of another chromosome, right below the site of an antibody gene. Leder and Croce theorized that the chromosomal mixup liberated the *myc* gene from its normal constraints. Having lost the DNA switch that kept it in check, *myc* fell under the sway of the antibody switch, a powerful promoter that's often in the "on" position. The result: *myc* was on all the time, rather than intermittently, as is normally the case. The work of Leder and Croce suggested that the mobility of the *myc* gene, like the point mutation in the *ras* gene, was a primary event in human carcinogenesis. To molecular biologists, the unfolding tale of *myc* was every bit as riveting as the latest news on *ras.*

"I thought *myc* was really hot," said Luis, "and I was looking for some way to work on it."

As work on the *myc* gene progressed, scientists discovered that the gene synthesized a protein which operates in the cell nucleus. They couldn't say what the protein was doing in its nuclear position, but, given the strong link between *myc* and human cancer, they thought that the protein could be a direct participant in DNA division.

Luis didn't have much self-assurance at that time. He cared about what people thought of him, and he didn't want to say anything stupid. Yet in reading so much about *myc,* he couldn't help putting the pieces together. The polyoma virus produces two proteins that cooperate to trigger cancer. One protein is like the *ras* protein and lives in the cytoplasm. The other protein, large-T, sits in the nucleus — just like the *myc* protein. Leder and Croce had implicated *myc* in human lymphomas. Weinberg and other scientists had detected *ras* mutations in a broad array of human tumors. *Myc* and *ras, myc* and *ras*: the combination had a nice ring to it.

During the next group meeting, when Bob began scribbling models on the blackboard for "transformation" and "immortalization," Luis cried out, "The *myc* gene!"

What about the *myc* gene? Weinberg demanded.

Maybe the *myc* gene immortalizes cells, Luis suggested. It's nuclear, just like large-T. And Mark's already found that *myc* and *ras* are both mutated in human cancer cells. Maybe that's not a coincidence. Maybe both genes are necessary for cancer, or at least some kinds of cancer.

I've got an idea, Luis continued. Hucky should try putting the *myc* gene together with the *ras* gene into rat embryo fibroblasts and see if the two human oncogenes will cooperate to cause full cancer. Hucky can do a simultaneous transfection of the two genes into the same set of cells, and then score for foci.

"Bob sort of nodded and smiled at that idea," said Luis. "He said, 'Luis, that's a very interesting idea.' I think he was sort of impressed that I'd come up with it."

But not profoundly impressed. Weinberg didn't do anything with the idea right away. When Cliff Tabin returned from a long stint in California, where he'd presented a series of seminars about the *ras* mutation, he told Weinberg about Lee Hood's work on *myc* at Cal Tech. That's all they're talking about on the West Coast, said Cliff. *Myc, myc, myc.* Unaware of Luis's latest notion, Cliff asked Weinberg whether the lab was planning to investigate the fashionable oncogene. Bob narrowed his eyes and asked Cliff why he was suddenly so gabby about *myc.*

I don't know, Cliff said. Why not?

Well, Luis thinks Hucky should try putting *myc* and *ras* together into primary tissue to see if we get full transformation, Weinberg explained.

That's a *great* idea, Cliff said.

Why is it so great? asked Weinberg.

The *myc* protein is nuclear, said Cliff. And so is large-T.

Weinberg shrugged. There are lots of nuclear proteins around, he replied.

"But I knew instantly that the idea would work," Cliff told me. "There was absolutely no doubt in my mind that it was right. It was probably one of the most important intellectual contributions I heard the whole time I was in the lab." The next time Cliff and Luis went to work out together on Nautilus machines, Cliff told his friend how excited he was about the idea, but Luis did not yet feel entitled to strut. He knew that Weinberg was still sitting on his intellectual contribution. He knew that clever little ideas were as cheap as bicarbonate of soda and that he didn't have the technical wherewithal to attempt the experiment on his own.

Finally, his confidence bolstered by his friend's confidence in him, Luis decided to take the initiative. He called Mike Bishop at the University of California at San Francisco to request a copy of the *myc* gene, which the Bishop lab had cloned. Then Luis asked Hucky Land whether he could collaborate with him. Hucky agreed, but first they had to work out the terms of the joint venture. Hucky was the postdoc, the senior member of the team; he had dedicated much time and effort to establishing the experimental system; he had defined the effects of *ras* on primary cells. He would be first author. Luis had had the idea about *myc*. He would be second author. Assuming the idea worked, of course. "I thought, okay, that seems fair," said Luis. "It was my idea, but Hucky had worked so hard, and I never could have done this thing on my own. I liked Hucky; he was a fun-loving guy — like me — and I wanted to collaborate with him." Okay for *now*.

At that stage, Hucky's fetal rat system had reached peak condition, and the first experiments proceeded rapidly. Familiar with the protocols, Hucky did the work himself. He simultaneously transfected cloned *ras* DNA and cloned *myc* DNA into freshly prepared fibroblasts. If the notion of cooperating oncogenes was correct, both genes would integrate into the DNA of the recipient cells. The *myc* gene would propagate a protein that infiltrated the fibroblast nucleus, allowing the cells to overleap the threshold of cell death and continue to divide and divide and divide again. The *ras* protein, master of the grotesque, would drift to its station at the inner face of the cell membrane and begin resculpting the cell morphology. Together, the genes would cause the cells to pile up into foci.

Within a month or two, Hucky was lumbering around the lab,

slapping people on the back and practically knocking the wind out of them as he announced that he had the "best bummer" of his career. His rat embryo fibroblasts need never be stained again. Through the synchronicity of *myc* and *ras*, they had swelled into the most flamboyant foci he'd ever seen. The experiment had worked. It worked when Hucky repeated the operation. It worked when he injected the doubly transfected cells into mice: this time, the rodents obliged him by growing tumors the size of unshelled peanuts.

The lab had contrived an *in vitro* imitation of multistep cancer. In human beings, cells must simmer for decades, acquiring enough mutations to wrest free of the body's intricate feedback control system and immune defenses. In cultured rat embryo fibroblasts, the cells must be massaged with a minimum of two oncogenic segments of DNA. Hucky's was a fast-forward mime of cancer, from initiation to progression. The lab was on its way to picking apart the molecular events that underlie our slow-growing tumors.

Weinberg was delighted. He couldn't believe his great good fortune. It was 1983, only a year or so since his previous success in identifying the *ras* lesion; now his lab had taken yet another stride in the study of cancer. Even better, his model had turned out to be correct. Strong-armed into action by his critics, he'd insisted on a transfection control that he thought would give him ammunition for the *ras* gene. Instead, the critics had turned out to be right; NIH 3T3 cells were not "normal" and could not serve as a fail-safe test for the malignancy of a gene. But Bob hadn't fallen into despondency. He'd exploited the opportunity to consider the larger problem of multistep carcinogenesis, and once again he had reduced a "muddled" problem to a few simple components. He'd seen that *ras* alone could not transform fresh tissue into cancerous tissue, and he'd postulated why it couldn't. He'd stressed the concept of immortalization. He was proud of his insight. "I'd seeded all these ideas by talking about the implications of the Kamen work," he said. "I'd suggested that *ras* needed something else, Luis thought of *myc*, and *myc* it was. Conceptually, it was a very important experiment."

So conceptually important was the discovery that before he could sort out why it was important, Weinberg worried that other people were probably thinking along the same lines. Everybody studying the *ras* gene had been using NIH 3T3 cells as receptacles for tumorigenic DNA; NIH cells are quick and amenable to transfection. Yet complaints about the validity of NIH 3T3 cells were common knowledge. So Wigler, Cooper, Barbacid, and a legion of others were likely to test their clones in primary tissue, discover that *ras* wasn't enough, and start tossing in other oncogenes to find co-conspirators of *ras*. Nothing seemed to happen outside the context of a race.

After Hucky's initial result, then, the Weinberg scientists concerned themselves almost exclusively with throwing together enough controls to buttress the result. Later, the Weinberg group would sit back and chin-stroke over grander themes. The paper on *myc* plus *ras* was sent to *Nature* on June 20, 1983. So impressed were the editors in Britain that they accepted the report on sight, without bothering to solicit the opinions of outside reviewers. Yet Weinberg had been prescient. His lab was not alone. *Nature* held on to the paper for several weeks, long enough for a similar report to reach the journal's in box. On August 18, 1983, the Weinberg paper about the cooperation of *myc* and *ras* was published together with the results of Earl Ruley, an unknown young scientist at Cold Spring Harbor.

But in this instance, Weinberg didn't mind the loss of an exclusive. Between Hucky Land's and Earl Ruley's papers, the Land work clearly was the sexier. Earl Ruley had also demonstrated that *ras* couldn't turn primary rat cells cancerous, and he too had transfected in a second "immortalizing" gene to do the trick. For his synergizing gene, however, Ruley had chosen a viral gene, isolated from the adenovirus. Like the polyoma pathogen, the adenovirus causes cancer in mammals other than humans. Ruley had taken one human cancer gene and one viral gene, transfected them into a crop of prenatal rat cells, and shown that the two cooperate to provoke malignancy. Hucky Land had used two human cancer genes, each a known factor in human tumors.

"I give the guy a lot of credit," said Cliff Tabin. "We had a huge lab behind us. Without Bob, we wouldn't have started thinking about the polyoma work. Without Hucky, the whole REF system might not have gotten started. Luis brought up the *myc* connection. Without all these contributions, we wouldn't have gotten anywhere. But Earl Ruley was working by himself. He did the whole thing on his own, and I think that's amazing."

Earl Ruley, though, didn't have a chance. When the two papers were published in *Nature,* all hell broke loose, and most of the publicity descended on the Weinberg lab. Ruley's name simply lacked the Weinberg cachet. Bob Weinberg already was famous. Not only did the popular press turn to the Weinberg coterie for quotes and speculations; many members of the scientific community also viewed the latest discovery as another Weinbergian stroke of genius.

"For me, the results were exciting enough," said Ruley, now a biologist at MIT. "They exceeded my wildest expectations. They were clean, and the significance of them was immediately obvious. I was glad the two papers were able to be published together, because they complemented each other. Of course, people may have preferred one paper over another, often for reasons that weren't entirely scientific. Some people may have preferred his paper because he was talking about two cellular oncogenes, two human oncogenes. I think I may have had the edge over him in controls. I was young and a relative newcomer, and I was so worried about ruling out alternative explanations that I spent months on my controls. As for who was going to be the spokesman for the two-hit model, well, the opportunities may have been there for me, but I didn't necessarily have the skills to articulate my views in a marketable way, the way Weinberg does. So when the National Cancer Institute invited me to a conference to talk to journalists about recent developments in oncology, I didn't bother going."

Beyond the size of his lab and his eloquence, Weinberg had another publicity advantage over Ruley: his talent for generalizing his results. Weinberg began to consider the big picture of multistep carcinogenesis. His previous attempts at model building seemed like the work of a boy with Popsicle sticks and Elmer's Glue-All. Now he was a committed hobbyist, constructing detailed three-rigged sailing vessels in bottles. What did it mean, that *ras* and *myc* colluded to precipitate foci in rat cells? By its simplest definition, cancer is a loss of growth control. It would have been easy to think that all genes capable of becoming oncogenes perform the same growth-related tasks in the cell — that *ras* does what *myc* does what *src* does. If the proteins all worked in lock-step biochemical fashion, multistage carcinogenesis would be a process of randomness and redundancy, of mutating any of the cell's growth-regulating genes that happened to stand in the crossfire of a carcinogen. The results that Hucky and Luis got indicated otherwise. "Until this work, people may have thought that all oncogenes were functionally similar and that they do very similar things in the cell," Weinberg explained. "But now we were compelled to consider oncogene heterogeneity. We had to think about the discrete biochemical changes that conspire to effect full transformation of a cell." Oncogenes are specialists, and each one confers a novel and presumably indispensable trait on the gestating cancer cell.

Weinberg considered a possible scenario for the birth and evolution of a tumor. He thought about the *myc* gene. Phil Leder and others had discovered that *myc* can be mutated by chromosomal rearrangement: in some types of lymphoma, the part of the chromosome holding the *myc* gene breaks off and moves to the bottom of another chromosome, where it falls prey to the overzealous influence of an immunoglobulin gene. That, Weinberg hypothesized, could be an initiating event in cancer. The *myc* gene moves to some chromosomal spot where it doesn't belong, and it either loses the DNA signals that keep its activity in check, or it picks up new DNA signals that it shouldn't have. The gene begins to produce too much *myc* protein, which lends the cell harboring the rearrangement a slight growth advantage over neighboring cells. The cell grows a little more aggressively than it did before, and it doesn't die after a few hundred divisions.

Years pass, and the original mutant cell is now a cluster of mutant cells, presenting a sizable target for a second mutational event. A *ras* gene in one of those cells gets hit. The twelfth codon of the gene is changed from a glycine to a valine, and chaos erupts. The cell excretes growth-promoting hormones; its membrane gets shiny and slippery and the cell gradually loses its figure. It starts ignoring the tempering touch of the surrounding matrix of cells. And as it becomes uglier and more antisocial, the cell continues to divide — faster now, because it's producing hormones that further stimulate malignancy. The cell is now an immortal Quasimodo — a cancer cell. Additional mutational changes probably occur, but Weinberg speculated that it's the first two, the immortalizing and transforming lesions, that are the foundation of cancer.

Weinberg could as easily have dreamed up a script in which a *ras* mutation was the initiating event, but he realized that such speculation could not readily be proved. He began to think about how he could expand the scope of his cooperation model in an experimentally reproducible manner. He wanted to set up a paradigm that he could publish. He thought about the qualitative distinctions between the *myc* protein and the *ras* protein. One protein is located in the nucleus and encourages cell immortality. Another protein acts in the cytoplasm and reconfigures cell shape. He wondered whether cancer always required the joint activity of cytoplasmic and nuclear proteins. Could the other known oncogenes be categorized by the position and presumed function of their proteins? If an oncogene wasn't *ras,* could it nevertheless be *ras*-like — an oncogene that altered cell morphology? If it wasn't *ras*-like, could it be *myc*-like, an immortalizing agent? His lab had a system for testing the thesis. Hucky could put various clones of oncogenes together with either *myc* or *ras* and screen for foci. "Our question was whether it would be helpful in understanding multistep carcinogenesis to attempt a classification of existing oncogenes, to give a framework to our current state of knowledge," said Weinberg. "I believed that it would."

Weinberg got out his Rolodex and began cadging clones from his colleagues. He acquired oncogenes that had been implicated in or at least detected in cat sarcoma, mouse leukemia, turkey leukemia, monkey sarcoma, human brain cancer, human breast cancer, and a dozen other cancers. Hucky was to transfect every one into his rat embryo fibroplasts. As often happened when an experiment showed promise, Weinberg diverted a large fraction of his lab's resources to the project. Hucky was given ample technical assistance, and even Luis had help, a rare privilege for a graduate student.

The Weinberg lab made two lists of the oncogenes that worked in the cooperation assay: *myc*-like and *ras*-like. The protein products of the various oncogenes were not always well understood, but the lab assumed that if a gene could be fitted into the *myc* immortalization category, the protein synthesized by the gene was probably based in the nucleus; if the oncogene conjoined with *myc* to transform fetal rat cells, its protein most likely lay in the cytoplasm. The lists grew until there were five or six oncogenes in the *ras* column and an equal number under the heading of *myc. Ras* and *myc* were no longer just fragments of DNA. In the Weinberg lab, they became family escutcheons.

Some oncogenes, however, did not fit on either list. When Hucky popped those genes into cells together with *myc* or *ras,* he could not foster foci. The negative results were mildly troubling. Among the clones that defied categorization was the *src* oncogene, an old favorite among molecular biologists. Weinberg didn't know what to make of the balky genes. He didn't know whether to blame the failures on the limitations of Hucky's assay or on the limitations of his model. But he still believed that his basic thesis of cooperation was secure and, in its fashion, profound.

Throughout 1983 and 1984, Weinberg spoke frequently in public on the subject of *myc* and *ras* cooperation and the classification of oncogenes, and many scientists were convinced. His work gave their field a more cosmopolitan view of cancer. For half a decade, scores of researchers had been living, breathing, eating, pacing, and napping on thoughts of *ras*. They'd tried to link *ras* to human cancer. They'd tried to determine when and how often the *ras* gene was mutated in tumor tissue and whether that mutation influenced the behavior of the *ras* protein. Oncogeneticists wanted to understand the molecular basis of cancer, and the *ras* gene seemed to be the dénouement of the story. From 1982 to 1983, for example, *Nature* published some twenty papers on the subject of *ras*: on the retroviral version of the gene, on the mammalian version of the gene, on the chromosomal location of the gene, on the detection of mutant *ras* genes in a panoply of animal and human tumors.

Running parallel to the *ras* fixation was the *myc* fad, which commanded an equal number of devotees. During that same period, 1982 to 1983, eighteen *myc* reports appeared in *Nature*. There were scatterings of *Nature* papers that discussed eight or so other oncogenes. Any scientific journal directly or distally concerned with molecular biology proferred scores of papers about oncogenes.

Regardless of their chosen clone, oncogene researchers tended to be reductionist. And to practicing medical people, the scientists seemed reductionist *ad absurdum*. Physicians never forgot that no single gene or single mutation could explain the intricate disorder they treated every day. Oncologists were waiting for basic researchers to say something that made sense about human cancer.

Weinberg's work on *myc* and *ras* made sense. He inspired his peers to begin thinking about cancer as oncologists do: in stages. He prodded them to consider the nature of the molecular steps behind the genesis and progression of cancer. His work was reminiscent of the celebrated promoter and initiator experiments of the 1940s.

"Those of us who had been working in chemical carcinogenesis were extremely relieved," said Seymour Garte, a molecular oncologist at New York University Medical School in Manhattan. "We'd all been skeptical about the point mutation in *ras*, because we knew that cancer couldn't be explained by a one-hit event. We wanted to feel that oncogenes had something useful to say about cancer, but we were worried that they would disappear, that they were just another fad. So the fact that Weinberg was talking about two oncogenes cooperating to transform primary cells made us happy. It put oncogenes back into our historical understanding of the real biology of cancer."

The work on *myc* and *ras*, said William Hayward of Memorial

Sloan-Kettering, "was one of the landmarks in cancer research." It bridged the gap, he explained, between the work that had been based on the contributions of virologists — and Weinberg *had* been a virologist — and those who had been working in the realms of human cancer.

Weinberg was now as famous as a basic researcher could be. He had more splashy successes to his credit than most of his peers and was the cynosure wherever he went. "I remember a Cold Spring Harbor meeting," said a young scientist in Mike Wigler's lab. "They were getting ready to take a picture of a group of the scientists, but then they had to stop, because Weinberg wasn't around. People started calling, 'Where's Bob? Where's Bob?' Suddenly he came running over to the group, and the whole crowd sort of opened to give him a place in the middle. It was like the parting of the Red Sea."

In light of Weinberg's glory, the tenor and composition of his lab changed dramatically and perhaps irrevocably. He stopped attracting graduate students, and the number of postdoctoral applicants skyrocketed. The stakes for the resident staff mounted higher. No longer were the scientists competing only to best other scientists. Now they were competing for newspaper space.

"At that point, the lab split into two camps," said Cliff Tabin. "There were those of us who had had front-page, *New York Times* successes and those who worked just as hard or harder but didn't have anything spectacular to show for it. As might be predicted, the have-nots were jealous of the haves." Some of the Weinberg researchers were studying the fine details of the *ras* gene. They were sequencing the end sections of the gene, or they were mutagenizing the twelfth codon of the gene in every possible way, to see which types of mutations turned the gene cancerous. They were doing the sort of work that Weinberg has described as "the dotting of *i*'s and crossing of *t*'s" — the sort of work he often finds dull. Weinberg was out of town more and more frequently, leaving him that much less time to dedicate to the smaller projects in his lab. His lab had become the "*myc*-plus-*ras*" lab, and *myc* plus *ras* was the project to which he felt the most pressing commitment.

Hucky and Luis paid a price for their stardom. Hucky worked so hard that nobody could claim he didn't deserve his success, but some members of the lab wondered whether he wasn't becoming too self-important. Because he spent his nights at the bench, he'd stagger into the weekly group meetings barely awake, then sprawl across the floor in a corner and seem to pay no attention to the presentations of others. He had a way of blocking out distractions that his sensitive peers interpreted as arrogance. The drop in his popularity index could be measured by what happened at parties. Soon after arriving at the lab — before he'd done anything noteworthy — Hucky had thrown a party in his apartment, and it was packed with members of the Weinberg group: Hucky was beloved by all. A year later, in the wake of his professional triumph, he held a second soirée. Almost nobody from the lab bothered to show up.

Luis Parada came under much more intense intramural sniping. He'd already had one success — the identification of the homology between *ras* and EJ — and it didn't seem right that he should be so lucky once again. "Some people thought he'd just been opportunistic, jumping aboard Hucky's bandwagon because he knew a hot project when he saw one," said Cori Bargmann. Luis had extraordinary social poise, and he enjoyed talking to the press. "Luis would sit around giving interviews while Hucky was breaking his ass doing experiments," said Alan Schechter. "I've always liked Luis, but some people thought he was being obnoxious." Most inexcusable of all, in the eyes of many, was that Luis was not an especially hard benchworker. He liked playing sports, and he liked going out with friends. Sometimes he left it up to the technician Jay Morgenstern to look after his experiments. Jay had been a graduate student with Harold Varmus in California, but emotional problems had prompted him to drop out. Jay knew at least as much molecular biology as Luis, and, like Hucky, Jay was a compulsive worker. Although he could blame none but himself for his decision to leave the Varmus lab, Jay resented being Luis's subordinate. "He'd take off to play squash," said Jay, "and I'd have to clean up all his scutwork after him." If Jay was in an especially foul mood, he wouldn't do as Luis had instructed. Luis's experiment would be ruined, and the two would teeter on the verge of a fistfight.

As the pressures of the project intensified, the Land-Parada team, which had been a model of an effective collaboration, began to crumble. Weinberg is forgetful, and he sometimes forgot whose idea the *myc* part of the project was. He'd walk around the lab extolling Hucky's work. He'd come up to Luis and say, Have you heard about Hucky's latest result? Luis would remind Weinberg that it wasn't just *Hucky's* result. Luis began to worry that he'd bargained away too much up front. The MIT thesis committee demanded that he take charge of a project and carry it through from beginning to end, and Luis wanted to assume responsibility for a freestanding part of the *myc* and *ras* experiment. Hucky Land, however, had his own career to navigate. Hucky had long been working much harder than Luis — they both knew that — and he wasn't going to sacrifice a piece of his system at the eleventh hour. Rebuffed, Luis would go to the gym with Cliff and bang away at the Nautilus machines, complaining about Hucky. Cliff stopped listening.

Robert Weinberg was the most famous figure in his field. During the previous three years, he'd produced a river of breakthrough papers. And now his lab was heaving and buckling like the Tacoma Narrows Bridge. The Weinberg researchers who weren't successful felt neglected, and they resented the stars. The people who were successful resented each other. Nobody was going to anybody's parties. Everybody resented Bob for not being around to patch things up. Eventually, a couple of the scientists got together and went in to lecture their boss. You're traveling too much, they told him. You're out of touch with your lab. Some of the people are floundering, and they don't know what to do. We need your help.

Weinberg felt terrible about the accusations. He promised to pay more attention to lab politics. He promised to cut down on his travels.

Yet Weinberg had scientific as well as administrative worries. The delight with his cooperation scenario was gone; his colleagues were beginning to complain that it was naïve and, in parts, downright wrong. They clucked that he was trying to cram all of cancer biology onto a cheap Chinese menu: you choose one gene from column A and one from column B, and you've got a tumor. They complained that Weinberg was underplaying the importance of column C — the *src*-like oncogenes that didn't cooperate with either *myc* or *ras*. Those oncogenes were known participants in tumorigenesis, said the critics. Why didn't the Weinberg model accommodate them? Scientists also attacked the premise that *myc* and *ras* could be compartmentalized. Their proteins may be located in different spots, but the *ras* protein seemed to have a slight immortalizing effect, and the *myc* enzyme probably impinged subtly on cell morphology.

"The genes may have worked in that assay, but the simple idea that *myc* and *ras* always complemented each other wasn't necessarily right," said Harold Varmus. "There were ways that *myc*-like oncogenes did what *ras*-like oncogenes did, and vice versa. The functions couldn't be so neatly segregated." Or, as I. B. Weinstein of Columbia University told a congress of cancer researchers in 1987, any very complicated question has a very simple answer, and that answer is always wrong.

Weinberg stoutly defended himself. He was not trying to argue that *myc, ras,* and their willing accomplices were the whole story behind every tumor known to science. He'd never made any such bombastic claim. "I think that many people overinterpreted what we said and attributed to us speculations that we never made — for example, that all oncogenes are either *ras*-like or *myc*-like," he said. "That's an idea that's implicitly attributed to us and then struck down as a straw man." Weinberg acknowledged that the *src*-like genes worked in some still unfathomed manner to cause cancer. Indeed, he said, the real trouble with his model was that it was conceived in the face of an astonishing degree of ignorance. Molecular biologists didn't know what any of the oncogenic proteins actually did in the cell. They didn't understand the biochemistry of the *ras* protein, the *myc* protein, *src, fms, fps,* or all the other three-letter words in oncogene research. That was the next big challenge of molecular biology, he said: to learn how the proteins work and how they send their malignant signal to the DNA. So of course Weinberg's cooperation scenario was incomplete, and he admitted it. Of course he couldn't say exactly what immortalization or transformation meant. Until researchers understood the biochemistry behind cancer, neither he nor anyone else could define the precise contributions of the individual cancer proteins. He was simply attempting to put the genes in some sort of context.

Weinberg was stubborn enough, though, to make a prediction. When scientists did divulge the complete biochemistry of cancer, his model would be upheld. He knew that his instincts were right: *myc* and *ras* were distinct creatures, and *myc* and *ras* were at the crux of cancer.

"The work was conceptually important, and I'm proud to have my name on it," he said. "I think that the lesson we taught or learned from these results will be seen to be correct. As we begin to learn the biochemistry of how the proteins work, these classifications will be vindicated."

Weinberg would not really have to wait for the biochemistry of cancer to be divined before his model was vindicated or at least accepted with reservation. Weinberg's cooperation theory took the same twisty, treacherous path that most innovative theses in science must hazard: widespread enthusiasm followed by widespread denunciation followed by tempered acceptance. Biologists realized that some aspects of the theory were legitimate and that it was up to them to distinguish insight from gimcrack.

Three years after the publication of Hucky and Luis's original paper, the *myc-ras* model had become a fixture in the cosmology of the field. Biologists still were struggling to understand how and when the oncogenes conspired to cause cancer, but they acknowledged that mutations in the two genes spelled calamity for cells — and for the animals that owned the cells. At a major oncogene conference in the summer of 1986, for example, an entire session of sixty-eight papers was devoted to the topic of *myc* and *ras* cooperation. One group from the National Cancer Institute reported that when they examined the tumor cells of eighteen patients with breast cancer, they detected evidence of lesions in both the *myc* and *ras* genes. A team from New York University announced that when they subjected rats to bolts of ionizing radiation, the rats developed skin tumors; and within those tumor cells were mutant *myc* and *ras* genes. In May of 1987, *Cell* featured a spectacular cover story about work on the allied genes by Philip Leder's lab at Harvard Medical School. The Leder scientists had injected specially doctored clones of either the *myc* gene or the *ras* gene into individual day-old embryos of mice and allowed the mice to mature. The scientists then had mated together the genetically engineered animals, obtaining offspring whose DNA contained both oncogenic clones. At three weeks of age, the *myc*-plus-*ras* rodents developed whopping tumors of the breast and salivary glands. The tumors grew earlier and were bigger and far more widespread than anything achieved by the embryonal insertion of either a *myc* or *ras* clone alone. Leder couldn't say what either of the oncogenic proteins was doing within the breast and salivary tissue of the maturing mice, but the two proteins evidently were cross-feeding each other's malignant potential. "In essence, our work is the Land-Parada experiment done in the context of living organisms," Leder explained. "We've shown that, even in living mice, the two oncogenes are highly synergistic. They're much more carcinogenic together than apart. It confirms the idea that cancer is a multi-hit process, and that the original tissue culture results were very important." At the same time, Leder added, his lab's experiments prove that *myc* and *ras* together are *not* sufficient to cause cancer. "If the two genes were enough," he said, "we would have seen masses of tumors in every cell that contained the two genes. But we don't. There are many cells in these mice that don't develop into tumors, which means that additional hits beyond *myc* and *ras* are necessary for tumorigenesis. What we're trying to do now is to identify those other hits, to get at the last events that result in cancer. But it's certainly true that *myc* and *ras* are a very powerful team of genes."

In the Weinberg lab, the researchers began to drift away from the *myc-ras* project. Hucky embarked on a job hunt and eventually landed an independent position at the Imperial Cancer Research Fund. Luis found a thesis project in a recently discovered oncogene called p53. Weinberg didn't have any great ideas about how to delve further into the biochemistry of cooperating oncogenes, so he set aside the *myc-ras* work. Until he could think of a way to spin off a new paradigm from the old, he'd let others worry about experimental minutiae.

In racing obsessively, Weinberg, Wigler, Barbacid, and the others at the forefront of oncogene research had overhauled the study of cancer biology. By introducing the transfection assay, they gave scientists a technique for separating mutant genes from the tens of thousands of normal genes in tumor cells. Through cloning the gene, they were able to analyze a component of cancer in unprecedented detail. The discovery of the point mutation marked the first time that lab people could see what happened when a gene took the Faustian path to cancer. The cooperation model encouraged biologists to think about multistep carcinogenesis. The vanguard scientists had established subspecialties within specialties, suggesting leads that tempted others to swarm in and pick up where the paradigm makers had gotten bored and left off. They were the Picassos of science. As Picasso told Gertrude Stein, when he did something new artistically, he was so concerned with making the new thing that he didn't care whether it came out ugly. Other people can worry, he said, about taking the new thing and making it beautiful.

There were many ugly problems left to be tackled in cancer research, and they were harder problems than any confronted to date. Scientists had to learn how tumor cells metastasized and how the cells evaded the guardians and the garbage collectors of the immune system. They had to understand how the cancer proteins worked.

Weinberg wanted to keep making the new. He wanted to continue hurtling forward. He hated hearing about the jealousies and the fights that were rending his lab, and he wanted everybody to be working on important, cream-skimming experiments.

As the cooperation project wound to a close, David Baltimore offered Weinberg the opportunity for a fresh start. He had invited Weinberg to join him at the Whitehead Institute, and by mid 1984 the building was ready for occupation. Everything about the Whitehead Institute was new. The six-story structure on Main Street in Cambridge — just down the block from the Cancer Center — was new. The lab equipment and furniture were new. And the arrangement between the Whitehead and MIT was unprecedented in the annals of American education.

The institute was the gift of Edwin (Jack) Whitehead, a sixty-ish industrial mogul with a deep tan and a salty tongue, who'd earned a fortune as co-founder and president of Technicon, a medical instruments firm. In the 1970s, Whitehead's accountants warned him that he needed a tax shelter to preserve his millions for his progeny, and they suggested that he donate a large sum of money to a university for the purposes of biomedical research. Although Whitehead liked the idea, he didn't want to take the standard philanthropic route and just write out a check. He'd never graduated from college, he didn't have any ties to a particular university, and he viewed many institutions of higher education as wasteful and inefficient. "If I gave the money to a university and told them to use it as they please, it would disappear into a black hole of bureaucracy and b.s.," he said. "I wanted to know how the hell my money was going to be spent." Whitehead decided that he wanted to make a grander statement, an enduring legacy. He wanted to establish an autonomous research center. But that center, he said, had to be associated with a sterling university to lend it prestige.

Over the years, Whitehead approached a number of universities to discuss his scheme, among them Stanford, Cal Tech, Harvard, and Duke. Duke University readily agreed to the terms of Whitehead's gift, going so far as to print up pamphlets that trumpeted the coming of a new medical research facility to the campus. The deal collapsed, however, when Whitehead discovered that he couldn't woo a Nobel laureate–caliber scientist to direct the institute. No great scientist was willing to move to the Southern provinces, where he'd be isolated from the scientific mainstream, and Whitehead demanded that a visionary scientist spearhead his venture.

In 1981, on the advice of an ad hoc committee of wise men, Whitehead asked David Baltimore whether he'd be interested in heading an independent institute at an unspecified location. Baltimore initially was suspicious of Whitehead's motives. As an academic scientist at MIT, Baltimore had a vague distrust of industrial barons, and he worried that Whitehead would expect any institute he bankrolled to turn out patentable biotechnology products. Whitehead assured Baltimore that he wanted only to advance the cause of pure research. The institute would be entirely in Baltimore's hands, he said. Baltimore would determine the theme of the institute; he would be in charge of hiring faculty; he and his staff would be free to chase after the most abstruse and unmarketable problems. It would be David Baltimore's playground.

Baltimore mulled over the offer. He'd already clambered up to the summit of his profession, having won a Nobel at the unusually young age of thirty-seven. Now he was in his mid forties, and he doubted that if he were to remain a research scientist, he would ever match the sublime moment when he'd paved the way for his Nobel by discovering reverse transcriptase. "Starting up something like the Whitehead Institute was another sort of challenge altogether," he said. "I didn't know if I was up to it, but I was damned well ready to give it a shot." Baltimore replied to Whitehead that he would be honored to serve as director, but first he wanted MIT to agree to an intimate affiliation with the institute. He needed his university's staunchest commitment to the enterprise, he said. Any scientist who came to work at the new center would have to be a full faculty member of the MIT biology department, and the researchers at the institute had to have access to MIT's pool of graduate students. "To Baltimore, MIT is the center of the universe. It's Nirvana," said Whitehead. "He didn't just want a loose association with MIT. He wanted a marriage."

In the spring of 1981, Whitehead and Baltimore presented a proposal to the MIT corporation, outlining the terms of an arrangement between the university and the institute. Jack Whitehead would spend $120 million to construct and endow the institute, which would be closely affiliated with the university. He would pay the salaries of the fifteen to twenty faculty members who worked there, and the scientists could be joint professors of the Whitehead Institute and MIT. Further, Whitehead would give $7.5 to MIT to spend on subjects related to the biomedical research interests of the institute.

Francis Low, the provost of MIT, embraced the proposition, and in July he distributed a memo to the members of the MIT biology department, announcing the relationship between the MIT corporation and the Whitehead Institute for Biomedical Research. "There will be opportunity for discussion by the Faculty at the September meeting, of course," he wrote. "However, since I hope that an agreement will have been drafted by then, I thought it best to communicate with you directly at this time." It was summer recess, and few of the faculty members were around to receive the memo, but many of those who did were outraged. "It was presented as a fait accompli," said John Buchanan, a professor of biochemistry. "They'd picked the site for the building; they knew how big it was going to be. They never even bothered to consult us beforehand."

Buchanan joined forces with other MIT biology professors, among them Jonathan King, Eugene Bell, and Sheldon Penman, Weinberg's old thesis adviser. They were bitterly opposed to the stipulations of the contract and to the unorthodox nature of the affiliation between the institute and MIT. In memos and at meetings, they disseminated their views throughout the biology department. They noted that Whitehead recently had sold Technicon to Revlon for $300 million, which made him the largest shareholder of the world's largest cosmetics corporation. Among Revlon's many subsidiaries was a biotechnology firm that manufactured interferon. Would research at the new institute really be "pure" and "basic," asked Jack Buchanan. Or was the Whitehead to be another example of the corporate hijacking of university resources? "You can't tell me," said Buchanan, "that Jack Whitehead doesn't have a vested interest in recombinant DNA research." The professors wanted to know who would own any patents that might stem from research at the Whitehead Institute: MIT or Jack Whitehead?

The dissenters had doubts about those new faculty slots, as well. Twenty positions would constitute about a third of the MIT biology department, and they would be filled in an entirely anomalous manner. As far as they could tell, Baltimore would select the lot, and the department would have no greater power over his decisions than the right of rejecting a candidate. How would that system affect the balance of research interests and the educational needs of the university? All aspects of work in biology would bear the unmistakable stamp of David Baltimore. Nobody else — not the head of the department, not even the president of the university — had that sort of power, they said. It would be much fairer, Jonathan King pointed out, to designate Whitehead employees adjunct professors of MIT rather than full, voting faculty. Finally, what was MIT really getting out of the arrangement? A mere $7.5 million. The rest of Whitehead's "gift" would go toward an institute over which he would exercise a vast degree of influence. Other philanthropists, like Cecil and Ida Green and Margaret Whitaker, had donated $20 million, $30 million to MIT without demanding such authority. All in all, said detractors, Mr. Whitehead was purchasing Lord & Taylor prestige at Filene's basement prices.

Baltimore fought back with equal vigor. He didn't like the insinuation that he was a pawn of a conniving power broker. He pointed out that he had resisted Whitehead's courtship until he had assured himself that the industrialist's motives were sound and that the institute would be dedicated to basic research, nothing more. He couldn't understand the objections to what would be the largest gift ever bestowed on any university at any time. Why shouldn't Mr. Whitehead participate in delineating the overall scope of his donation? As for the suggestion that Whitehead faculty be adjunct members of MIT, Baltimore refused to consider it. Such a concession would seriously undermine his ability to recruit blue ribbon scientists.

Over the ensuing months, the dispute grew more brackish, and it spread beyond the confines of the biology department. The opponents drafted a position paper, stating the reasons that they were against the institute: the conflict of interest between the needs of a wealthy businessman and the needs of a public university; the regal powers that would be vested in David Baltimore; the unfair advantages that anybody working at the Whitehead would have over members of the regular MIT biology department — more money, better equipment, larger labs, possibly a lightened teaching load. The opponents pleaded with the faculty to attend a meeting that Low had planned for November 18, when all would be asked to vote on whether MIT should accept Whitehead's offer, and they asked that thumbs be turned floorward.

At the same time, Francis Low mounted a pro-Whitehead campaign, circulating letters that emphasized how the president, provost, and corporation "strongly support the proposed affiliation." The national press picked up the story, and most articles began with a variation on the same bemused theme: "Edwin Whitehead found it easier to earn $100 million than to give it away."

With uncommon civic zeal, an enormous percentage of the MIT faculty showed up to cast their ballots at the interdepartmental meeting. "I hadn't seen such a packed house since we voted on whether or not to oppose the Vietnam War," said one of the critics of the Whitehead Institute. The floor debate was gladiatorial, but in the end the pro-Whitehead forces won the vote by a sizable majority. The MIT faculty could not justify turning down so much money, particularly not in an era of tight government budgets. From the faculty's perspective, the line between academia and industry already had been blurred: many of the MIT biology professors either had launched small biotechnology firms on the side or moonlighted as consultants to industry. The Whitehead Institute seemed a relatively innocuous way to accommodate private investment in higher education. And Baltimore was an imposing figure in the MIT hierarchy. His colleagues generally respected his judgment and commitment to superlative basic research; if he swore that Jack Whitehead would keep his nose out of institute affairs, they saw no reason to doubt him. They were willing to give him a chance to demonstrate his integrity.

The new building was dedicated in December of 1984. Lewis Thomas — physician, researcher, and author — presented a poetic keynote address; an appealing portrait of Jack Whitehead seated before the picture window of his Connecticut mansion was hung on the second floor; and a plaque honoring Whitehead was installed in the lobby. Baltimore got his Whitehead Institute and, with it, the opportunity to influence the future of the life sciences.

The major area of research at the institute is developmental biology — the study of how a single fertilized egg becomes a complex multicellular animal — but such a rubric is broad enough to cover any area of molecular and cell biology that piques Baltimore's interest. The $100 million gift makes the little institute one of the wealthiest of its size. Its political structure is unique. The Whitehead board of directors is independent, able to initiate ventures without seeking university approval. Yet arrayed behind the little institute is the vastness of MIT — its resources, graduate students, history, reputation. David Baltimore and the people who work with him are simultaneously members of the landed gentry and the nouveaux riches.

"I've got the best of both worlds here," said Baltimore. "On the one hand, the institute is just another part of MIT, and that's how I want it to be perceived. We do science here exactly as people do anywhere else in the university. Our standards are exactly the same. On the other hand, the institute is small, and it's autonomous in some ways. I hope that I can give it a focus, a common purpose, that the MIT biology department doesn't always have. But only time can put us to the test."

Although he was one of Baltimore's confidants, Bob Weinberg had more or less kept his distance from the Whitehead wars. "Academicians love to fight," he said. "Give them the least excuse, and they'll be at each other's throats." But he was glad when his lab finally transferred from the overcrowded Cancer Center to the extravagant quarters at Nine Cambridge Center. *Nature* published a little column about the move. It reported that Bob Weinberg was last seen in the back of a van, surrounded like a wood nymph by his beloved potted plants.

The building was new, the machinery was new, there was more money to throw around than ever before. Weinberg was primed to take on the bigger and harder problems of cancer research. He too was ready to be put to the test.

Yet in the midst of wealth, commonness of purpose, bays so new there was still sawdust in the corners and snips of masking tape on the benches, the great Weinberg lab mysteriously collapsed. Its hands and feet were bound with bandages, its face tied up with a cloth, and it looked like Lazarus laid out in his cave.

7 Desert Days

THE WHITEHEAD INSTITUTE sits on Main Street in Cambridge, but there's nothing of small-town Main Street about this thoroughfare — no frisky collies, no freckle-cheeked children, no white clapboard church. More appropriate is the rather stiff name of the neighborhood, Cambridge Center. On the south side of the center lies the Massachusetts Institute of Technology. To the north are the various bioengineering and computer firms launched by faculty members of MIT (and, occasionally, of Harvard). There's Biogen, a genetic engineering firm founded by the Nobel Prize–winning chemist Walter Gilbert, who was unceremoniously dumped when the company's fortunes flagged; Symbolics, Inc., a computer software developer; and the tiny Applied bioTechnology, also devoted to recombinant DNA products and cofounded by none other than Robert Weinberg.

On the east side of Cambridge Center, abutting the Charles River, is a canker of about six new, ungainly office buildings, clustered around a Marriott Hotel. All are made of the same red brick, as though the employment of the traditional Boston-Cambridge construction material could compensate for the buildings' hostile scale. "I'm afraid to walk by here from one week to the next," Weinberg said one day, as we stepped gingerly around the potholes, stray cables, and mounds of sand and wet cement that lay at the foot of the complex. "You never know what new surprises may have sprung up overnight."

And on the west end of the center, nested in a private lot, is the Whitehead Institute. It's a charming, idiosyncratic little building and a postmodernist potpourri. Prim concrete columns slice the right side of the façade into neat bays, evoking the stately nineteenth-century town houses that line Boston's Beacon Hill. In contrast to such modest historicism, the left side is punctuated vertically by rounded false balconies, each sheathed in glossy, cranberry-colored tile, like a row of fruit Life Savers. To complete the design concept, the ground floor balloons out into one enormous shining purple hoop skirt, the external projection of the Whitehead cafeteria. When the weather is fine, the building glows with a sense of humor, and it's worth a detour across the street just to admire it whole. If the sky is drizzly or you're feeling less generous, you can't help wondering, Now why did the architects give their building that fat lip at the bottom?

The lobby inside looks less like the portal to a basic research institute than a corporate headquarters in Dallas or Minneapolis. A receptionist sits behind a high, dark-wood desk, and the room is dominated by a busy green and black painting by Frank Stella titled *Pillars and Cones, 1984*. Most of the lobby furniture was chosen to match that painting. There are kelly green modular couches, green and black plastic triangular coffee tables that can be shuffled around into star or polygon configurations, and a beige rug patterned with, yes, a green triangle. "You don't like the furniture?" said Cori Bargmann. "I don't see why not; it's found in all the best Marriott Hotels." Only one small feature gives away the building's true nature: down a hall off to the right is a bright vending machine filled with Cheez Doodles, Sugar Babies, and M&M's, the preferred fare of any laboratory nightshift.

Yet none of these design elements matters once you behold the upper floors, where the real work gets done. If *Town & Country* were to feature a biomedical research facility between spreads of Loire Valley châteaux and Manhattan pieds-à-terre, the Whitehead would be it. Most university labs look like basements even when they're on the twelfth floor; as one postdoc from Columbia University put it, "Peeling paint and UV light are considered good for the soul." At the Whitehead, sunshine tumbles through the huge windows that are everywhere. Velvety gray carpeting covers the front area of each floor, and a broad spiral staircase punches through the third and fourth stories, opening up the core space to more light still. Whenever I needed to write out notes, I would sit on the caterpillar-shaped couch that curved against the base of the staircase, feeling like a lady in an opera house lounge. The only other people who used the couch regularly were two young American scientists who smoked cigarettes with the ferocity of old Europeans.

The institute's elegance is not an accident of aesthetics. It's part of David Baltimore's philosophy. When the interior of the building was being designed, he oversaw the blueprints every sketch of the way, suggesting general changes and describing what he called his vision of "an atmosphere for creative thinking." He wanted high ceilings, contemplative colors, carpets, inspiring artwork, sun. He wanted windows wherever walls could bear them. "Look, I'm not going to exaggerate what a good physical environment can do for science," he told me. "Obviously some great science has been done in some pretty ugly buildings. But for most of the people who work here, this isn't a home away from home — this is home. So I wanted to give them someplace they would look forward to coming to every day. I wanted them to enjoy spending many long hours here."

What he gave them, then, was a gilded cage. He gave them labs as bright as beaches and wide individual workbenches that would accommodate two students at other institutes. He gave them new cold rooms for protein experiments, warm rooms for bacterial work, darkrooms to develop autoradiographs, computer rooms, enough ultracentrifuges to satisfy all. He gave them art by Man Ray, Berenice Abbott, Robert Rauschenberg (which soon inspired some of the postdocs and graduate students to display their own creations: a mobile of test tubes filled with beer, for example, and a metal trash can brimming over with hardened yellow styrofoam, as though oozing garbage). He gave them an auditorium where a lecturer can manipulate lighting, screen, and slides from a little control box on stage — "toys for grownups," he said — and where the seats are as comfortable as La-Z-Boy recliners. He gave them a library where the latest issues of scientific journals may not survive a day before winding up in somebody's knapsack, but where a bored scientist at least can flip through *The New Yorker* and the sports pages. And as his greatest gift to physical contentment, he gave them the Whitehead cafeteria and Lynn the Head Chef. Lynn's cooking was so good that after a while Baltimore said he needed a bouncer to keep random Cambridge lunchtimers from straying in off the street.

Less obvious but every bit as palpable, David Baltimore gave the institute David Baltimore. Although he is an elegant and sophisticated man, with a trim Elizabethan beard and a compact figure, who likes painting, music, literature, wine, sailboarding — all the elements of an attractive "personals" ad — Baltimore is first and last a scientist, and by every account a brilliant one. "I have met only four truly creative scientists in my life," said Salvador Luria. "David Baltimore is one of them." For all his administrative duties as the Whitehead director, Baltimore retains a lab of some thirty visiting scientists, postdocs, graduate students, and technicians. The work is divided into three main disciplines: the reproductive cycle of the polio virus; retroviruses (including the AIDS virus); and immunology. If Baltimore disappears to conferences for several weeks at a stretch, he needs only a brief tour of his quarters to learn exactly what is going on and decide how floundering experiments may be improved. "He doesn't waste words," said one of his senior graduate students. "He'll only make a comment or two about what you're doing, but those comments will cut straight to the heart of things."

Baltimore has a ferocious capacity for work. During a weekend at home he will read and write comments on several papers others have submitted for publication, dash off a sweeping summary of a field for a scientific journal like *Cell* or *Nature,* and touch up the textbook that he and two other biologists are writing. Whatever it is, he tackles it with his tough, street-smart-kid-from-New York aggressiveness. One night at a cocktail party we talked about his textbook, which I'd read in part and admired for its breezy clarity. I then praised the classic text *Molecular Biology of the Cell,* by Bruce Alberts and others, for being the kind of book that novices like me can understand and professional biologists refer to again and again. Suddenly Baltimore's smile dropped, and he got this funny, shaded look in his eyes. "Our book is going to be better," he said. "We're going to be the first ones to beat out Alberts." I was tempted to kid him about his reaction: winning a Nobel Prize and directing one of the nation's finest research institutes aren't enough? Can't you ever relax? But Baltimore, unlike Weinberg, is not a man you can tease.

Baltimore's dark impatience and competitiveness float freely through the Whitehead corridors, setting the tone and the pace. MIT has one of the best and most difficult biology departments in the world; the Whitehead Institute is MIT on an imploded scale. Baltimore expects more than hard work of his scientists. He expects their lives. "People have asked me if there are any characteristics that are common to great scientists," he said. "I've thought about that quite a bit, and I'd have to say yes and no. You can be eccentric or you can be conventional. You can come to work in a tie or in a monkey suit. But if there's one thing that I believe all great scientists share, it's an obsession with science. There's no getting around it. You have to be obsessed." You dream about science, he said. You wake up thinking about science. You take a shower and find yourself thinking about science. You marry a scientist so that you can talk about science at home.

Baltimore's choice of staff reflects his convictions. There are thirteen full, associate, and assistant professors at the Whitehead, and slots for an additional seven. "All of them are second to none in their fields," said Baltimore, "or are expected to become so." Of course, the Whitehead's affiliation with MIT has given Baltimore great leverage in hand-picking the best, but the power of his personality and reputation in attracting people must not be underestimated. "David is one helluva good reason for bothering to come here," said Gerald Fink, a Whitehead professor. "The only other scientist I know who could pull off a directorship with such aplomb is Jim Watson." Fink, who helped boost yeast from the Dark Ages, when it was considered a "stupid" organism to study, to its present status as biology's favorite tool, forsook a full professorship and a thriving lab at Cornell to join the Whitehead. Rudolf Jaenisch, a pioneer in the technique of inserting foreign genes into the embryos of mice, abandoned his West German fatherland and a prominent position at the University of Hamburg. For a couple of years, Baltimore had trouble recruiting a woman scientist to the institute — and the absence of a senior-level woman, he said in 1985, was "a real embarrassment" to him — but in 1986, he persuaded Terry Orr-Weaver to leave the Carnegie Institution in Baltimore and set up a fruit fly laboratory at the Whitehead.

For his dependable bests, Baltimore raided Nirvana — MIT. In addition to Weinberg, two other MIT professors joined the Whitehead staff: Harvey Lodish, an international authority in membrane proteins, and Richard Mulligan, the youngest senior scientist at the institute and among the most famous. Mulligan (who came to MIT with the understanding that he would go to the Whitehead) is perhaps the world's greatest designer of retroviral vectors, recombinantly doctored viruses used to pop alien genes into chromosomes. If and when human beings are treated for inherited diseases by the introduction of healthy genes into their bone marrow cells, Mulligan's viral constructs are likely to be the conveyances of choice. Because of his connection with a medically sexy topic like human gene therapy, his handsome, bearded, satanic face has been splashed across the popular press. He was a recipient of a MacArthur "genius" award, and in 1985 he was listed in *Esquire*'s under-forty Register of Achievers. Without bothering to ask his permission, Toyota used Mulligan's name in its auto ads — until Mulligan and his lawyer politely threatened to sue. ("I figured the least I should get out of it," said Mulligan, "is a new car.") But the place where Mulligan's face appears most frequently is at the Plough and Stars, a raucous, literary, and garishly lit Hibernian bar a mile from the Whitehead, where MIT biologists congregate nightly to drink beer and badmouth their competitors.

It would be difficult to exaggerate the competitiveness of the Whitehead Institute. Even those people who cultivate an air of hip nonchalance and confidence in their cleverness can be as scratchy as emery boards. Richard Mulligan, for example, hates the monomania of science — and according to David Baltimore, he's one of the few superlative scientists who can get away without being obsessed. "He's so good," said Baltimore, "that he can do whatever he wants." "Mulligan will go to great lengths to avoid talking about science," said Cori Bargmann, who deeply admires him. "If he's at a conference and somebody asks him about his work, he'll change the subject. If the person persists, he'll walk away." Yet Mulligan expects unqualified devotion from the researchers who work for him, and he is pitiless in his criticism of other scientists. Once, when a prominent biochemist, pleading jet lag, gave a disorganized and rambling keynote address at a Whitehead-sponsored symposium, Mulligan was scathing. "The guy's done two really major things in oncogenes," he said to René Bernards, "but after that talk, I wonder about him. It wasn't just that the talk was bad. Anybody can give a bad talk. But he didn't sound very bright. In fact, now that I think of it, he may be stupid."

David Baltimore has eased the mechanics of doing science. At the Whitehead Institute there are abundant beakers, pipette tips, buffer solvents, comfortable furniture. If a new technique is invented in molecular biology, a Whitehead scientist will be among the first to learn it and to pass it on to others. Luxuries, however, do not come cheap. Although the Whitehead Institute is new, Baltimore is like a California winegrower: he wants his charge to age in a hurry. "Nothing would make me happier than if this place were regarded as an extension of MIT," he said. "I want it to look and act like any other first-rate research institute." At the same time, he said, he strives to give it a cohesiveness that MIT may lack — a David Baltimore imprimatur. Young scientists are expected to produce and to perform, and if, through the vagaries of nature — which doesn't always cooperate with an experimentalist — they fail to produce, they'd better have a good rap. Every week, the individual senior scientists hold two-hour group meetings with the members of their lab, during which several scientists must discuss their work-in-progress. Group meetings are common enough in science, but Baltimore demands more. He, Weinberg, and others also instituted weekly floor meetings back at the Cancer Center; in the new setting of the Whitehead Institute, the floor meetings are as formal as scientific conferences. Three or more labs gather in the institute's Carnegie Hall–like auditorium at the same time, and a designated speaker from each of those labs presents a talk of twenty minutes, complete with slides and diagrams. The senior scientists cluster in the front row, sipping coffee or slurping soup, and offer pointed criticisms of the presentations. ("I hate the way they do that," said one Weinberg postdoc. "They remind me of the gorilla judges in *Planet of the Apes*.") Young researchers scheme bravely to skip their turn on stage, but sooner or later their excuses wear thin. David Baltimore is a consummate public speaker, and he believes that conveying data persuasively is at least as important as obtaining the data in the first place. "What good is a great result," he said, "if you're the only one who knows about it?"

The Whitehead's junior scientists are not yet second to none in their fields, but they're expected to become so. The problem is that there is only so much room for the David Baltimores, the Bob Weinbergs, the Richard Mulligans. Several postdocs and graduate students described to me the feeling they had when they went to an important scientific meeting. "You see them together, the big-name scientists," said Bargmann. "They usually sit together, talk together, eat lunch together. And all around them, fanning out as far as the eye can see, are hundreds or thousands of graduate students and postdocs. We all want to be where the honchos are, sitting up front, projecting charisma. I think the really successful scientists have a special aura about them; you can tell they're important even if you don't recognize their faces. And that's what we dream about being. But look, science is a pyramid, and you can fit only so many people on that pinnacle."

Postdoctoral fellows and graduate students at the Whitehead Institute have a better chance of succeeding than do young researchers at, say, Ohio State University. They'll almost surely publish more first-author papers in *Nature* and *Cell*; they'll get job interviews at more, and better, universities. But their models of success are also larger, and thus their opportunities for "failure," on a relative scale, are the same as anywhere else, if not greater. A Ph.D. candidate at a state university may easily envision being a lab director at a state university; a Ph.D. candidate at the Whitehead has a harder time imagining being David Baltimore. "How many of us are going to win a Nobel Prize?" said the Weinberg postdoc David Stern. "All of us, of course. Well, maybe only a couple of us. Well, maybe none of us."

The Whitehead Institute is the best and the worst place to be. Best because scientists-in-training have access not only to advanced lab equipment and to the minds of the resident second-to-nones, but also to the wider galaxy of scientists. The senior scientists at the Whitehead know everybody: after all, everybody wants to know them. Any time a visiting scientist from another city or nation is in town, he or she will stop by the Whitehead; sometimes Weinberg's office looks like a corridor at the United Nations. And when a Weinberg postdoc needs a particular clone to perform an experiment, Bob can call a friend at whatever lab has the clone — an operation known as "cloning by phone." "I can do things so much faster here than I could in Holland," said René Bernards. "All it takes is one little call by Bob and, boom, the delivery man is at the door." Better still, the scientific grapevine twines thickly around the Main Street building. Nothing important or scandalous can happen in molecular biology without somebody at the Whitehead hearing about it the next day. Hearing about an important result long before the official announcement means that a scientist can shift gears in midexperiment to incorporate the new data, saving time and staying ahead of the less enlightened researchers at smaller institutions. Hearing delicious rumors — of fakery, infighting, a famous scientist who's involved in a paternity suit — is simply a way of being entertained through the long days and nights. "If you want good gossip," said Raymond White, a biologist at the University of Utah, "you've got to be in Cambridge."

But the Whitehead is the worst place to be when a person is floundering — when experiments refuse to work, when results can't be reproduced, when months flutter by and there are no publications in sight. It's the worst because it's easy enough to feel dimwitted there under the happiest of circumstances. Worst because everybody works so hard, and few things are more painful than working hard when nothing is working. Worst because the rest of the scientific community is watching. "There are lots of people out there," said

Michael Gilman, a Weinberg postdoc, "who would love to see Bob fail." Scientists know that most of science is a story of failure, but when the director is determined to make his institute the world's best, the first thing to go in the face of failure is perspective. Problems become grotesquely magnified as though under a scanning electron microscope, put-downs become funnier and more withering, the lab supervisor becomes the enemy, trips to the country become more frequent.

Ironically, though, it is during periods of failure that a scientist's reasons for being in science are clearest. Anybody can understand the why of doing science in the wake of a big discovery. What's harder to fathom is why anybody would be in such a risky, low-paying, spine-breaking business when so many years can pass with nothing to show for them. And not only do people continue to slog away at their experiments; most retain a love of science. They continue to be obsessed. At that point, the point of failure, motivations other than simple competitiveness — the drive to be first and best — come into play. As all happy families are alike, so all triumphant scientists sound goofily alike. (If you don't believe me, just go back through the newspapers of the last few years and read the responses of Nobel laureates to learning of their prizes.) It is the stress of failure that lends scientists their character, their humanity. Failure makes them interesting.

That, in any case, is what I told myself during the months I spent at the Whitehead Institute. For, during much of 1985, the Weinberg lab was in its deepest slump ever. Morale was poor, everybody was complaining about everybody else, Weinberg wondered whether he had "shot his wad." After having heard so much about the glory days of the Weinberg lab, I theorized that great labs have a limited shelf life and that I'd caught this lab just when the fungus had begun to bloom. Not that the Weinberg lab was alone in its moldering: after three or four headline years, the entire field of oncogenes had stagnated. Many oncogenes had been cloned, and three of them, the *ras, myc,* and *src* genes, had been analyzed in exhausting, but not exhaustive, detail. Still the big questions remained, circling round and round like madly grinning horses on a carousel. What do these genes do? What is their normal role in cells, and how do the mutant versions cause cancer? How many genes must be disrupted to initiate a human tumor? Observing that the oncogene field seemed unproductive, a few critics were even saying that oncogenes might not be so important to human cancer after all.

Weinberg and his lab workers found little comfort in the frustrations of others. "I can't very well send my people out onto the job market with résumés that say, 'Well, everybody was having trouble; therefore I didn't publish during my tenure with the Weinberg lab.' It just won't wash," said Weinberg. "My people must publish." His determination wasn't enough. The torrent of papers from the Weinberg lab had slowed to a dribble, and for some members of his lab the dribble had parched to a drought. In lieu of results, the scientists lived on a strange and subtle sort of fluid, an elaborate wellspring of motivations that I didn't understand until much, much later. Oh, they had a few of the standard traits of the persevering heroes. They had energy; they worked and worked. But after a while, it didn't look like fist-waving, we'll-lick-'em-yet energy; it was more like the surreal energy you get when you haven't slept enough. And they had hope. Some of the scientists were working on big projects, which, if they panned out, would amply repay the agony. But hope was not really the motivation — at least, not hope that any particular grandiose project would work. Rather, they had hope that sooner or later something would work. Nature is there to be known. There were defined problems to be solved, and because the problems were defined, they could be solved. That is the advantage which scientists have over many of us: the assurance that the object of desire exists. Despite writers' talk of receiving a complete literary masterpiece from their muse, a novel doesn't exist until it's on paper, and nobody notices if a given book never materializes. Although advertising and public relations executives may work themselves to the marrow, a market for a product does not necessarily exist. But there are cells and genes and cancer. Human cancer is no figment, and it can be understood.

In seeing how the individual scientists responded to repeated setbacks — whether they wept, shrugged, considered alternative careers, blamed themselves, fate, Weinberg, or those around them, improvised, philosophized, or simply continued to work — I learned about the unpleasant, unglamorous part of science, which approaches closer to scientific truth than does the rarer success of transfection or of discovering the point mutation. If I hadn't seen failure, if by some freak the Weinberg lab had presented only an unbroken string of triumphs, I would have had an inaccurate view of science. I had to keep reminding myself that failure is fundamental to science, because, frankly, it looked and smelled rotten.

And yet, if I hadn't witnessed so much frustration, I might not have appreciated the sweet triumphs that came later. I certainly wouldn't have appreciated the intricate and contradictory characteristics that make Weinberg a genuinely great scientist. David Baltimore acts like a great scientist; his brilliance barrels before him. Weinberg doesn't act brilliant. Sometimes he's pompous, sometimes he's narcoleptic, sometimes he seems unaware of the goings-on around him. Sometimes he even acts silly. So I might have assumed that much of his past success had been the result of great good fortune, aided by his stubbornness.

I would have been worse than wrong. I would have been stupid.

The epoch of the setback crept up slowly on the Weinberg lab, and it came from behind. In fact, it initially looked like success. Bob Weinberg had an accidental tradition of publishing a major paper every eighteen months (although between 1982 and 1983 he published three big reports — on the cloning of the EJ gene, the comparison between EJ and the viral *ras* gene, and the point mutation at the twelfth codon). In the spring of 1985, eighteen months after Hucky Land's paper on the cooperation between oncogenes, Shelly Bernstein broke the fast. He announced in the *Proceedings of the National Academy of Sciences* that he had identified a DNA fragment associated with one of the most lethal features of a cancer cell: its ability to spread throughout the body. Shelly Bernstein was closing in on a gene for metastasis.

Although the discovery of the metastasis gene was not widely publicized, as previous Weinberg coups had been, Weinberg considered the advance to be as important as all his work on oncogenes to date. When a person dies of cancer, he noted, it is almost always metastasis that kills. A single tumor can be carved out with a scalpel or disintegrated with radiation, but once a few cells have been launched on an interorgan odyssey, there is usually little that a doctor can do except struggle to slow the inevitable spread.

I was delighted to have arrived at the Whitehead Institute in the wake of a big event. I called Shelly Bernstein and arranged to meet him the following afternoon.

Shelly came to our interview dressed with an elegance unusual for a postdoctoral fellow: a crisply ironed cotton shirt and a cheerful plaid tie. He explained that he had just finished his clinical rounds at Children's Hospital, a teaching affiliate of Harvard Medical School. Shelly is an M.D.–Ph.D., and for the M.D. half of his title he serves as a pediatric oncologist and hematologist, treating children with cancer or severe blood disorders. "I have clinic one day a week, and the rest of the time I work here," he said, indicating the Whitehead. "Once a week is enough. When I was doing my residency and spending all my time at the hospital, I'd sometimes be depressed by the intensity of it. This way, I know I have the lab to come to."

Shelly Bernstein, who's in his mid thirties, is one of the slimmest and most wiry men I've ever seen; he looks as though he were strung together with pencils and rubber bands and then snapped loose. He has big, romantic, sea-blue eyes set off by a fringe of thick black lashes, a black mustache, and black hair cut in a bowl-shaped bush. He talks slowly and with measured kindness; he talks like a doctor accustomed to talking to very young and very sick children.

Shelly's prospects, however, seemed anything but measured. Not only had he just published a big paper, but he'd recently become the second Whitehead scientist to win a MacArthur Foundation Award, which would give him something like $35,000 a year for five years, tax- and obligation-free. By the nature of the award, Shelly didn't even have to pretend he was looking for cancer's cure, nor was he under any obligation to give talks or to demonstrate progress in his research; MacArthur fellows are encouraged merely to relax and let their creativity rummage at random. For Shelly, the prize removed the strain of being chronically underpaid — as all scientists-in-training are — and also certified his scientific potential as exceptional, an honor that, when you're at the Whitehead and surrounded by uniformly smart people, any ego would be grateful for.

To top it off, said Shelly, he had some new and promising experimental results. The *PNAS* paper had reported that he'd identified a fragment of human DNA that seemed to be associated with metastasis. Now he was pretty sure he'd cloned the gene proper. "So what we're doing is testing to see that the clone is active — that it can cause nonmetastatic tumor cells to metastasize in mice," he explained. "Then we'll know we have the gene."

That Shelly had found a segment related to metastasis was a manifold victory. The previous year, like many oncologists, he'd lost 50 percent of his patients. They died because their malignancies had metastasized, and Shelly was eager to do something that might eventually result in improved diagnosis, prognosis, treatment — anything. Discovering the metastasis-linked fragment was the first step toward intervention. What's more, he'd been cloning the gene on his own. When he came to the Whitehead, he knew almost no molecular biology. Over a year and a half, he'd mastered a strategy to search for the metastasis gene, and he'd taught himself recombinant DNA and related techniques: transfection, DNA probing, cloning. Most gratifying, he'd succeeded with a project that many people in the lab had dismissed as a dog.

The notion that there might be a single gene responsible for malignant imperialism originated largely with Weinberg, and it sprang from his unwavering faith in biological simplicity. At first glance, metastasis seems to be the most ornate event in the progress of human cancer. Even more than a primary tumor cell, a metastasizing cell overturns all rules — rules of place, of function, finally of sanity. A liver cell does not normally decide to prick the tender rosy-brown underside of the kidney. A lung cell is a lung cell because it takes part in a matrix of spongy cells shaped like tiny Nerf balls; if it were an outer epidermal skin cell, it would be firm and square. Metastases are the body's resident suicide terrorists.

Weinberg considered the facts known about what a cell must do to wrench free of familiar and like-minded companions. It must secrete a protease, an enzyme that chews up the proteins binding the cell to its surroundings, and then burrow out through layers of fats, connective tissues, and blood vessels. After floating through the blood or lymph system to a different organ, it must reverse the process, penetrating through other layers of tissue, fats, and vessels, and clasping on to its new home like a barnacle — or like Anne Baxter in *All About Eve*.

In other words, the cell must first pull away from its origin and then hook back up again to some other spot. Given that the two cellular maneuvers were contradictory, thought Weinberg, they might involve the contribution of two or more proteins — hence of two or more genes. He realized that if metastasis was a multistep operation, his lab would find it excessively difficult to identify the molecular basis of cancer spread. Recombinant DNA is a potent tool for isolating single genes, but it stumbles when it comes to tracing genetic networks.

The second possibility, thought Weinberg, was that one powerful and versatile enzyme can open the cage of a cancer cell, allowing it to roam at will. Perhaps a gene involved in tissue migration during embryogenesis — when cells disperse up and down the central neutral crest column to become constituents of developing organs — can be reactivated in adulthood under extraordinary circumstances. Perhaps the hysteria of a growing tumor flicks on the migratory gene, causing a malignant cell to break away from the original neoplasm, and perhaps the gene is switched off once the cell reaches a more sedate part of the body. After all, Weinberg reasoned, most cancers eventually metastasize, and that argues for molecular simplicity: not several genes turned on in coordinated, stepwise fashion — an unlikely accident to occur so repeatedly — but the ignition of a single gene.

Weinberg preferred the idea of a solo gene to that of a genetic network, and he liked the problem of metastasis. He liked its conceptual breadth. Metastasis is not one quirk of one cancer; it is a fundamental aspect of Cancer. And that conceptual bigness appealed to him. He'd found the *ras* gene, a participant in at least one fifth of all human malignancies. Now he wanted to find another ambient cancer gene, the gene that allows a tumor to kill.

A sheaf of clinical papers on the subject stuffed under his arm, he approached several people in his lab, among them David Stern. "He came over to me, papers spilling everywhere," Stern said. "He talked in an excited voice, as though he were about to give me the biggest present of my career. He said, 'David, how about working on metastasis? It's great, it's important, it's the obvious next step.' But I told him to forget it. I thought it was ridiculous. It's too complicated, and I didn't think there would be only one gene for metastasis." Luis Parada also turned down the proposal. He wanted to finish up his doctorate, and looking for the metastasis gene was not likely to be an overnight affair.

When Shelly arrived at the lab, he happily adopted the project. As a physician, he believed he had a better feel for how cancer expands throughout the body than might a basic researcher who'd never treated a patient. He understood the course of metastasis, its pain, its ugly drive. Shelly didn't know much about designing molecular biology experiments, but with Weinberg's aid he helped to perfect a very clever assay for seeking the metastasis gene.

The key to finding the gene, said Weinberg, was somehow to discriminate between tumorigenic cells that were immobile and those which wandered. When the cancerous *ras* gene is transfected into NIH 3T3 cells and those cells are injected into mice, the mice develop tumors at the point of inoculation but nowhere else. Well, then, he suggested, why not try adding the DNA from human metastases to *ras*-transformed cells? If the human DNA succeeded in turning the stationary tumor cells into metastatic tumor cells, Shelly would know that he'd inserted an important fragment of DNA into his mouse cells. It was the same scheme that the Weinberg lab had used to isolate the original oncogenes: add fractionated DNA to cells until something interesting happens — in this case, until the cells begin metastasizing. Weinberg even recommended that Shelly try an old Chiaho Shih trick to clone the gene. Shelly could tweeze the human gene from the metastatic rodent cells by screening for *Alu* sequences.

Shelly riffled through a catalogue of human tumor cells for purchase and ordered nine types of human metastases, including a uterine cervical cancer that had spread to the fatty tissue adjoining the intestine, a pancreatic malignancy that had infiltrated the liver, a kidney tumor that had penetrated a lymph node. He isolated the DNA from the metastatic cells, inserted it into mouse cells that had been transfected beforehand with the malignant *ras* gene, and injected the doubly damaged cells into mice.

After six weeks or so, when the mice appeared to be near death, he killed them and cut them open. Of the 108 inoculated animals, 107 were metastasis-free: they had tumors below the skin where the needle had been inserted, but the rest of their bodies looked clean. One mouse, however, had a lone tumor in its lung. The primary, subcutaneous tumor had metastasized. "It was only one mouse and one tumor," said Shelly, "but we were pretty excited about getting any positive results at all." The stricken rodent was among a dozen mice that had received cells harboring the DNA of the cervical metastasis.

The next question was: Did the mouse metastasis arise spontaneously, or could it truly be ascribed to a human gene? To get the answer, Shelly isolated the DNA from the animal's lung tumor, once again snapped it into nonmetastatic 3T3 cells, and injected another batch of animals. This time, the effect was amplified beautifully. Four of six mice spawned lung metastases. More exhilarating still, when Shelly probed the purified DNA of those four lung tumors, he found an identical pattern of human *Alu* sequences in each one. This must be it, he thought. This must be the fragment behind metastasis; it had been sent from one round to the next. The human sequence he saw was large, measuring 23,000 chemical bases in length; it might contain extraneous DNA that was not part of the gene he was interested in. But the project was almost there, and the results were good enough to write up for publication. He'd bugle his initial success to the world and then proceed to the hard part: cloning the gene.

When I first talked with Shelly, two weeks after the release of his paper in the *Proceedings of the National Academy of Sciences,* he explained his prior experiments and the progress he'd made since publication. "We think we've got the right fragment," he said. I have a few reservations, and I know Bob has reservations. I'm concerned that so few of the original tumor cells gave us metastases. But I believe we have something important nonetheless. I'm not saying this gene is the only gene required for metastasis. I think that's unlikely. What I do think is that in this case, with these cells, a single gene leads to metastatic conversion, and that in itself is significant. It's fair to say I'm hopeful."

There was another reason to hope that Shelly's metastasis project was going to soar. After eighteen months of toiling on his own, he now had the assistance of a new postdoctoral fellow in the lab, the ace molecular biologist from Holland, René Bernards. A long-time gene jockey, René could be expected to accelerate the pace of the metastasis work considerably. At least, that's how Bob Weinberg saw it when he suggested the collaboration. Shelly said he was optimistic that the new partnership would be a productive one, although he admitted he felt a little strange, after having worked alone for so long.

Toward the end of our conversation, I asked Shelly whether I could accompany him someday on his rounds at the Jimmy Fund Clinic. He replied that he would be happy to take me there once he'd cleared the visit with the hospital, his fellow physicians, and his patients. For the moment, Shelly's schedule was tight and he was feeling harried, because he wanted to complete the isolation of the metastasis gene.

I was still unsure about what stage the cloning was at. I was also curious to see how the new collaboration between Shelly and René Bernards would pan out. I'd met René on my initial exploratory visit to the Whitehead months earlier, and he'd struck me as forceful and driven; I couldn't imagine him playing second fiddle to anybody's virtuoso violin. Nor did Shelly seem to be especially laid back. Indeed, for all his courteousness, I was a little intimidated by him.

When the time was right, I'd meet some of Shelly's young patients. But the time would not be right for a long time to come.

Later that afternoon, I ran into René Bernards, who was walking around with his left arm in a cast and sling, the result of a recent skiing accident in Killington, Vermont. The disability was particularly hard for René, because he is left-handed. Indeed, he was carrying a bottle of bleach in his right hand: he'd just made a slop at his workbench of ethidium bromide, a highly carcinogenic chemical used to stain the slippery gels of DNA, and now he had to clean up the spill with bleach, a neutralizing agent. "So I hear you're helping Shelly with the metastasis gene," I said. "He told me that he's pretty close to having the clone."

"Oh, he's got it," he replied. "It cuts in all the right places."

"Really?" I said, confused. "Does Shelly know? He told me he was still waiting to be sure."

"I'm pretty sure he's got it," said René. "Here, let me show you." He took me back to his desk to sketch out a restriction enzyme map of the gene. Drawing awkwardly with his right hand, he showed me where on the putative clone of the metastasis gene different enzymes nicked the DNA. In his excellent if heavily accented English, he described how the map was made and how Shelly had figured out the position and orientation of the enzyme-generated fractions of the gene. I was amazed by the lucidity of his explanation. Apart from Weinberg, René was the best expositor of science I'd heard. "It's fun, isn't it?" he said. "It's like doing a puzzle."

I asked René how he'd gotten involved with the metastasis project. I recalled that during a lunch Weinberg had invited me to eavesdrop on, he and René had discussed what René wanted to work on when he came to the Whitehead, and none of René's ideas had had to do with metastasis.

"Well, I guess he wasn't very impressed with anything I suggested," said René. "Besides, he needed somebody else to work on metastasis. And because I was new, I was handy."

I asked how René and Shelly had divvied up the project. "I was very careful about that," said René. "I said to Shelly, I don't want to interfere with your experiments, but Weinberg thinks that maybe it's better if some more people start working on this, and I agree. Because if this takes off it will be so incredibly important we'd better put some manpower into this." The two men drew up clearly bounded nation-states. Shelly would continue studying the gene, and René would attempt to define physical properties of metastatic cells: whether the cells secreted any kinds of unusual proteins, for example, and how the membrane of a migrant cell differed from that of a primary tumor cell. René planned to compare the mouse lung tumor cells with the nonmetastatic but tumorigenic NIH 3T3 cells. "That was a completely different area from what Shelly was doing, and really more difficult, because you're working with whole cells instead of extracted DNA," said René. "So Shelly agreed that he had no time whatsoever to do that kind of thing, and told me it was fine with him if I started on it."

A division of labor was not the only reason that the Bernstein-Bernards collaboration began so auspiciously. Around the same time that René undertook the metastasis work, he began a passel of other experiments, which he hoped would prevent him from getting restless. Most of the people in Weinberg's lab found one or two projects more than sufficient to consume their days and nights. René was like Shiva, the Hindu deity whose multiple arms splay in every direction. René usually had three, four, six experiments simmering simultaneously. "I think it's silly to put all your effort into a single thing," he said. "What if it doesn't work out? Then you're stuck; you've got nothing. This way, I figure at least one of my experiments will hit pay dirt." In addition to his work on metastasis, René was trying to develop an anticancer vaccine. He was studying a member of the family of *myc* genes. He was examining the interaction between the immune system and cancer. And whenever one of his projects failed to yield anything after a few weeks or a few months, he dropped the experiment and picked up something else.

"Don't you ever get confused?" I asked. "Don't you ever forget what you're supposed to be doing?"

"You don't forget things if you concentrate," he said. "How much time out of the day does any single experiment take? Twenty minutes here to feed cells, ten minutes there to check up on the mice. What am I supposed to do for the rest of the day? Sit around and dream up great Weinbergian models?"

It's true that no matter how many projects he took on, René never seemed to bobble or even to misplace an occasional petri dish. For one thing, he worked extremely hard, often staying in the lab until well after midnight, six or seven days a week — but then, most scientists work hard. More to the point, he loves science, with an almost romantic sort of ardor. He could be describing to me the most elementary concepts of molecular biology, yet he'd sound excited, as though he were learning the thing for the first time himself. And he's so bright that I sometimes wished his English were worse, which might have taken the edge off his intelligence. "René Bernards is brilliant," Bob told me, in a hushed, confiding voice that indicated "Don't tell him I said so or he'll get cocky."

Others apparently concurred with Weinberg's judgment. René had come to the United States on an extravagant fellowship that made him one of the wealthiest postdocs in the lab. The Dutch government provided him with $27,000 a year for five years, entirely tax-free on both sides of the Atlantic. The average postdoctoral fellowship is $8000 or $9000 less than that and runs two to three years. In addition, René was given extra money for going to scientific meetings, and the government would help him set up his own oncogene laboratory when he returned home. It was a fellowship that several hundred graduate students in the tiny Netherlands had applied for but only five received.

The award was a particularly gratifying vindication. As a high school student, René had performed so poorly that his teachers warned his parents he would never make it to college and was barely fit for a job as an auto mechanic or a train conductor. "I think they thought I was slightly retarded," said René. "I was a *very* late bloomer." The fellowship was also a triumph in that enduring scrabble known as sibling rivalry. René's brother André, who is a year older and had gone into molecular biology first, had applied for the government fellowship the same time René did, only to become one of the disappointeds. Nevertheless, by an amusing twist of coincidence, André had ended up doing his postdoc, on a more modest fellowship, in David Baltimore's lab, right across the hall from René. "I want you to know that I wrote my letter of request to Weinberg first," René declared to me as the three of us sat in the cafeteria. "You can check the date on it." "Oh, yeah?" said André. "Well, I have my letter to Baltimore, and it's written on papyrus."

I couldn't quite figure out how those two felt about each other. They shared a two-story house in nearby Brookline, along with André's girl friend, Marcy MacDonald, a postdoc working at Harvard Medical School on Huntington's chorea, the genetic disorder that killed Woody Guthrie; and René's lovely girl friend (and eventually his wife), Sophie, a psychologist and the daughter of Holland's most famous biologist, Piet Borst. René had purchased the first household car, an outsize Mercury they nicknamed the Guzzler, and he usually let André have it. Yet René insisted they weren't very close. When I jokingly told him that I thought André was wittier than he was, he never forgave me and from then on strove to convince me of my error.

There must be a gene for competitiveness, because René has inherited a mutant copy of it that never shuts up. Sure, he said, André is older, André was the family's first scientist, André is an inch taller than René's six feet four, and *some* people may think he's wittier — but isn't he, René, more handsome? (The answer is yes. With his high, sculpted cheekbones, tousled blond hair, and a jaw as strong as the side of a desk, René looks like a Leonardo sketch of a young horseman, or at least a contemporary interpretation by Calvin Klein.) René was always watching to see what other people were doing. He knew the status of any given experiment in the lab and had an opinion about it, often negative but invariably clever. Because he worked so hard and thought so quickly, he was intolerant of others' slowness. If I sometimes forgot an experiment or concept that he'd explained to me in the past, he'd seem disgusted, as though I were lazy and not doing my job. He would glare with his eerie blue eyes, which were his most noticeable physical imperfection: his left optic muscle was weak and he couldn't focus both eyes in precisely the same direction.

"You're not paying attention," he hounded me one evening when admittedly I wasn't. "You're as bad as Weinberg. Half the time when I'm trying to tell him something, he'll stop listening and start waving to someone across the room."

"Christ, René," I said in despair, "I can only absorb so much of this at one time. It's really hard."

"Ah, you're right," he said, his humor returning. "Sometimes I have trouble paying attention to myself."

As the metastasis project coalesced, Weinberg plowed ever more of the lab's resources into it. René was assigned his personal technician, an unusual privilege for a new postdoc even at the ample Whitehead. Shelly Bernstein had a kind of miniature metastasis lab: one full-time and one part-time technician and a college student worked under him. As a clinical doctor, Shelly was accustomed to supervising a pool of nurses, medical students, and interns — as well as his patients — and he seemed to enjoy organizing assignments and just generally being in charge. I could tell that he was proficient at giving orders by listening to him talk on the lab phone. He often received calls from Children's Hospital and would respond to whoever was on the other end in a firm voice that bounced all over the room.

"Tell her she's got to take her medication," he said during one conversation. "Tell her if she doesn't take her medication, she's going to die."

Despite the evident effort and Shelly's brisk managerial style, Weinberg was eager to nudge progress along. The metastasis project, he believed, was the most vital experiment in his lab, and he wanted to secure the data for the second paper, one that would announce: "We have cloned a human metastasis gene. This is what the gene looks like, here's its chemical sequence, and X times out of Y the clone causes nonmetastatic tumor cells to spread." Weinberg often stopped by Shelly's workbench to offer advice or support, and whenever he called in from out of town, Shelly was one of the people he'd ask to speak to.

Swept up by the accelerating excitement, I kept my ears cupped forward, like satellite dishes, for news of the final breakthrough, the definitive cloning of a metastasis gene. One lunch hour, I saw Shelly and Bob standing ahead of me in the cafeteria line. I crept up behind them to hear their conversation.

"If you're pretty sure you've got the clone, I think you should try transfecting it into REF's," Bob said to Shelly. The rat embryo fibroblast is considered closer to the cells of a living animal than the mouse cell line that Shelly had used in his experiments. "That shouldn't be too hard to do on the side," said Weinberg. "I think it would be a good thing to include in the next paper."

"Mmm, hmm," said Shelly, noncommittally.

"Well, let's talk about it when you've nailed down the clone," Weinberg said. And they scurried up ahead to the cash register. I was pleased that Weinberg sounded so optimistic about the cloning of the metastasis gene.

That afternoon, I wandered by to talk to Shelly again. He was washing up at the sink that gushed forth hot and cold water when you step on floor pedals. I had my maroon leather briefcase with me, and I set it down beside our feet.

"Hi, Shelly," I said.

"Hello," he replied, distractedly. He wasn't looking at me, and it crossed my mind that he might not want to be disturbed, but I felt too awkward to stop the conversation.

"So, are you going to transfect your gene into REF's?" I asked.

"What?" he said.

"Rat embryo fibroblasts," I said foolishly. "I heard Bob talking to you about it at the cafeteria today."

The question seemed to bother him, because he looked down at my briefcase to indicate that it was in his way. I scooped it up in embarrassment. "Sorry," I said.

"I don't know about that yet," he said, wiping his hands on a paper towel brusquely. "Look, I'd like to talk now, but I have to run a gel."

I couldn't blame him if he were peeved. I'd be peeved, too, if some snoop monitored my conversations. He seemed annoyed the next day when I joined him for lunch in the cafeteria, and I didn't improve matters during the conversation. After brief badinage about his wife, Nancy, who is also a scientist, and the effects of budget cutbacks on basic research, I asked Shelly whether he was close to finishing up the cloning of the metastasis gene.

"Yes, it's going well," he said. "We have some positives, and we're testing them now to be sure they're active."

Not pausing to think, I said that René had told me he thought Shelly had the right clone. Shelly put down his soup spoon and said sharply that he did not appreciate another scientist discussing his new results before consulting him. I apologized, but without conviction. After he'd left the table to return to work, I sat there angrily and pettishly shoving the lettuce back and forth in my salad bowl. Scientists could be so touchy and defensive, I thought. Did Shelly really believe that somebody would clone the metastasis gene out from under him?

René Bernards was bored. His half of the Bernstein-Bernards collaboration wasn't going anywhere. He had rummaged through the scientific literature, seeking out reports on the behavior of metastatic cells that might spark a few ideas for experiments. There were very few papers, but that paucity wasn't the worst of it. Lately he'd begun to realize that his original idea — comparing the characteristics of Shelly's lung tumor cells with those of the nonmetastatic cells — was flawed. The problem, he decided, lay with the mouse intermediary. Those cells which had spread to the animal's lungs were a highly select collection, chosen by the mouse's body to succeed. The putative metastasis gene may have been responsible, but so might a thousand other factors well beyond experimental scrutiny. If the lung cells secreted strange proteins, René couldn't be sure whether the secretion was the result of the metastasis gene or of something else.

There was only one sure way to prove cause and effect between the gene and the appearance of metastatic cells. René needed a clone of the metastasis gene. That would enable him to transfect the gene into cultured cells and *then* check for oddities like a spurt in novel proteins. But Shelly alone could provide René with the cloned gene, and Shelly was moving too slowly. In an attempt to hasten progress, René asked Shelly whether he could do some preliminary work on the RNA transcript of the gene — the message encrypted by the 23,000-base-pair fragment. René reasoned that if he characterized the RNA of the metastasis gene, the lab would be in a better position to run with the full gene clone once Shelly had it in hand. René also hoped to compare the expression level of a key secreted protein with that of the metastasis gene. He had to request Shelly's permission, however, because RNA, as the flip side of DNA, fell within Shelly's territory in the carefully gerrymandered project. Shelly said fine, if that was what René wanted to do. Sure, René could characterize the central transcript of the gene — if that was all he wanted to do. Shelly gave René a copy of the bacteriophage that contained the sprawling 23,000-base-pair metastasis fragment. Shelly assumed René would use the bacteriophage to obtain the RNA transcript, and leave it at that. Soon afterward, Shelly was called out of town to attend to his father, who'd had a heart attack.

René, however, was not content to work with the full-size cloning bacteriophage and its ungainly metastasis gene. While Shelly was away from the lab, René decided that it would be easier to divide the lengthy segment into smaller pieces, or subclones, and to insert each fraction into plasmids, little circles of DNA and protein. René fashioned one plasmid that incorporated the central RNA transcript of the metastasis gene and other plasmids that held surrounding pieces of the 23,000-base segment. He characterized all the transcripts in great detail, preparing restriction enzyme maps of the subclones. No longer was René working on the physical traits of a metastatic cell. In essence, he was working on the metastasis fragment. Shelly's fragment.

When Shelly returned to the Whitehead, René told him of some of his RNA work and that he had made a subclone of the central transcript. The phage you gave me was too big, he said, so I snapped the transcript into a plasmid. You can have it if you want. You'll probably need it for your experiments.

His nerves already tested by his father's illness, Shelly reacted with anger. In subcloning the metastasis fragment and carrying out such extensive RNA studies, René was encroaching too far onto Shelly's turf. Shelly had been on the project for eighteen months, perfecting the mouse assays, doing the cloning, riding out the dips and bumps — all on his own. He was under the impression that he'd be permitted to retain control of the metastasis gene, but now it looked as though René wanted to take over his half of the experiments.

Shelly thought that René had agreed to work with the bacteriophage, and he wanted to know why René was performing so many operations that concerned the metastasis gene proper.

Anyway, it's done, said René sharply. Do you want the subclone or not?

Shelly took the plasmid that encased the central metastasis transcript, but he made it plain to René that he was unhappy about the course the project was taking. Obviously the lines were not as clearly drawn as I thought, he said. The two men retired to the conference room to try drawing the lines anew.

René, however, was so nettled by Shelly's reaction that he didn't even mention the other subclones he'd designed. René knew that Shelly could use those subclones as well, but to hell with it. If Shelly was upset, thought René, I'm not going to make more trouble. René threw away the rest of his subclones. Later, Jay Morgenstern told René that Shelly had asked him to prepare subclones of the remaining portions of the metastasis fragment.

Look in the garbage, Jay, René said. You'll probably find everything you need there.

"I know what Shelly's doing," René told me angrily. "He doesn't want me to be an author on his paper, and he figures that Jay can do something without being a threat to him. I'm a postdoc. Jay's only a technician. Shit. If Shelly had to make the subclones himself, he'd be sweating for two weeks to do something I can do in a day."

René was beginning to feel scorn for Shelly. Not only did their personalities clash, but René was unimpressed by Shelly's scientific training. Shelly was an M.D.–Ph.D., and to date he'd dedicated more time to medicine than he had to science. René had spent eleven years earning his doctorate. Who was Shelly to think that he could master real research in half that time or less? Like other pure Ph.D.'s in the lab, René had a lofty image of Science, and he doubted that any human being could excel at both his field *and* medicine. Richard Mulligan — a purist if there ever was one — told me during a Whitehead beer hour that he believed the very notion of a combined M.D.–Ph.D. program was ridiculous.

Everything that Shelly had to do for these experiments, thought René, took four times longer than it should. Occasionally René could not suppress his anger, and cast doubt on Shelly's right to a MacArthur award. When Weinberg asked René how he and Shelly were doing, René said that they weren't getting along at all, and that he didn't know how much he was contributing to the collaboration.

I don't know what's gone wrong, he said, but I'm miserable. Maybe it would be better if I was taken off the project.

Please, Weinberg said. Don't you realize how important metastasis is in human cancer? We can't let personal differences interfere. Can't you just try to work together?

René felt ashamed. "Bob was right," he told me later. "We were being trivial. He put it very nicely. He didn't say we were acting like children, but that's what he meant." He vowed never again to get mired down in bickering, and he more or less set aside his part of the metastasis effort until Shelly came up with the cloned gene.

But René's contribution to Shelly's project was not quite finished. The week after René's discussion with Bob, Weinberg was elected to the National Academy of Sciences, the researchers' equivalent of the 400 Club. It was a great honor, in some ways a deeper recognition of his scientific merit than a Nobel Prize, and everybody was happy. Amy gave Bob a huge salmon-colored azalea plant to add to his office jungle. David Baltimore held a champagne pour, and Robert Gallo, the famed AIDS researcher from the National Institutes of Health in Washington and an old friend of Weinberg's, stopped by to tip a glass. "It's a big deal because your peers elect you to the academy, not just a handful of fogies in Stockholm," said David Stern. "Bob was long overdue for this one."

Yet I sensed, beyond the bonhomie, that something was up. As I strolled through the lab corridors in the evening, the air felt tight and prickly. René was talking to André in the equipment room that separates the Baltimore corridor from the Weinberg hall, and everything about their bodies said gossip. Not the usual joky insults, but big work-related, can-you-believe-it gossip. They leaned against the blue metal top of a scintillation counter, their blond heads bowed together, their voices low yet excited, René's eyes darting this way and that, and his smile very near a smirk.

"So did you hear the news?" he asked me later.

"You mean about Bob being elected to the National Academy of Sciences?" I said, puzzled. Had René missed the champagne pour?

"No-o-o," he replied, doodling in his lab notebook.

"Then what?"

He smacked his thigh happily. "Well, at last I know something that's going on that you don't."

"René, I'm supposed to know what's going on — that's my job," I insisted, getting a big·strung up. "So what is it now? Is it good news or bad?"

He hesitated. "If you don't know, I don't think it's my place to tell you."

"Wait a minute!" I cried. "Who else knows about this? Does Bob know?"

He nodded. "But not many other people do. Not yet, anyway. Don't worry — you'll find out soon enough. And, hey, you're lucky. This is a good time for you to be around the lab." He picked up his pipette and jabbed it into a box of plastic tips, returning to his precise business of shredding DNA with enzymes. I wheedled and blandished for a few minutes, but his resistance stood. "Okay, forget it," I said. "I just hope I don't read about it on the front page of the *New York Times* tomorrow."

Nothing pertinent appeared in the paper the next day. The cure for cancer hadn't been discovered, and I hadn't been scooped. Instead, halfway through the Weinberg Friday afternoon group meeting another visitor and I were asked to leave the room. It seemed very unlike Bob to make such a request; he enjoyed a hush-hush atmosphere, but he normally liked to have as many colluders as possible. I skulked around right outside the conference room, bantering nervously with the other barred visitor and barely able to keep from pressing my ear against the closed door.

When the meeting was over, nobody would tell me what was going on — at least, not immediately. It took a few days, but eventually I heard various shards of the story. Shelly Bernstein's metastasis gene, the proud, filigreed project that he had singlehandedly nurtured for eighteen months and that he had hoped to coax into something that would buoy his career for years, was in big trouble. And perhaps ironically it was René Bernards who'd knocked Shelly and his gene to the ground.

Scientists make ridiculous mistakes in the lab all the time. They stir together the wrong biological compounds. They set their pipettes in the wrong position. They use enzymes that, unknown to them, have been sitting on the shelf a little too long and have lost their potency — or crapped out, to use the scientific jargon. They pick a dish of cells that has been contaminated by bacteria or viruses. Even the simple little experiments that I tried my hand at involved fifteen or sixteen steps, each requiring tiny, exacting measurements and great concentration. After two or three hours of benchwork, the fingers creak, the eyes smart, the mind meanders, and you make sloppy mistakes.

What is more perilous for experienced researchers who don't tire so easily is that all biology projects are prey to gremlins known as artifacts: results that, on first glance, look real but turn out to be a consequence of the experimental system. An artifact is a human-made result, not an inherent property of the cell. For example, when a scientist wants to see whether fruit fly DNA has a gene homologous to a given human gene, the scientist will prepare a Southern blot, screening the fly DNA with a radioactive copy of the human gene. Under certain experimental conditions, however, genes can hybridize promiscuously: the probe sticks to a gene that it is not truly related to, leaving behind a deceptive dot of radioactivity. Such a result is an artifact. Only by rehybridizing the genes under more demanding conditions (higher temperature and more salt in the solution), or by sequencing the fruit fly gene, will the scientist unveil the mistake. "Most of the time, when you get a result that's entirely unexpected or crazy, it's not because you've discovered a new biological phenomenon," says Mike Wigler, "but because you've screwed up the experiment."

The majority of ridiculous mistakes or artifacts never make it past the scientist's bench, let alone into the dark of ink. Scientists may err as often as the rest of us, but they're more careful in the long run. They know that a lot of people are going to be picking over their papers with an eye for any stupidities, so they do a procedure over and over until they get it right. "One thing you can never mind as a scientist is repetition," said Stephen Friend, a Weinberg postdoc. "I'm not saying you have to love it, but you can't mind it or get impatient with it. If you do, you're gone."

The second reason for the relatively modest number of published snafus is controls. Scientists are fanatical about controls. A picture of an autoradiograph in a scientific paper is likely to be one part result and four parts control. Those controls will be of two classes: negative and positive. As its name implies, a negative control should be naught, blank, mute, empty. Bob Weinberg had asked Luis Parada to provide him with a necessary negative control for the EJ gene to show that the gene was not related to any of the known viral oncogenes. Luis's "negative control," of course, turned out to be a major discovery in itself. A positive control is something that should be happening, something mundane and well understood. Even if the rest of a scientist's model was misconceived, and nothing that the researcher wanted to find could be found, the positive control should be active. The scientist may, for example, demonstrate that a probe is biologically sound by daubing it onto a sample of DNA where an analogue of the probe is known to exist. Thus, when the researcher fails to detect the presence of an active gene in the DNA of a cancer cell, he or she can't ascribe the failure to a rotten probe; the positive control speaks otherwise.

Through negative controls, positive controls, and the occasional single- or double-blind controls, a scientist hopes to obviate the mortification of a published error. Yet mistakes are printed. The Letters editor of the journal *Science* estimated that she receives at least five complaints a week about published errors. Most of those errors are niggling: a mislabeled diagram, an incorrect reference to another scientist's paper, a deviant decimal point in one ancillary equation. Many errors are never even spotted, and when they are, they rarely merit mention in a journal's corrigenda.

Every so often, however, a terrible mistake is published, a mistake that calls gene A gene B or an artifact a discovery, a mistake that destroys the thesis of the paper, endangering a scientist's reputation, ego, and equanimity. In 1984, Geoffrey Cooper, Weinberg's archrival, published a ballyhooed paper in *Cell*. Cooper claimed that the *ras* protein interacts with the protein receptor for transferrin, an important iron-based molecule. The finding appeared to be a watershed: the iron receptor previously had been linked to cell division, and by adding *ras* to the package, Cooper offered a plausible function for the oncogenic protein. Several months later, Joe Harford of NIH wrote a letter to *Nature* refuting Cooper's results. Harford had attempted to duplicate the experiment and failed. In his letter, he carefully pointed out when during the purification of the *ras* and transferrin molecules Cooper had, he believed, erred. The "apparent" association between *ras* and the iron receptor, wrote Harford to the British journal, "is an artefact." Thus challenged, Cooper's lab people repeated their work, realized that their initial findings had indeed been wrong, and were forced to confess, in response to Harford's letter, "We concur with Harford that our findings are explained by the artefact he describes and, therefore, do not indicate an interaction between transferrin receptor and *ras* proteins" — much to the merriment of the molecular biology community. Scientists have an admirable respect for accuracy — a greater respect, I think, than do people in any other profession — and when, through carelessness, a lack of proper controls, or simple misfortune, that accuracy is breached, it's open season on the erring scientist.

The summer of 1985 was a hard, open season on Shelly Bernstein. Bob Weinberg, according to the people who have worked for him, is not as nutty about controls as are some scientists; he's very careful, but he's not anal compulsive. Sometimes, when he's eager to get out a paper, he may decide that the lab should forgo a particularly time-consuming control. David Steffen recalled one experiment he did on the viral *myc* gene. Toward the end of the experiment, he told Weinberg that the results looked good but that he wanted to do one more control. Weinberg said that the control was unnecessary and that the paper should be sent off as is. "We had a loud and prolonged argument about it, and finally I submitted the paper without the control," said Steffen. "The reviewers wrote back that it was a very nice paper, a very interesting paper, but that we needed to do one more control — just the control I'd been planning to do. At that point, Bob and I were both sick of the whole thing, and the paper never did get published."

Because Shelly's approach to metastasis was relatively risky, however, Weinberg and Shelly had agreed that the experiment would require a haystack of controls. Shelly injected batches of mice with all sorts of placebos: normal cells that had received no foreign DNA, NIH 3T3 cells that had the cancerous — and presumably nonmetastatic — *ras* gene alone, even pure saline solution. He wanted to be sure that metastasis wasn't a random event that could arise spontaneously as a result of some funny factor in the rodent's body. In addition, Shelly needed to guarantee that the metastasis fragment he had found was a novel length of DNA and not a sequence that had been cloned by other biologists in the past. He screened the metastasis segment with probes of several known oncogenes: three types of *ras* genes and the *myc* gene.

Shelly believed he had been diligent about his controls, and, satisfied with the answers the controls had given him, he published the paper, claiming that the metastasis-linked fragment was not related to any of his oncogene probes. Once he thought he had cloned the gene, however, he decided to repeat and extend the controls. He incubated the isolated gene with *ras, src, myc,* and fifteen or so additional probes, checking to make sure that the gene was not sullied by fragments of other oncogenes. He set up the incubation experiment to run overnight. Shelly liked to be home by dinnertime, and there was no rush to see the results of the controls. He'd look at the hybridizations in the morning. As he was getting ready to leave, however, René came over to his desk. Shelly, said René, there's something I have to show you. From the expression on René's face, Shelly wondered whether he was about to lose his appetite.

René is hard on people, and sometimes he even takes a soft dig at himself. Rather than claim to have a photographic memory, he'll say, "My mind is cluttered up with all these useless details." Details such as most of the papers he's read, the talks he's heard, and the characteristics of every gene he's worked with.

A couple of weeks after René quietly decided to drop out of his collaboration with Shelly, he was looking over the data he'd obtained while preparing the ill-fated subclones. He was trying to figure out some way to study metastasis that would not again cross Shelly's path. As René stared at the restriction enzyme map of the metastasis gene, he had a disturbing sensation. The map looked awfully familiar. Yes, René had seen the metastasis data before, in recent times, but *this* map was an old, old friend. Two EcoRI sites, 23 kilobases between them; a HindIII site here, a BamHI site there . . . Why did the data points ring so many bells and whistles?

The *ras* gene. The restriction map of the metastasis gene was identical with that of the *ras* gene. René had studied the *ras* gene in Holland. It hadn't been his specialty, but he knew it well enough. Racing down to the Whitehead library, on the second floor, René located an old Chiaho Shih paper that had a map of the EJ-*ras* gene. Sure enough, the restriction enzyme map of the *ras* gene was the same as that for Shelly's metastasis gene. René knew that the chances of the two genes coincidentally having the same maps were virtually nil. He knew immediately what had happened. Shelly hadn't isolated a novel gene. He hadn't even isolated a gene involved in metastasis. The gene that Shelly had reported as the metastasis gene was none other than *ras*, the America of oncogenes, the gene that kept getting rediscovered. Somehow, Shelly Bernstein had profoundly screwed up his experiment.

René took the two maps over to Shelly and laid them side by side on his desk. René offered Shelly his estimate of the odds that the resemblance of the maps was a coincidence. Shelly stared at the maps. He looked from one to the other. His face turned as white as rice. It's true, he said to René. It can't be a coincidence.

That night Shelly hardly slept. The next day he looked at his hybridization results, and they confirmed the terrible truth. The "metastasis" gene was *ras*. He went into Weinberg's office to break the news.

During the closed-door group session, the Weinberg lab thrashed out the chronology of the crisis. According to the accounts of various people who attended the meeting, Shelly didn't know exactly how the mistake had happened, but he had a general notion of what had gone wrong. Somebody had thawed out a plate of cells. It may have been his fault, he said, or it may have been a technician's, but in either case the mistake had occurred early in the experiments — at least a year ago, he estimated. Stored in the laboratory freezer were two types of cells that had been established by another researcher long before Shelly began his experiments. One was a cancerous, nonmetastatic mouse cell line that harbored the human *ras* gene; and the *ras* gene, said Shelly, normally has an *Alu* sequence, as do most human genes. The second frozen mouse cell line also was cancerous and also contained the human *ras* gene, but — and it was important — this *ras* segment had been slivered with a restriction enzyme ahead of time to separate the oncogene from its known *Alu* sequence.

Shelly had meant to work with the second, enzyme-treated cell line. Those were the cells he'd been planning to use as recipients for his transfections. When he later screened the DNA from the metastatic cells in search of the human metastasis gene, he didn't want to worry about his *Alu* probe picking up signs of the *ras* gene.

Shelly thought he had been working all along with the proper enzyme-treated cells; he thought that when he detected a conserved *Alu* sequence in his four mouse lung tumors, it belonged to the human metastasis gene. But obviously, he said, the wrong cells had been taken from the freezer. He had begun the metastasis part of the project with the group of mouse cells that had a *ras* gene with its *Alu* sequence intact. As the months passed, that first little error — a wrong set of petri dishes — had bloated into a major disaster. From one round of transfection and injection to the next, Shelly had been chasing after an *Alu* sequence in the *ras* gene rather than in the metastasis gene.

The Weinberg group agreed that it was a sad mistake, but one that any of them could have made. "I'm sure we've all, at one time or another, pulled out the wrong dish of cells," Cori Bargmann said to me later. "Usually it wouldn't turn into such a big deal, but you can see how it would happen."

What Shelly initially couldn't understand is why his negative control had failed to reveal the mistake. He had probed the presumed metastasis DNA with the *ras* gene. Why hadn't the original hybridization worked when he'd blotted the metastasis DNA with *ras* and *myc*? Shelly displayed the autoradiograph of the negative-control blot that had been published in the *PNAS* paper.

Cori Bargmann softly suggested that Shelly had flipped his DNA blot around and published it in the wrong direction. If it were turned around the other way, she said, so that what was now marked Lanes

A, B, and C were instead interpreted as Lanes C, B, and A, Shelly's mistake would have been clear. He would have noticed that there was *ras* where *ras* should not have been.

Upset and flustered, Shelly at first could not respond to her suggestion. He didn't believe that he'd flipped his blot, and later, through additional checking, he became convinced that the error had been committed elsewhere, on another blot. Shelly was sure that he had screened the wrong cell line at the wrong time. The other scientists in the lab could not be absolutely sure why the negative control had failed, but they concurred that it was an unfortunate failure indeed.

Maintaining a calm demeanor, Weinberg wanted to press ahead. How bad was the mistake? Was the metastasis fragment a complete figment, and would the lab have to retract the paper? Poring over Shelly's data, the group decided that all was not lost. There were other conserved *Alu* bands on the autoradiographs, any one of which might be the true metastasis gene. Shelly would have to go after these fragments, testing each for its metastatic potential. For now, the lab would make no official confession, and Weinberg requested that the people in the lab keep quiet about the affair. Shelly deserves a chance to work this out, he said, without the whole world knowing about it.

The scientific grapevine, however, is long and leafless, and there's no place to hide a succulent piece of gossip. Biologists soon heard that there was something wrong with the metastasis gene, although nobody knew exactly what the trouble was. Thomas Maniatis, a scientist at Harvard, reportedly said to Richard Mulligan that he'd been told the metastasis gene was really the herpes virus. The connection seemed logical: in some animals, the herpes virus has been associated with metastatic tumors. Pretty soon, the nature of the error had been well publicized, to the point where a Weinberg scientist attending a meeting in Frederick, Maryland, and overhearing two people discussing a paper displayed on a poster board, caught them saying that the authors of the paper had confused a couple of cell lines — and adding, cruelly, that the confusion was the Weinberg syndrome.

Shelly's travails would not cease for months to come. He tried many different cloning techniques, but for unknown reasons could not pull out the metastasis gene. The gene probably exists — the mouse experiments performed by Shelly and other scientists demonstrate that the tumor cells have some sort of metastatic potential, which is amplified from one round of transfections to the next — but the gene proper remained elusive; as of the fall of 1987 it still had not been cloned. Shelly's lab mates began to feel sorry for him, and even René became his good friend. "He's a really nice guy," said René. "He still has a lot to learn about molecular biology, but he's a nice guy."

True to his word, Shelly eventually took me to the Jimmy Fund Clinic at Children's Hospital, where I met, among other patients, a seven-year-old boy named Kevin. Although Kevin had a cancer of the nervous system, and although he was in the hospital to receive blood platelets that would help him recover from the devastations of chemotherapy, Kevin looked remarkably healthy. His skin was fresh and rosy, he wasn't too thin, and he had a full head of glossy brown hair. But Shelly told me that Kevin's cancer had spread throughout his body, and that his prospects were very bleak. Kevin died the following year.

By the end of my visit to the clinic, I admired Shelly. He confronted the incomparable horror of terminally ill children, week in and week out; I was at the Jimmy Fund for only a few hours, but I was desperate for the moment when I could bolt through the hospital door. I understood why he wanted any sort of molecular handle on metastasis, and why he was so unhappy about his troubles with the metastasis gene.

Shelly was not alone in his disappointment, though. The metastasis debacle seemed to cast a pall over the lab. Projects just weren't working. Bob Weinberg, for whom ideas had always cascaded down as though from an astral river, started to grope. Something would work out, he thought. Something always had in the past. The allure of science was that you could never predict when or how the next big break would come. The misery of science was the possibility that the big events would break in another lab. And nothing repelled Weinberg more than the thought of doing *i*-dotting, *t*-crossing, small-time, boring science.

8 Down the Garden Pathway

BOB WEINBERG had his first great notion the way scientists are supposed to have great notions: romantically, poetically, with the imposing Longfellow Bridge beneath his boots, the snow and the idea swirling around and blinding him. Contrapuntally, Weinberg had what he thought was his next great idea in a conventional, starchy setting — during a scientific meeting in the middle of August. This second idea was no surprise package, as the *aha!* of transfection had been. When Weinberg decided he knew the function of the *ras* protein, he had long been truffling about for inspiration. Cliff Tabin and the postdoc Alan Schechter had spent more than a year pursuing a tantalizing lead on the biochemical pathway of *ras*. For much of that time, they were sure they were near a breakthrough. "It looked as if we had a result. A fabulous result," said Schechter. "We were manic about those experiments. But it all turned out to be an artifact."

Weinberg was dismayed by the setback; he'd come so close to the supreme victory. He'd found the *ras* gene, he'd cloned it, he'd discovered that it *was* the *ras* gene, and he'd pinpointed its mutation. But to reveal the gene's function, to discover the biochemical basis of cancer, was ultimately the only problem that counted. The difficulties of biochemistry notwithstanding, Weinberg wasn't about to drop what he'd begun six years earlier. Just as Woody Allen hears anti-Semitism even in innocuous remarks, so Weinberg couldn't learn anything new about the cell without trying to shape the information into a theory about the function of *ras*.

It was nearing the end of a conference in Saxtons River, Vermont. The weather was so-so, the food was bad, and the talks were dull. Weinberg had spent most of his time chatting with his colleagues, trying to tease forth news without giving away too many tidbits of his own. He was looking forward to going home, but he decided to attend one last seminar on a subject he knew relatively little about: phospholipid turnover in the cell membrane. A scientist was presenting the new and intriguing work of Michael Berridge, a biochemist at the University of Cambridge in England. And as the scientist spoke, Bob Weinberg's mental ratchets began to clatter and purr. "An idea occurred to me, as I'm sure it occurred to eighty other people sitting in the audience," he said. "At that moment I thought I understood the biochemistry of *ras*."

Phospholipids, as Weinberg already knew, are large, greasy molecules composed of a core of fatty acid and one or more arms of phosphate molecules. About 50 percent of a cell membrane is made up of phospholipids, and the other half of protein. Phospholipids help lend the membrane its structural integrity, yet they're not static molecules. They're continually synthesized, altered, and degraded in response to external stimulation, as when, for instance, a hormone receptor — a protein that pokes through the surface of the cell — catches a passing unit of hormone. The receptor then turns around to knead the phospholipids embedded in the membrane. As the phospholipids shift or dissolve, they affect other compounds within the cell.

Weinberg knew, in a hazy way, about phospholipid verve. What he didn't know were the details of the phospholipid under discussion. The seminar speaker was not talking about any or all greasy molecules in the membrane. He focused on the rise and fall of a phospholipid with the stumbling name of phosphatidyl inositol, or PI. When cells are stimulated, phosphatidyl inositol in the cell membrane is broken down into two compounds — diacyl glycerol and inositol trisphosphate. Both substances are capable of sparking great changes in the cell. Inositol trisphosphate induces the release of calcium from little storage pockets in the cytoplasm; the calcium goes on to hot-wire a number of enzymes. At the same time, diacyl glycerol, the second product of the bifurcated phospholipid, ignites the multipurpose enzyme protein kinase C, which starts prodding yet another string of cellular enzymes. Eventually, all that kicking and banging of enzymes reaches the nucleus, and the cell divides.

Phosphatidyl inositol, thought Weinberg, lies in the cell membrane. *Ras* lies in the cell membrane.

Diacyl glycerol detonates protein kinase C. *Ras* is believed to have some connection to protein kinase C.

The *ras* protein, Weinberg concluded, must work through the phospholipid pathway. *Ras* must modulate the breakdown of PI into its two active components. The oncogenic *ras* protein must indirectly tear apart the slippery molecule too often, releasing into the cytoplasm a continuous supply of not one but two pharmacologically powerful molecules.

Weinberg was ablaze with his new idea. The ramifications were unlimited. If he could determine the pathway of *ras*, he could start thinking about ways to block the pathway. He could start thinking about new treatments for human cancer. Not yet, of course; Weinberg was too rational to get carried away by fancies. Yet once the notion of phospholipid turnover had infiltrated his mind, Weinberg couldn't help considering the possibility of new therapies for cancer. When he returned to the Whitehead Institute, he was certain that his scientists would be able to fit his treasured *ras* oncogene into the phospholipid pathway.

Weinberg excitedly told the people in the lab about his latest model. This is it, he said. This is the biochemical basis of cancer. Weinberg's subordinates reacted as they often did to their leader's models: with jaded incredulity. Sure, *ras* could act through the PI pathway, but it could act through twenty other cellular pathways, too. "They were reserved," Weinberg recalled. "They thought it sounded too neat, too easy." Well, not quite easy: phospholipid biochemistry, they reminded Weinberg, is the most difficult biochemistry of all. They refused to work on it.

"Of course, I couldn't force people to do something against their will," said Weinberg. "But I had a new postdoc coming, a young woman named Snezna Rogelj. Being new to the lab, she was at my mercy."

Let's give it a try, he said to Snezna. If the idea works, your career is set for life. If nothing promising comes out of it within a few months, we'll cut bait and move on.

This time I agreed with Weinberg's assessment. Snezna Rogelj was at his mercy.

Meeting Snezna convinced me that Bob Weinberg is waging a private war against the myth of the science nerd. You know the myth. It's as persistent and unidimensional as America's vision of Communism. In movies and on TV, scientists are portrayed as eggheaded misfits. Doctors can be handsome and dashing, but research scientists are awkward and asexual. Male scientists have acne, wear glasses that are taped at the nosepiece, and keep a plastic pack of pens in their shirt pocket. Female scientists also wear glasses, are either chubby or completely fleshless, and talk in loud, whining voices. Genuine science nerds exist in real life. Bob Weinberg, when he's plodding around the Whitehead in his Earth Shoes and old sweater, his hair flying every which way, is a winsome sort of nerd. Yet a large percentage of the people who work for him are unusually fetching, which led me to wonder whether Weinberg didn't subconsciously choose researchers partly on the basis of their physical attractiveness. "Bob has something of the lech about him," said David Stern. "He'd never act on it, but he's a geezer. When you walk along the street with him, he turns around to gawk at every pretty woman who passes." (And if you're with Weinberg in a restaurant, said René, "forget about talking science. Bob is too busy watching the waitresses.")

If he has no inclination to act on his fantasies, Weinberg at least can surround himself with fair representatives of them. And the fairest of all, in the opinion of many men at the Whitehead, is Snezna Rogelj. Snezna is from Yugoslavia, and it took me weeks to learn to say her name; a gross approximation of the pronunciation is "Snezhna Roe-gull." Snezna isn't just pretty; she's dazzling. Her hair is a rich shade of auburn, the kind that glows in the sun like new pennies. Sometimes she piles it on her head Gibson Girl style, with loose locks falling around her face. Her eyes are wide and green, her cheekbones high, and her complexion as clear as morning. The only feature that she complains about is her nose: it has a slight hook to it, which gives her character and a more intelligent beauty, but she occasionally grabs and waggles it in playful dismay.

Snezna also has an excellent figure. For a while she kept a piece of paper on the wall of her desk area that she had once stuck by a telephone in the lab. It said, "Snezna Rogelj's extension is now 5175." Somebody had crossed out her first name with a black felt pen and written "Sexna" above it. She dresses very creatively, in a striped jumpsuit or in a short leather skirt that she had made from Indian purses. "When I first came to the lab, I wore high heels, because all the women in Yugoslavia wear them," she said. "But people would watch me walking around, carrying a bucket of dry ice, and tell me I was nuts to wear shoes like that. So I switched to these" — and she lifted her foot to show a flat-heeled shoe. "They drive me crazy. They don't support the foot at all." After a while, I noticed that she had switched back to wearing high heels most of the time.

Snezna was the beautiful lab den mother. She worried about people being happy. She took me under her wing. "You need a woman friend around here," she said. "You need a sponsor." When people in the lab began bad-mouthing a new postdoctoral fellow from Israel, calling her slow, Snezna came to me to praise her lavishly. "She's really amazing," said Snezna. "She was trained as an immunologist, and I think she's made tremendous progress in learning molecular biology. She's done so much. And all that while being a mother to two small children and keeping an Orthodox home."

I asked Snezna whether she was telling me this because she'd guessed I might have heard the insults.

"Oh, I know," she said in exasperation. "People around here are always quick to pass judgment. They think they can't look good unless they make other people look bad."

Alan Schechter, who was one of the quick judges, declared that Snezna was the only nice person at the Whitehead.

Yet Snezna is not treacly. She has her edges. When we attended a shockingly bad conference at Harvard Medical School and saw a woman give a presentation as though she were a TV game show hostess slinking seductively around a refrigerator, I couldn't wait to see what Snezna would make of the performance. As soon as we left the conference room, Snezna launched into a flawless imitation of the woman, cooing, flirting, and flouncing. "It's hard enough being a woman in science," said Snezna, "without worrying about bunnies like her."

Snezna also likes to argue, a taste that she claimed to have picked up from Vivek Dhawan, her husband. Vivek is an astrophysicist from Bangalore, India, where his father until recently was the head of the Indian Space Agency. The son is slim and graceful and almond-brown, every bit as good-looking as Snezna. At home or in company the two of them fought like stray cats, and always about the most high-minded subjects: whether economics is as legitimate a science as biology and physics, or who should decide how information about AIDS is distributed. "It's only since I've known Vivek," said Snezna, "that I've realized somebody can fight with you and love you at the same time."

There was one person, however, with whom Snezna would not argue: Bob Weinberg. She would never talk to him the way David Smotkin did when Smotkin felt put upon; she would never say, "Oh, you're being silly." She wouldn't say, "I can't do this," "I refuse to do that," or "I need help." Instead, she worked on the phospholipid pathway in utter isolation. She was the only person in the lab attempting to do difficult biochemistry, analyzing reactions between the *ras* protein and the various species of PI in the cell membrane. Whenever an experiment failed, she had only herself to consult. "Bob does his best to make suggestions, but he's no biochemist," she said. "Nobody really understands what I'm doing." Her colleagues wouldn't deny it.

"Just thinking about her project gives me a headache," said the postdoc Mike Gilman.

"Every time I hear about the PI pathway, I get a headache," said Bob Finkelstein, one of Weinberg's graduate students.

Quite frankly, Snezna did little to soothe their migrainous fog. She hated discussing her experiments, and that dislike showed when she was compelled to talk at group meetings. She'd scribble a diagram on the blackboard, sigh, mutter a barely audible explanation of what she'd drawn, qualify what she'd just said, draw another diagram, go back to the original figure to fill in additional complexities. Whenever I stopped by her workbench to see how her project was going, she'd sigh, pull out her latest autoradiographs reluctantly, start to explain the data, and become so flustered that I'd worry that she might start throwing her film around the room. She preferred to talk about anything else — Vivek, me, clothes, leftist politics. She was a lousy saleswoman of her own work.

And yet I never stopped being enthusiastic about Snezna's project, partly because, as a nonscientist, I could ignore contradictory or messy details, and partly because Weinberg insisted that her experiments were the most important in the lab. Once I began to understand it, I thought that the system she'd developed was lovely. It had clarity and logic. This isn't so hard, I thought. Maybe the cell really is as simple as Weinberg says.

The basic thrust of Snezna's experiments was to examine PI turnover in cells that differed only in their kind and apportionment of *ras* genes. Snezna had cultivated three types of cells: healthy NIH 3T3 cells, which had the standard complement of two indigenous *ras* genes; mouse cells that had been given extra copies of the normal, noncancerous *ras* gene; and mouse cells transfected with the tumorigenic *ras* gene. She then injected the cells with radioactive guanine triphosphate, or GTP. There were two reasons for the addition of GTP. The compound is a source of phosphate molecules, and phospholipids, as they shift, tumble, break down, or build up, require a steady supply of new phosphate molecules. What's more, the *ras* protein is known to need GTP to be active. Snezna wanted to measure the flux of phosphatidyl inositol in response to stimulation by growth proteins: whether and how the radioactive phosphate groups were absorbed by the phospholipids. She theorized that if the *ras* protein communicated with membrane phospholipids, and if the cancerous *ras* protein was an especially enterprising modulator of PI turnover, the phospholipids in the mutant *ras* membranes should consume more phosphate molecules when stimulated than either the healthy NIH cells or the cells with the surplus copies of the normal *ras* gene.

After exciting the membranes with growth-promoting proteins, she solubilized the membranes in large glass tanks and dipped chemically treated sheets of plastic into the solution. The phospholipid molecules from the dissolved membranes crawled up the sheets and stopped at a point determined by their molecular weight. An x-ray snapshot of the sheet revealed whether there was any difference in the radioactivity of the PI in the three classes of mouse cells — whether *ras* had any effect on how many phosphate molecules the phospholipids absorbed. She measured other parameters as well. Sometimes she looked for an increase in the amount of free inositol trisphosphate, one of the two active byproducts of PI. Sometimes she examined the condition of protein kinase C, the enzyme thwacked by inositol trisphosphate, or of phospholipase C, yet another enzyme situated along the PI pathway. Always she searched for discrepancies among her three categories of mouse cells.

Snezna found discrepancies and she lost discrepancies. One day she'd see a fiftyfold increase in PI stimulation in her cancerous *ras* membranes. The next day she'd repeat the experiment, and the increase would drop to two- or threefold. She'd repeat the experiment again, and her healthy NIH cells would seem to take up more radioactive phosphate. Sometimes she thought *ras* stimulated phospholipase C; sometimes she thought it communicated with protein kinase C. Sometimes she thought the oncogenic *ras* protein caused PI to be degraded more rapidly into its active subunits. More often, she decided that *ras* boosted the creation of new phospholipids on the other end of the equation.

In sum, nothing Snezna did was reproducible. Most of the experiments took only a day or two to complete, but somehow days aggregated into weeks, weeks into months, months approached a year, and still she couldn't get the same results twice.

"Do you see those notebooks?" she said, pointing to a shelf crowded with black plastic looseleaf books. "Those notebooks are the history of my postdoc. I have so many data. I have reams and reams of data. I've learned so much about lipid metabolism. I'm really glad I've done these experiments. But nothing is reproducible!"

I asked whether she couldn't cobble together into a good paper the information she had gathered thus far.

"And say what?" she replied. "That I haven't been able to reproduce any of my results?"

"I don't know. Your observations?" I suggested lamely.

She shook her head. "It's so frustrating. I believe that *ras* does act somewhere along this pathway. But I can't make head or tail of what's going on. It could be a primary interaction, secondary, tertiary — who knows? Why should it be that one day I get one result, and the next day, when I repeat the exact same experiment, under the exact same conditions, I get nothing, or I get the opposite result?

"We could be on the completely wrong track. That's how little we know about this signaling pathway stuff. Let's say there's a slight increase in PI turnover in *ras*-transformed cells. What does that tell us? So many things happen when cells are stimulated. So many enzymes are activated, so many changes in the membrane. It's almost impossible to separate out one signal from another, to know if the effects you're seeing are primary or secondary or tertiary. I could try working with isolated components *in vitro*, but purifying lipids is a bitch. Besides, when you're doing *in vitro* biochemistry you invariably leave something behind. If we want to find the pathway of *ras*, we can't just start picking out enzymes at random. I want to know what really happens in the cell."

("Talk to any biochemist about the irreproducibility of biochemistry," Mike Wigler said to me. "They'll always say the same thing: how hard it is to reproduce even the simplest things. In the nineteenth century, German scientists had a nice term for it. They called biochemistry 'dirt chemistry.' And it still is. It's dirt chemistry.")

"Are you worried?" I asked Snezna. "Are you worried because you're a postdoc and you haven't published any papers yet?"

"I don't care about papers," she said. She jabbed her pipette into a box of plastic tips. "You should stand back a little. These tubes are really hot."

Hot means radioactive. I took several steps backward.

"I don't care about papers, and I don't care about making a name for myself," she continued. "That's not important. What's important is that I prove to myself that I can do this." She jabbed her pipette again. "I have to prove to myself that I'm a scientist, that I can design a difficult experiment and just do it. I have this little voice in my head, telling me I can't do it, and I have to shut that voice up."

I believed Snezna when she said she didn't care about papers and fame, and I think her indifference stems in part from her background. Cori Bargmann once told me that most research scientists come from middle- or upper-middle-class families. "Scientists need strong egos to go into something as risky and anonymous as basic research," she said. (And it's true that few scientists are the children of plumbers or elevator operators; truer still that almost no molecular biologists are black or Hispanic. Of the hundreds of biologists I heard present seminars at various conferences, only one speaker was black.) Snezna's upbringing was unstable and occasionally impoverished. She spent her girlhood in Ljubljana, Yugoslavia, where her father often was unemployed and Snezna and her brother sometimes went hungry; today, she always keeps something to eat — an apple, a candy bar — in her desk or pocketbook. As a young teenager, she moved to her uncle's house in Columbus, Ohio, and her life improved. Yet Snezna's ego remains frail. She worries about not being smart enough or good enough or sufficiently respected.

"Tell me something," she said to me at a party. She looked embarrassed. "What do people in the lab think of me? What do they say about me?"

"Alan Schechter thinks you're the nicest person in the lab," I replied.

"I mean as a scientist. Do they think I'm a good scientist?"

I hesitated. I'd heard less about Snezna's scientific ability than I had about her personality and her appearance. People seemed unable to judge the merits of her work. "They say they don't understand your project," I replied. "They say just thinking about it gives them a headache." I was relieved to see that my answer made her smile.

Snezna's low self-esteem tugged her in paradoxical directions. On the one hand, she wanted to be respected as a scientist; on the other, she denied the importance of the standard parameters of success: number of papers published, prizes, fame. Nor did it help her that Weinberg took such a keen interest in her experiments. Snezna would try to explain to him the difficulties of the approach. He'd make suggestions. Why not use antibodies against the *ras* protein? he'd say. If you could demonstrate that by blocking the activity of *ras* you blocked PI turnover, that would be compelling evidence that the *ras* protein worked directly on phospholipids. She tried anything he or she could dream up. Her whole-cell system proved too complicated. All the reactions that concerned her took place in the cell membrane, so she designed a system by throwing away everything except the membranes. She tried working with a form of GTP that couldn't be dissolved by water. She varied the time points of her experiments. She was too proud to complain to Weinberg of her unhappiness in any but the most oblique manner.

When I came and sat by Snezna's bench and watched her clicking the button on her pipette or stretching her arms around a plastic shield that protected her from radiation, I thought that her entire body looked depressed; even her hair and fingernails looked depressed. She sniffled a lot. She often seemed to have a cold.

"If nothing comes up soon," said Weinberg, "we'll have to cut bait. I can't sacrifice her career on the altar of my ideas."

But then something provocative would come up to reconsecrate Weinberg's altar. Returning to the lab after salmon dinner one night, Snezna performed an experiment that gave her the best results ever. She showed me the autoradiograph the following morning, and for the first time since I'd met her she sounded optimistic — guarded, but optimistic. During the reaction, the level of PI turnover had increased incrementally, starting at a low basal level in the NIH 3T3 cells, rising in the cells with the extra copies of normal *ras,* and exploding in the *ras*-transformed cells — just the results she'd expect if the *ras*-phospholipid model was correct. "It must have been the salmon," she said. "It was great salmon." When she repeated the experiment that day, she got somewhat less dramatic results, but at least they followed a similar pattern of incremental rise from normal to cancerous cells. Weinberg asked Snezna to talk about her findings at the weekly group meeting.

"I'm very encouraged by this," he told the lab. "These are the first positive results we've gotten in almost a year. I think Snezna may be on to something." After the meeting he wagged his finger and said to me, "There's something to this. I'll be vindicated yet. I can smell it."

Within a few days, however, the encouraging results had collapsed into their old irreproducibility, and observers began to deride the whole approach. "Phospholipid turnover is a red herring," said David Baltimore. "It obviously has something to do with transformation, but I'd guess that it was a consequence of transformation rather than a primary event."

"If Bob was serious about the PI project," said Cori, "he'd set something up along the lines of Alfred Gilman's lab at the University of Texas. Gilman is studying G proteins, period. He has twenty or thirty people who do nothing but study G proteins. They purify the catalytic subunits of G proteins, the inhibitory subunits of G proteins. They identify the G proteins' targets. It's a philosophy of science that goes against the grain of MIT, but Gilman is an excellent biochemist. His lab is picking apart one kind of protein, piece by piece. Bob won't do that. He says, 'let's look at these phospholipids in the membrane and hope that *ras* fits in somewhere.' He says to Snezna, 'You do it. You solve, by yourself, the sort of biochemical problem that other labs have dedicated twenty people to solving.'"

"I don't know why she let herself get talked into this," said René. "If Bob had tried to convince me to work on something like phospholipids when I first came to the lab, I would have laughed. No way. It's like signing your own death certificate."

I asked Snezna whether she'd be happier if Weinberg put more people on the project. "I'd be happier not so much because I want help doing the experiments," she said, "but because it would be nice to have somebody to talk to about my work. I feel really isolated here."

Bob was reluctant to assign extra hands to the effort. For one thing, the resident scientists hadn't changed their minds about the viability of the project, and Weinberg knew that none could be persuaded to dump what he or she was doing to help Snezna; for another, it was a Weinberg tradition that a single, lonely researcher ply a risky project until the results were promising enough to warrant additional lab resources. Smotkin had begun transfections; Mitch Goldfarb and then Chiaho Shih had transfected oncogenes; Shelly Bernstein had searched for the metastasis gene.

"What are you going to do about Snezna?" I asked Weinberg as her project approached the one-year mark.

"We may have to think of something else for her to work on," he said, "but I hope that doesn't happen."

I asked him whether he thought Snezna needed help.

"That wouldn't be fair to her," he said. "She's been working on this by herself for so long. If she succeeds, why should anybody share the glory with her?"

"But you still believe in phospholipids?" I asked. "Baltimore says that he thinks they're a red herring."

Weinberg grew churlish at mention of the criticism, which I knew he'd heard from Baltimore before. "Baltimore is entitled to his opinions," he said. "But they're just that — opinions. He may be right, or he may be wrong. He has no proof for saying such a thing, one way or another. I happen to believe that he's wrong, although I don't have proof to support my conviction, either."

As the weeks wore on, Snezna spent less time in the lab. She didn't come in on the weekends, and often she didn't stay late during the week. She and Vivek took a vacation in India, and when she returned, Snezna had made a decision: she wasn't going to stay in cancer biology. She wanted to apply the tools of genetic engineering to parasitology or infectious disease — some field that would be of concern to the majority of the world's people. "In Western countries, everybody is scared of cancer, so that's where all the research money goes," she said. "Nobody in the U.S. or Europe cares about malaria or hookworms or tropical diseases. Nobody builds a Whitehead Institute for Parasitology. Sure, there are a few groups working on a malaria vaccine, but that's only because the recombinant DNA technology involved is flashy. If a disease isn't flashy or sexy, who cares?

"But most people in the world don't live long enough to get cancer. Cancer would be a luxury to them. If they survive childhood malnutrition, they die at twenty-four of parasites, or tse-tse flies, tuberculosis, cholera. I'd like to do something that has a more direct impact on people's lives, even if that means I'll be an obscure scientist in an obscure little clinic working on a shoestring budget. What's science for? For ourselves? Is it some sort of fancy masturbation?" When Vivek finished his doctorate, she said, they would move to India. "Shiv thinks I'm crazy," she said, referring to Shiv Pillai, a David Baltimore postdoc from Bombay. "He said, 'Snezna, that will be the end of your scientific career. You don't realize how primitive Indian science still is.' That's why he came here, because he couldn't do what he wanted back home. But I'm sure I'll find something worthwhile to do. If not, I'll start up something of my own. Who knows? But I'm sure oncogene research can survive without me. The future of cancer biology doesn't depend on Snezna Rogelj."

When she talked about phosphatidyl inositol, Snezna drooped like an old scarf. When she talked about Third World poverty, about the woman she befriended in India whose monthly salary wouldn't pay for a day's worth of one enzyme at the Whitehead, Snezna became a firebrand. Dreaming about her future, however, didn't ease the present. She was still at the Whitehead, and her experiments weren't going anywhere. Snezna wanted to do something else, but Weinberg was reluctant to admit defeat. He believed that his theory remained viable; he couldn't shake the sweet memory of Snezna's post-salmon success. Other labs were working on the connection between oncogenes and PI turnover, and the idea seemed to be gaining momentum. Weinberg refused to allow his lab to drop out of the *ras* race. As the midway mark of her three-year fellowship approached, Snezna grew desperate. So she did what many postdocs do when confronted with a hated project and an implacable patriarch. She eloped.

The Whitehead Institute is a center for advanced research, but as an affiliate of MIT, it's also a university; people are there to learn. Weinberg considers himself as much a teacher as a scientist. He teaches one or two graduate-level courses at MIT each year, and he teaches the people who work for him how to do and think about science. As part of the in-house adult education program, Weinberg often invites scientists from other universities to speak at group meetings.

In the summer of 1985, Michael Klagsbrun, a cell biologist at Harvard Medical School, talked to the Weinberg lab about his research on the blood vessels of tumors. Judah Folkman, a colleague of Klagsbrun's, had discovered that a neoplasm can grow to about a millimeter in diameter on the amount of blood supplied by the blood vessels in normal tissue. To continue swelling, however, a tumor needs more blood and more blood vessels. How to stimulate new vessel formation? Tumor cells, said Klagsbrun, craftily exploit the body's built-in mechanism to seed fresh capillaries. They secrete a protein called angiogenesis factor — *angio* from the Greek word meaning vessel, *genesis* meaning creation. When stimulated by angiogenesis factor, nearby blood vessels begin branching off into little buds. The venule buds lengthen into thin tubes, converging on the small colony of tumor cells from all directions and finally penetrating it. The tumor turns pink, a rush of blood-borne nutrients pours in, and wastes are speedily removed. Fed by the new capillaries, the malignancy rapidly grows in volume to a cubic centimeter or more.

The synthesis of angiogenesis factor normally is a rare and specific cellular event. During pregnancy, said Klagsbrun, the mother's uterus releases angiogenesis factor to form the extra blood vessels needed to supply a developing fetus with oxygen and nutrients. When a person is wounded, the capillaries surrounding the injury secrete angiogenesis factor as part of the healing process. Tumor cells simply make a rare event commonplace, said Klagsbrun. All solid tumors, in all organs, begin secreting the blood vessel factor once they've bulged beyond the size of a pinhead. If deprived of the factor, a tumor stops growing and eventually dies on the vine. Significantly, said Klagsbrun, human malignancies almost never metastasize until they're larger than a millimeter, the size when a tumor begins secreting angiogenesis factor. In other words, cancer spread and blood vessel formation proceed hand in hand.

The Weinberg scientists were mesmerized by Klagsbrun's talk. They hadn't heard much about angiogenesis factor, and they certainly didn't know about its role in tumor growth. "The guy's got the cure for cancer right there," Alan Schechter said at lunch that afternoon. "All you have to do is block angiogenesis factor. You could either block the synthesis of the factor or block the angiogenesis receptors on the blood vessel cells. There wouldn't even be any side effects, unless you're a fetus or you've just had a heart attack. Jesus, forget about oncogenes. Just go after angiogenesis factor."

Everybody chattered about working with Michael Klagsbrun. Lacking skills in molecular biology, Klagsbrun hadn't yet cloned the gene for angiogenesis factor. Instead, he'd spent ten years purifying the protein from tumor and uterine tissue, a tiresome, painstaking procedure. Klagsbrun had asked the group, "Do you think there's any molecular biology that can be done with this protein?" and the scientists had laughed in disbelief. Can you do molecular biology? With the amount of purified protein that Klagsbrun had around, they said, a molecular biologist could work backward, designing DNA probes based on the amino acid sequence of the protein. A molecular biologist, they said, could clone the gene in his sleep.

Snezna sat at lunch, smiling to herself and glowing like a candle. She listened to Alan Schechter and Mike Gilman discuss which researchers Klagsbrun might get to help him clone the gene. She didn't talk much, but I could see that she was listening closely.

The next week, Snezna told Weinberg that she wanted to collaborate with Klagsbrun. Weinberg didn't respond immediately, but over the following months she made occasional forays to Harvard to talk with Klagsbrun. In the winter of 1986, Klagsbrun came on a year's sabbatical to work at the Weinberg lab. And the scientist he worked with was Snezna. They studied angiogenesis factor. They sought out related growth-promoting proteins that might specifically stimulate the generation of other organs in the body. Snezna discovered a new type of growth protein in mother's milk, which presumably has some beneficial effect on a breast-feeding infant. As she analyzed the peculiar growth factor, she gathered data that would likely result in a paper. She didn't exactly abandon the phospholipid project; she simply allowed her research with Klagsbrun to consume more and more of her time. I stopped seeing the glass tanks and plastic sheets on her bench.

"The PI project was so depressing. Even now, when I think or talk about it, I get depressed again," she said. "With this work, I feel as though I have a handle on something. I'm in control of the experiments, not being controlled by them. Oh, I know Bob doesn't care as much about angiogenesis factor or growth factors in milk. Maybe he's disappointed in me. I wanted to make him happy, but I also have to make myself happy." Snezna looked and acted happy. She returned to her old twelve-hour shifts, and she came in on the weekends. Her colds and sniffles disappeared. "I think she'll get an interesting paper out of her work," said Weinberg. "Nothing earth-shattering, but it could be interesting. The important thing is that Snezna should feel that she's being a productive member of the lab. She's working on something do-able. I don't know if that's a legitimate word — do-able? But you get my drift. Her career comes first."

Weinberg didn't discard the phospholipid idea. He put it aside for a little while. He wasn't a bully, and he wouldn't push people against their will — but he did not forget. Although growth factors in veins and in milk were fine, they were not conceptually profound. The study of angiogenesis factor would not radically change scientific understanding of the cell. Weinberg cared more about the big questions, the enzyme cascades, the proteins, lipids, and phosphates spinning and hurtling against one another, the coordination of tens or scores or hundreds of signals into one mammoth signal that tells the cell: Hurry up, please, it's time — to divide, to desist, to die. Weinberg believed, with the ferocious tenacity that had sustained him through the bedevilments and saharas of the past, that understanding the function of just one oncogene would blow all of biology wide open. If the *ras* problem remained for the moment intractable, perhaps some other oncogene might be understood.

"Mike Gilman," said Weinberg, "is doing one of the most potentially important experiments in my lab."

Mike Gilman has a wonderful facial expression that makes him look as though he's smiling even when he's disgruntled or despondent. His skin and hair are the color of wheat, he's forever pushing his glasses up his nose, and he's about ten or fifteen pounds plump. He's like a busy, friendly marmot.

His tone of voice is happy, too. It has a little lilt to it, a fringe of vocal apostrophes; he's always on the perimeter of a giggle. He enjoyed talking with me, and he was unfailingly generous with his time, but I was often stumped by his apparent cheeriness. I would scrutinize his wheat-pale face and listen to his singsong voice and try to collate what I saw and heard with the sense of his words. Was he feeling optimistic or fatalistic? Did he mean what he'd just said, or was he being ironic? Now was he happy, or was it the sunniness of the lab that made his strange smile beam brighter? Did he trust and admire his boss, or did he not give a Susan B. Anthony dollar for Weinberg's style of science? Whenever I was baffled, I would await David Stern's reaction to what Mike had said. David worked at a bench opposite Mike's, and he was Gilman's closest friend in the lab. David would smirk to himself or say to me, "That's bullshit," or "Be sure you write that down. He's whining again."

Mike Gilman did whine, but it was an amusing, self-deprecatory sort of whining. If he complained about being unappreciated, he acknowledged that he'd done little worthy of appreciation. When I met him, Mike had been in the lab for almost three years but claimed that he still felt like an interloper, an anomaly. It was amazing, he said, that he'd weaseled his way into Weinberg's group in the first place. As a graduate student at the University of California, Berkeley, Mike had studied gene expression in bacteria. He believed that Weinberg usually chose postdoctoral fellows who worked on subjects that intrigued him, and *Bacillus subtilis* was not among his preferred topics. Gilman felt that he didn't have much to offer Weinberg, but after reading Weinberg's early transfection papers, he was so excited by the new approach to cancer genetics that he figured he'd take a shot at it: he wrote a letter to Weinberg requesting a position as a postdoc.

"Bob wasn't such a big deal yet, but I guess his star was rising fast," said Mike. "So a couple of weeks later I called him and said, 'Hi, this is Mike Gilman.' There was a long pause on his end, and finally, 'Ye-e-s?' 'Well, I wrote you a letter asking if I could do a postdoc at your lab.' Another few moments of silence. 'Ye-e-s?' At that point I was really beginning to panic, thinking this guy doesn't know me from Adam. Finally I just decided to fly across the country to see him, and when I got here he showed me this stack of letters from people requesting postdocs. I got in just before the deluge. I think the only reason he took me was that now he had a face that he could connect to a name. If I were to apply today, I wouldn't have a chance."

"Oh, come on," I said. "Do you really believe that?"

"It's not that I think I'm less intelligent than other people here," said Mike. He twitched his pug nose to prevent his glasses from sliding down. "I think I'm as good a scientist as anybody else. It's just that the stuff I was working on wouldn't really have interested him if he'd stopped to think about it. I was a prokaryote jock. Maybe he was impressed because I'd read his papers and given them some thought, but I didn't have any particular expertise to offer him beyond my intellect. Today Bob must get two hundred letters a month from people, begging to come work for him." (The actual figure is closer to eight a month.) "Bob can look for intellect and specialized skills and a graduate résumé with fifty publications on it."

His anecdotal modesty, however, couldn't mask his sturdy ego. In contrast to Snezna, who's shy about voicing her opinions, Mike often spoke up at group meetings, interpreting others' results and proffering advice. He was secure enough to be scientifically supple. Many researchers, once they find an experimental protocol that works, stick to it long after the introduction of a better, quicker method; Mike frequently was the first in the Weinberg lab to adopt advances in recombinant DNA technology. For all his good nature, Mike could fight; he had sass. And he didn't much care about making Bob happy.

One afternoon, for example, Weinberg padded over to Gilman's bench to tell Mike about a result from Gerald Fink's lab. Fink had discovered a membrane protein in yeast cells that Weinberg thought Gilman should follow up on. "I think it might be helpful to your project," he said.

As Weinberg spoke, Mike didn't bother looking up from his experiment, and he didn't hesitate a moment before criticizing the suggestion. "I'm not really interested," he said. "At this point, I think it would be a waste of my time."

"Are you sure?" said Weinberg. He went into detail about what the Fink lab had done, but Mike showed no flicker of enthusiasm.

"The next time I have a free Saturday, Bob, I'll look into it," said Mike. "But I don't anticipate having too many free Saturdays."

Weinberg shrugged and wandered off. I was aghast. I imagined how one of my magazine editors would have reacted to such a retort. I recalled a job I'd had right after college; I had refused to answer the telephone because I thought I had more important things to do. Two weeks later, I was unemployed. "You didn't even fake it," I said to Mike.

"Why should I fake it?" Mike said. "If I were to take every one of Bob's ideas seriously, I'd never get any work done."

"I wonder if you guys realize how spoiled you are," I teased. "You wouldn't last more than a month at a corporation."

"You're probably right," said Mike. "In fact, I might not last as long as a month. But I doubt that Bob is going to fire me because I'm not jumping up and down over a yeast protein."

Job security is not the only way in which Mike and the other postdocs are pampered. Scientists can dress as they please and drift into the lab when they please. They're trained to be skeptical of what they're told, so they learn to say whatever pops into their head, however cruel or tactless; if they notice that somebody has gained weight or is breaking out, they'll say it. And their time is pure: they're beyond doing course work and taking exams, but they're not yet burdened by administrative inanities, grant writing, or teaching loads. "Postdocs tell me that they treasure the years of their fellowship, because they know that it's the last time they'll be able to concentrate totally on the work," said James Broach, a geneticist at Princeton. "They're not responsible to anybody but themselves."

"I'm in no hurry to leave Bob's lab," said David Stern. "I feel protected here. I'm taken care of. Bob looks after getting the money and keeping the lab going, and I can spend all my time at the bench, playing around."

But the freedom is in no way worry-free. A postdoctoral fellowship is the most critical and intense period of a scientist's career. It's the professional equivalent of the SAT's: it doesn't last long, but the results have unnerving resonance. "A postdoc can afford to blow the first year or eighteen months of his or her fellowship," said Weinberg. "It's expected that one will spend a certain amount of time experimenting with different ideas, and even floundering. But after that, a hiring committee at a university may begin to wonder, Does this person have sound scientific judgment? Does this person know when a problem is intractable, and can this person tell when it's time to move on? Good judgment is as important to a scientist's success as is tenacity or intuition."

From week to week, month to month, postdocs must assess their progress and weigh the riskiness of a project against its potential payoff. This is the postdoc's conundrum. Young researchers must publish. In science, publish or perish is not a cliché; it's book and word. A small paper on a small result is better than no paper on a cosmic undertaking.

But postdocs — particularly at a trendsetting center like the Whitehead — must distinguish themselves. They must do something original, something that will flower into a project capable of sustaining a whole new lab. Like capitalism, science must continue to grow in order to stay the same. Each postdoc must think beyond an immediate experiment to developing a "system" that will eventually keep three, four, ten people busy. A great experiment is great not only on its own terms; it's great because it has a future spin. "It opens doors that you hope others will follow you through," said Mike Wigler. Timid experiments may yield concrete results, but they're unlikely to establish paradigms. And paradigm makers are what the prestigious universities seek when selecting new faculty members.

Mike Gilman was trained as a small-scale scientist. He was an enzymologist. "I did clean, *in vitro* science," he said. "I mixed together two enzymes in a test tube, and I knew I'd get something out. I knew I'd have a result." How much impact the results would have was another matter: gene regulation in *Bacillus subtilis* is of concern only to a handful of microbe cognoscenti.

"It can be said with complete confidence," the late Sir Peter Medawar wrote in *Advice to a Young Scientist*, "that any scientist of any age who wants to make important discoveries must study important problems. Dull or piffling problems yield dull or piffling answers." Medawar cited a hypothetical example of a zoologist who decided to find out why 36 percent of sea urchin eggs have tiny spots on them. Declared Medawar, "This is not an interesting problem."

So when he moved to MIT, Gilman traded the clean assurance of "piffling" test tube reactions for what he would later call the black box of the cell. He didn't care about making Weinberg happy, but he quickly adjusted to Weinberg's philosophy of science. He took on a big, chancy project, one that would either raise him to fame and guarantee him a position at the university of his choice — or leave him bankrupt. He was unlikely to obtain any intermediary results; the experiment had to work as his model predicted, or he'd have nothing to say. In short, his project was exactly like Snezna's: it was an altar of an idea.

Snezna was working on the function of *ras*, attempting to trace the signal of the oncogene gene by studying the loops and whorls of proteins and lipids in living membranes. Mike too was studying oncogene signaling. He wanted to dissect the biochemical changes that are associated with normal and cancerous cell division, which is why

Weinberg described Mike's project as "one of the two most important" in his lab. Oncogene function was the outstanding problem in the field. Scientists were desperate to know how the genes that supposedly modulate some aspect of cell replication actually modulate cell replication. Until biologists understood the grand biochemical cascades of oncogenes, all their cloned genes were so many *chotchkes* of DNA. Researchers certainly couldn't hope to devise clinical interventions of cancer gene activity — to apply their knowledge to cancer treatment — before they knew the nature of that gene activity.

Weinberg wanted to understand oncogene function. And if he couldn't get at the function of *ras,* the workings of another oncogene would do nearly as well. Gilman went for broke, seeking the biochemical signals that controlled not one but two oncogenes: the *myc* gene, which had been implicated in several human leukemias and lymphomas (and which Hucky Land and Luis Parada had mixed together with the *ras* gene to turn fresh tissue cancerous), and the *fos* gene. First detected in a retrovirus that causes mouse bone cancer, the *fos* gene is another indispensable cellular gene that is vigorous in nearly all organs of the body. Like the *myc* gene, *fos* synthesizes a protein that works in the cell nucleus, presumably socializing directly with the DNA molecule. *Myc* and *fos* have more in common than the location of their proteins. When cells get ready to divide, *myc* and *fos* are among the first genes to flick on: they're front-line genes that a splitting cell cannot do without. They may also play a negative role in cell maturation. Scientists have found, for example, that the *myc* gene must shut down before primitive white blood cells can become T cells, B cells, macrophages, and other denizens of an adult immune system; without a *myc* shutdown, blood cells cannot differentiate, and thus retain a deadly capacity to divide with the recklessness of amorphous, fetal-like tissue. *Myc* and *fos* are obviously important genes with much to say about cell growth and maturation, signaling pathways and cancer.

Mike was looking at the expression of the *myc* and *fos* proto-oncogenes in normal rodent cells. He reasoned that if he could decipher the function of the healthy genes, he'd be a nose away from understanding the genetic transgressions that promote tumors. The project had several points in its favor. From a technical standpoint, it was easy. Unlike Snezna, Mike didn't have to soil his hands with biochemistry. He performed RNA analysis, checking to see when the two different oncogenes switched on and shut down in response to stimulation. I have a photograph of Mike Gilman leaning over the lab freezer, because that's where I often found him. I'd hear the crunch and hiss of dry ice being shoveled into a bucket, and I'd know that Mike was gearing up for another RNA experiment. Purified RNA disintegrates rapidly, and Mike had to keep his little tubes of RNA on ice until he was ready to throw everything onto an agarose gel and screen the RNA with radioactive probes. Moreover, Gilman liked the genes because he had a model for their behavior that might have suited Adolf Loos, the modernist architect who equated ornament with crime: this model was stripped down to skin and girders. It was a simple, testable, Weinbergian sort of model.

Mike set out to find the physiologic switches that stimulate *myc* and *fos*. He knew that the genes must be ignited before cell division, and he wanted to identify the cellular changes that turned the key. "It was a kind of ass-backward way of getting at an oncogenic signaling pathway," said Mike. "Rather than looking at a protein like *ras* and trying to figure out the sequence of all the target enzymes it affects, I started with the knowledge that *myc* and *fos* are activated and then tried to see how they got that way."

He considered a couple of knowns. First, he knew that the genes respond vigorously to a growth protein called platelet-derived growth factor (PDGF), a hormonelike substance carried by red blood cells to help heal wounds. When Mike applied PDGF to mouse skin cells, he detected abundant *myc* and *fos* messenger RNA bubbling through the cytoplasm, a sign that the genes were switched on and ready to be turned into protein.

A second, more telling clue was the divergent time courses of *myc* and *fos* activation. Although both genes respond dapperly to PDGF, Gilman saw that they do so in idiosyncratic patterns. *Fos* turns on first and shuts down first. *Myc* is more languorous, responding later and lasting longer. Gilman wondered what the wound-healing protein could be doing to the mouse cells to have such different effects on *myc* and *fos*. He wondered what sort of divergent signals PDGF was sending through the cytoplasm to the two genes nestled on the DNA molecule. Scanning journals old and new, he looked for reports of biochemical fluxes that might correspond to the time contours of *myc* and *fos* activation. Mike was hoping that somewhere a scientist had described cellular events that just happened to start and stop at the same moments *myc* and *fos* turned on and off.

Miraculously, he found two events that fitted. The *myc* connection came first. Cell biologists had shown that a cell's internal pH level — a measure of the concentration of hydrogen ions — changes as the cell prepares to divide. The shift in pH is slow but sure. Hydrogen ions begin drifting out of the cytoplasm an hour or so before division and then drift out again through the membrane right before the big moment of replication. Mike noticed that the time curve of pH rise and fall had the same topography as that of *myc* activation and deactivation. So pH rise was Mike's candidate for the biochemical event that switched on the *myc* gene. "You worry about placing your hopes on something like pH levels," he said. "It seems so obvious that you figure everybody studying *myc* would have already thought of it. They would have said, 'Oh, yeah, pH turns on *myc*,' gone ahead and done the experiment, and then come up negative. You'd never hear about the negative results. But then again, just because something is obvious doesn't necessarily mean it's been thought of before. Good ideas *always* seem obvious after you've thought of them."

A nominee for the *fos* switch soon followed, and it proved equally unexotic. Another response to growth proteins, Mike explained, is a jump in internal calcium levels. Cells hoard calcium in dense pockets scattered throughout the cytoplasm. When the cells are signaled by growth factors, the pockets burst, spraying calcium across the cytoplasm and leading to an overall spike in calcium concentration. Over the next several minutes, the freed calcium galvanizes a string of known and unknown enzymes presumably critical to cell division. Its task complete, the calcium is stuffed back into its storage purses. The reaction is quick and sharp — exactly like the activation of *fos*.

Mike Gilman's equation was thus complete. A dose of PDGF hikes the pH level and releases calcium stored in cells. The pH rise turns on *myc*, the freed calcium turns on *fos*, the cells divide. When he told me his model, I was delighted and grateful. Listening to Snezna had often left me in a muddle.

Even Gilman's efforts to test the model were blissfully straightforward. All he had to do was to raise the pH inside the cells through some clever artificial means and demonstrate that *myc* responded but *fos* did not; then he would pump up the calcium level in mouse cells to show a singular reaction by *fos*. "That wasn't hard," he said. "I had a few fancy tricks up my sleeve." By manipulating the conditions of the tissue culture serum that he fed to the cells, Mike was able to raise the pH levels within the cytoplasm. To elevate the calcium concentration, he sprinkled the dishes with calcium ionophores, little calcium transporters that can penetrate the cell membrane.

After subjecting the rodent cells to one or another treatment, he isolated the RNA and combed through the thousands of messages for signs of active *myc* and *fos*.

Mike showed me the autoradiographs on which he had screened the cellular RNA with hot *myc* and *fos* probes. I saw that pH did indeed boot *myc* into gear while leaving *fos* cold. I saw, too, that the calcium ionophore bath had the opposite effect. On the bottom of the film Mike had marked the date of the experiment: March 1984. A year earlier.

"Pretty convincing, huh?" he said.

"It's great," I replied, staring at the shadows of radioactivity on the film. "It looks real."

"Yeah, well, I thought so too," he said. "I thought it was just the grooviest thing. I repeated the experiment about five times, and it kept working. I thought I had it made. I thought, Great, I'm a first-year postdoc, and I've figured out the signaling pathways for *myc* and *fos*. I was getting down to the brass tacks of how these genes work in the cell. Hundreds of people are trying to do the same thing, and I'm the one who gets it."

He pulled out other autoradiographs from the same year. They looked real, they looked groovy, they looked as though Gilman had found the biochemical pathways for not one but two cancer-related genes: calcium-controlled *fos* and pH-controlled *myc*. I imagined how Mike felt when he first saw those blots, believing that he was on to something so big yet so easy. He had a model, he tested it, and it worked, five times in a row. He had found the biochemical basis for oncogene activation. His ass-backward approach to signaling pathways had twirled around and marched forward.

In the spring of 1984, everything about Mike's life was groovy. His daughter, Melanie, was born, and though he'd been apprehensive about fatherhood while his wife was pregnant, Mike greeted the new arrival with uninhibited joy. "The first time I held Melanie," he said, "was the best moment of my life." Mike doesn't seem much like a father. Through his dress and demeanor, he still pays noncommittal homage to the sixties, the way so many young — and unencumbered — scientists do. But Mike is the sort of father we grownup daughters wish we'd had. He loves Melanie, his beautiful girl, who has the dark button eyes and matching dark hair of her mother, Laurie. Whenever Mike talked about Melanie, I had no doubt of the sincerity of his smile and his joy. He talked about the first time she crawled, the first time she walked, her first words, her subsequent words. He talked about how she liked to fiddle around with his stereo equipment, as though she knew her father had once been a part-time disc jockey. He talked about her imitations of cow sounds and duck sounds.

"The other day, I was walking through Harvard Square. I used to do that a lot when I was at MIT as an undergraduate," he said. "And for a while I felt the same way I used to. I could look around at everybody and see that they all looked like me. We all looked like students. But suddenly it hit me. I'm not like most of these people. I'm a father. It was a relevation. I'm somebody's father."

Mike Gilman embraced paternity greedily, and he soon found that Melanie tempered his obsession with science. He didn't want to spend all night and every weekend in the lab. He wanted to go home and play with his daughter. He wanted to see her grow up. "Sometimes when I was at the lab," said Mike, "Laurie would call and say, 'Melanie just smiled for the first time,' or 'Melanie can drink from a glass by herself now,' and I'd be so jealous that I'd missed having seen for myself." Mike began to organize his work time so that he could leave the lab by 6:00 P.M. and have his weekends free.

"When we lived in Berkeley, I could never count on Mike to come home at any particular hour," said Laurie. "He'd wake up on Saturday and have this big debate with himself: 'Should I go into the lab? I probably should go into the lab.' And he usually did. Now he wants to spend time with his family."

I was eating dinner with the Gilmans at their home in nearby Arlington, a small, serene house that, despite a scattering of toys on the living room floor, was immaculate. Laurie had cooked the best Mexican food I'd ever tasted, and the only person not attacking the meal voraciously was Melanie; she was too busy squealing and singing and banging her spoon on her highchair. Laurie tried to calm her by picking her up, but Melanie stretched her arms toward her father and fussed until he took her on his lap.

"I think I'm more efficient now than I used to be," Mike insisted, jiggling his daughter up and down on his knee. "I think I can get as much done in eight or nine hours as other people get done in twelve or fourteen. I don't stop to goof off or bullshit. And I really believe that being with Melanie helps to replenish me. Whenever I get stressed out at the lab, I can come home and give Melanie a bath or read her a bedtime story. That relaxes me, and it makes me work better the next day. So I don't feel guilty if I leave a little earlier than some people."

As Mike spoke, however, I only half believed his claims. I think he did feel guilty. He kept a book of hours in his head. He knew who worked how long, and he knew where his schedule ranked. Shelly and Luis worked about the same amount; Cori, René, and David worked harder. Bob Finkelstein worked very hard, but maybe he wasn't very efficient. Weinberg — well, Weinberg could do whatever he wanted. Mike believed that the time spent with his daughter was worth much more than a few extra hours of shoveling dry ice and labeling tubes, but he seemed to worry nonetheless.

He had reason to worry — not so much about the number of hours he clocked in the lab, but about what those hours were yielding. His scientific career, he said, was falling apart. Soon after the Weinberg lab moved from the Cancer Center to the Whitehead, in the fall of 1984, his *myc* and *fos* experiments had stopped working. It was very mysterious, he said. They'd worked and worked before, and then, suddenly, zero. "I've been trying to get things going again for eight or nine months now," he said to me one day in the lab. "I've tried everything possible." He'd changed the serum conditions; he'd changed the incubation time; he'd changed his probes; he'd changed the hybridization solution that links the probes to the RNA; he'd measured the cells' internal pH level to make sure it really had risen as a result of the serum treatment. "I've been grasping at straws," he said. Still he was unable to recapitulate the pretty pH-*myc,* calcium-*fos* correlations. *Myc* would come on when *fos* was supposed to; *fos* would appear where *myc* should be signaling; both would blink on; both would be blank.

"I don't know why I'm showing you this," he said, "but here goes." He yanked out a stack of autoradiographs from beneath a pile of papers and catalogues on his desk. "This one is garbage," he said of the top autoradiograph, and threw it aside. "This one is crap. This one is so-so. This one is shit." The autorads whisked past like calendar leaves in an old movie. I couldn't see any of the data, but I got the idea.

"So now I'm going back to the very beginning. Instead of starting at some midway point where I thought the experiments began to fall apart, I'm doing the whole thing over. I'm growing new cells. I'll just pretend I'm back at the Cancer Center."

It takes about three weeks to cultivate a sizable batch of cells. Six weeks after starting over, Mike found that the experiments hadn't improved. If anything, they were getting worse. He kept at it. I often saw him squatting by the freezer to get dry ice. I noticed that the stack of autoradiographs on his desk was growing. He was trying — most of the time. At other times I'd catch him slumped in a chair by his desk, looking as though he never wanted to stand up again — and experiments are done by people standing up. "Hi," he'd say. "Are you here to see me just out of morbid curiosity, or did you actually think I might have something worth showing you?" Fridays were the worst days, because Mike no longer came to the lab on weekends. "This week has been such a bust," he'd say. "I should come in tomorrow, but for some reason I can't get inspired by the idea. Maybe I'll stop by here on Sunday afternoon." He wouldn't stop by on Sunday, though, and he didn't think his neglect mattered. From Monday to Friday, *myc* and *fos* were ungovernable, and Mike didn't see how weekend work would make his great model any more true.

I also knew that his work depressed him because he talked more than ever about Melanie. "If it weren't for her, I'd have gone crazy months ago," he said. "Did I show you the pictures I took of her when she was just born?" His whining stopped sounding amusing. For one thing, he complained about money. Laurie was staying at home to take care of Melanie, which left Mike to support the family on a fellowship of $18,000 or $19,000 a year, no benefits included. "We're really destitute," he said. "We pay for rent, food, clothing, medical insurance — and we barely have enough for that. We never go out. There's nothing but the essentials. I'm worse off than I was as a graduate student."

Nor were his complaints directed entirely at himself. He felt that Weinberg was ignoring him, allowing him to drift along without support. Mike hadn't published a paper in years, and he didn't think that Bob noticed it. He was beginning to nurse a somewhat frosty philosophy about the inevitable conflict between the people who work in a lab and the principal investigator who runs it, a philosophy that I was to hear voiced by several other Weinberg scientists.

"Bob has twelve or fifteen people working for him, and they're all doing different things," said Mike. "He doesn't have to worry about all those projects paying off. If only one or two things work, he's safe. His grants will keep coming in, and he'll still have a great reputation. And it's better for him to go after the brass ring, the signal pathway sort of thing, rather than have everybody doing things that are guaranteed to work. That's okay, that's his style, but what happens to the ten or twelve people whose projects don't work out?" Sometimes, he said, they get left out in the cold.

Mike was feeling the chill. Although he'd thought up many of the details of the *myc-fos* project, he still considered it a Weinberg type of endeavor. Mike's initial faith in his project had been supplanted by a deep atheism. He didn't care anymore about setting up paradigms or discovering the big cellular phenomenon that would make him stand head and neck above competing postdocs. He wanted something that would give him results, something he could comprehend. Thus he set forth the peculiar proposition that it is the young postdoc who must be scientifically conservative. Only a tenured professor with a solid reputation can afford radicalism.

Mike admired Weinberg's panache, but he was beginning to suspect that Bob grossly oversimplified the cell. Either that, or he came up with ideas that were impossible to test. Gilman waxed nostalgic about the old days, when he worked on gene regulation in bacteria. Maybe he'd hoped that he could crack some monolithic code of cancer, but apparently that was a pipe dream. He'd do better to do what he did best.

"I feel uncomfortable working with whole cells. Whole cells are a black box. You can't separate one component away from the cell and expect to get an idea of how the system works *in vivo*. You start with one of the most complex things in existence — the cell — you scrounge around for a model, and then you have to twist the data to make them fit that model. You have to ignore all the messy complications. But we know so little that we're just stabbing in the dark.

"I'd like to go back to studying gene regulation. I've been thinking about looking for proteins that bind to the *myc* or the *fos* promoter, the proteins that regulate gene expression. I think a project like that is more up my alley, but gene regulation is definitely not something Bob is interested in."

On the days when his experiments failed completely, when his results were uninterpretable and the whole idea of looking for an oncogenic signaling pathway seemed ludicrous, Mike's testiness grew extreme. One day while we were waiting on line at the cafeteria, I mentioned to Mike that a member of David Baltimore's lab who likes to predict future Nobelists had prophesied that Weinberg almost would surely be a recipient within the next five years.

"If Bob wins a Nobel Prize," said Mike, "that will convince me that anybody can win one."

I studied his face, trying to decide whether he was serious. "You don't think he deserves it?" I asked.

"My thesis adviser at Berkeley is the kind of scientist who deserves a Nobel," said Mike. "He's brilliant, and his science is superlative. He's discovered some very fundamental things about gene regulation. But because he's not working in some headline-grabbing field like oncogenes, he probably doesn't have a chance."

Gilman's feelings about Bob were not cut and dried, however. His ego wasn't so strong that he didn't crave Weinberg's approval. When I said to him that Weinberg had rated his project as one of the two most important in the lab, he looked surprised but pleased. "Well, I'm glad Bob feels that way," he said. "I just wish his enthusiasm could make the experiments work." One day Mike told me that he'd just had a "heart-to-heart" with Weinberg and that it had bolstered his confidence. "Bob said that I didn't deserve this. He didn't think I should be having all this grief," Mike said. "He told me that he wouldn't say that to everybody in the lab. Some people bring on their own troubles, he said, because they don't think things through. But he absolved me of all blame."

Indeed, Mike was not so much angered by Bob as he was hurt. Bob obviously found Mike's suggestions for alternative projects somewhat uninspiring. You want to identify the proteins that regulate *myc* and *fos* expression? said Bob. But why? If you find the proteins, what will you do with them? What will they tell you about the cell? Weinberg always encouraged his people to consider the broader implications of their experiments, to think about models; a result is worth something only if you know what you'll do with it when you discover it. "I try to get people to look beyond their immediate experiments," said Weinberg. "I feel that that is the most important lesson I have to offer them — to think about models."

I asked Weinberg whether he thought he succeeded.

"No," he said, frowning thoughtfully. "I don't believe I succeed very often at all."

Mike hated to think that Weinberg could be bored with his ideas. So he lashed out at Bob's scientific style. Gilman didn't like big, simple models; the cell wasn't so simple. He didn't like working with whole cells. He didn't like having to reduce everything to two or three parameters. Models were figments, fun to mull over but impossible to do. Bob didn't care what happened to him. Bob had a job, a reputation, a great salary, a fat pension. Mike had no money and he needed to start thinking about a real job at a university, but he didn't have anything to show for his postdoctoral fellowship.

Bob Weinberg sacrificed him to his myths, said Mike, and then he didn't even pay attention during the ritual dismemberment.

"You hear two complaints from postdocs and graduate students," said Cori Bargmann. "One is that the boss is overbearing and demanding, always nagging the person to get the work done. The other is that the boss is never around and doesn't care. It's been my experience that people will complain one way or another when their experiments aren't working. When they're not happy with their projects, they're not happy with the supervisor. When experiments work, the supervisor is great." (And in fairness to Weinberg and supervisors everywhere, they can hardly release a stream of dissatisfied postdocs into the world without soon finding that their pool of applicants has dried up. Weinberg's pool has done anything but.)

Life might have been more difficult for Gilman — he might have gotten sick, or taken long, desultory vacations — had it not been for his lab support system. Snezna was alone; Mike had a young-boy network. He shared a bay with two of his closest friends, David Stern and Bob Finkelstein, and the bay had a homogeneous spirit rare at the multinational Whitehead. All three men are Jewish and North American (Mike is from Canada), and all three are wiseasses. They almost always ate lunch together. They jointly purchased a tape deck for their bay, on which they played such old favorites as the Grateful Dead and the Rolling Stones, or soft-core New Wave bands like the Smiths and the Talking Heads. They took turns sporting beards and shaving beards. And they took turns buoying each other up during periods of protracted experimental poverty. Bob Finkelstein had been struggling for seven years to patch together a dissertation project that would please both Weinberg and the imperious MIT thesis committee. "I'll say to Bob, If I do this and this and this, will that be enough for a thesis project?" said Finkelstein. "And he'll say, Oh, yes, that will do it. Or if it doesn't, you'll be very close. So then I'll do this and this and this, and I'll get ambiguous results, and I'll be back to where I started — with no Ph.D."

David (who, much to his displeasure, was nicknamed Sterno) had spent the first eighteen months of his postdoc gathering data that he describes as "pure shit" and feeling like the lab leper. "I knew that Bob wasn't interested in my work," he said. "When he's thinking about what you're doing, he'll send you papers that he believes may apply to your experiments. For a while, all he sent to me were supply catalogues."

Sterno and Finkelstein could empathize with Mike and with failure. Perhaps more important, they could discuss his experiments with him. They'd examine one of his gels and throw out various interpretations. "These guys are my family away from family," said Mike. "Actually, I spend more time with Bob and David than I do with Laurie and Melanie, so maybe it's my real family that's my family away from family."

It was David who broke the news to me. I was sitting near Snezna's bench, scribbling notes, when he stopped by and asked me whether I had talked to Mike yet.

"No, why?" I asked. "Is something wrong?"

"You should talk to him," said David. "It works."

"'It?'"

"pH turns on *myc,* and calcium turns on *fos,*" he said. "But you should go talk to him."

He didn't have to tell me again.

Scurrying down to Mike Gilman's bay, I found him sitting at his desk, holding an autoradiograph to the light of the window.

"Good news?" I asked. "David said you had good news."

"Well, I got what I was looking for," he said, still squinting at the film. "I guess that should make me suspicious, having everything turn out the way I'd hoped, but . . ." He placed the film on his desk and pulled a second autorad from the pile so that I could get a peek at both. "You remember how I told about my idealized model, where pH should turn on *myc* but not *fos,* and where calcium should turn on *fos* but not *myc*?"

"Of course," I said.

"And do you remember how I told you that I'd gotten it to work eight months ago, but then the whole thing fell apart?"

I nodded. "You're really great at building up the dramatic tension," I said. "So did you get it to work again?"

He began pointing at the different lanes on the film where his radioactive probes had coupled with the RNA messages of activated *myc* and *fos.* He explained the conditions of the experiments, going over the shifting of his mouse cells from one type of serum to another, his use of calcium ionophores, and the length of time that he had allowed the cells to incubate before pulverizing them to remove the RNA. On one gel, he had screened the RNA with a *myc* probe at different time points after subjecting the cells to changes in both the pH and the calcium concentration. On another gel, he'd hybridized the RNA with a *fos* probe. I saw that the finickiness of the genes was just right — one preferred calcium, the other a massage of pH. The time courses were right, too: *fos* RNA was brightest five minutes after stimulation; *myc* responded best a half hour later. Everything was right.

"This is beautiful," I said, gazing at the gels.

"Yeah, well, the *myc* signal still isn't as strong as it was eight months ago," he said. "And I've only gotten this result once so far."

"But this is really great," I insisted. "How did you do it?"

"That's the funny part," he said. "After I'd eliminated all the usual suspects — the probes, the cells, the RNA, the serum — the last thing I was left with was the incubator. My troubles began right after we moved our lab to the Whitehead. At the Cancer Center I was using a shitty, old-fashioned incubator. Here we have this fancy thing with digital readout that controls the carbon dioxide levels inside by some sort of microprocessor." The carbon dioxide concentration of the incubator, he explained, affects the pH level of the cultured cells warming within. He hadn't known how to use the fancy digital incubator, so he'd turned up the dial on the carbon dioxide regulator to about twice what it should have been for his experiments. Once he'd realized what had happened, he turned down the carbon dioxide, and his *myc* signal reappeared.

He was somewhat less pleased by the ruse he'd been forced to employ to stimulate *fos*. With a fatalistic shrug, he'd finally dumped an abnormally high dose of calcium ionophores onto the cells. "My feeling was that I had nothing left to lose," said Mike. The calcium blast did indeed spark a peak of *fos*, but now Mike worried that such a megadose of ionophores might be having a multitude of effects on the cells rather than the discrete, *fos*-only stimulation he would have liked. Still, the high dose of ionophores did the trick, and, most heartening, it left *myc* cold. "I have to repeat all this," he said. "I have to try all sorts of things. I have to try another kind of ionophore. And then I have to measure the internal pH of the cells to see if my bicarbonate-free stuff is doing what I think it is." He told me several other "tries" he'd have to attend to.

"But you're happy, aren't you?" I pressed.

"Yeah, I guess I am. This is the first positive reinforcement I've had since last fall," he said. "It came just in time, too. I have to give a floor meeting next week, and without these data I'd just embarrass myself and bore everybody else."

For the next few days Mike puttered around, gathering slides, scrawling out notes, and discussing his upcoming speech with his bay mates. Sequestered in his bedroom at night, he rehearsed the talk in front of a mirror. "I could hear his voice rising and falling as he went over it again and again," said Laurie. "But mostly what I heard was him clearing his throat and coughing." Mike also crowed about his results to anyone who would listen. The day before the floor meeting, he had lunch with Paolo Dotto, a Weinberg postdoc from Italy, and Alan Schechter, neither of whom had heard the news yet. "pH turns on *myc* but not *fos*; calcium turns on *fos* but not *myc*," said Mike, and then elaborated on his experimental techniques. "That's very good," said Paolo, sounding genuinely impressed; Paolo is a quiet and highly critical scientist who isn't impressed very often. Alan asked a few questions about Mike's experimental procedures.

"You're giving a floor talk tomorrow, aren't you?" he said.

"Yeah, why?" asked Mike.

"You're not going to talk about this stuff, are you?" Alan said.

"I was planning to," Mike replied. "Why shouldn't I?"

"I wouldn't if I were you," said Alan. "These meetings are pretty open. They're not supposed to be, but you can never be sure who's in the audience. There could be somebody from Brent Cochran's lab sitting there taking notes." Brent Cochran, a molecular biologist at Harvard University, was working on the activation of different oncogenes during the course of the cell cycle. As such, he was a competitor of Mike Gilman's. "You'd probably be safer standing on Mass Ave announcing your results through a loudspeaker."

Mike Gilman looked confused. The blood vanished from his face. Alan added that he too was giving a floor talk the next day, but that he wasn't going to mention his latest and best results; he'd talk about data that nobody would bother stealing. "I'm not like you, Alan," Mike said, his voice getting squeaky. "I don't think about things like keeping secrets or how many spies may be looking over my shoulder."

"Suit yourself," said Alan. "But I don't know, I think you'd be crazy to. It's not as though you're using any special reagents or techniques. Your experiments would be very easy for somebody to reproduce once you gave them the idea. Look at how big Cochran's lab is. They'd do it in no time."

Alan Schechter is tall and dark-haired and brutally attractive. He wears black leather boots with square toes and rolls his shirt sleeves up past his biceps. (I kept expecting to find a pack of Marlboros tucked into one of those shirt cuffs, but the only time he smoked was when he could bum a cigarette from Snezna.) Adding to his tough handsomeness is his unusual nose: the bridge of it isn't quite there, and though I didn't dare ask him, I imagined that he'd broken his nose badly during some boyhood brawl. More noticeable than his fashions and his features, however, are his opinions. He describes himself as opinionated, and you can't argue with him there. You can't argue with him about anything. Even when his ideas were so patently silly that I knew he was disagreeing just to be obstreperous, I never heard Alan Schechter lose an argument.

In this case, he sounded eminently if aggressively reasonable. Anybody from any lab could wander into the Whitehead auditorium. (David Baltimore reserved his bouncer's services for the cafeteria.) Mike Gilman didn't argue with Alan. He just panicked for the rest of the day about what Alan had said. He solicited the advice of other people in Weinberg's lab, and most agreed with Alan: it would be better to button up about it. He consulted Bob Weinberg. Weinberg granted that Alan Schechter's point was valid, but he told Mike that it was up to him: he'd accept Mike's decision either way. "So I thought, Fuck it, I'm going to talk about it," said Mike. "Scientists are inherently competitive and overachieving, but I don't want to acknowledge that level of paranoia. It's fun to talk about your results. It's not fun to worry about who might know it."

Mike Gilman talked about his *myc* and *fos* results the next morning at the floor meeting, and he wasn't the only one to have fun. His presentation was trenchant and vivid and crackly with jokes about his eight months of plagues and the piddling insignificance of an experiment that had come out right only once. Afterward, several people told him that it was the best talk they'd ever heard.

With the success of his performance, Mike's anger faded. He still wasn't willing to stake his future on a model of the cell, but his vigorous ego had revived. So much so that the following year, when Mike visited Cold Spring Harbor in search of a staff position (which, with Weinberg's highest recommendation, he eventually got), several of the scientists there guessed his academic origins. "Only a person from MIT," a Cold Spring Harbor postdoc noted, "could act as though he knew just how the world worked."

Mike Gilman later told me that I'd met him at one of the lowest points in his life, but his low was not the lowest of the lab. A blacker moment came when half the lab converged on a single, fragile gene and chopped it to pieces. It was a gene that Chiaho Shih had identified half a decade earlier without knowing what he'd found. It was a gene that for a long time seemed barely worth the energies of one researcher — yet it commanded the attention of four postdocs, one graduate student, and two or three technicians. It's called the *neu* gene, and eventually it would live up to its homonym and help usher in the renaissance of the lab. It would burst alive, a dawn of flame after a dark, fitful night. Through studying the gene, the lab would discover an extraordinary mechanism by which a normal gene could become cancerous — a mechanism that would buttress Weinberg's belief in the horrible simplicity of cancer. And another lab would use the gene to help in the diagnosis of human breast cancer, one of the major cancer killers.

But all that came later. First were the months upon months when the *neu* gene seemed tired and tattered and hateful.

9 Something Borrowed, Something Blue

BOB WEINBERG is a creative scientist who likes big, unhomogenized problems, and he doesn't hide his distaste for mopping up after himself. In the past, his emphasis on paradigms had helped bisect his lab into those who had results fit for the *New York Times* and those who didn't.

By the mid 1980s the Weinberg lab had grown desperate for something to do. Yet even as the researchers suffered together, they were divided into camps. There were the big have-nots, the little have-nots, and the maybes. The big have-nots, like Snezna Rogelj and Mike Gilman, had big models and no results. The little have-nots had no models, or minor models, and no results. The maybes had no models, but they had the possibility of results. For young scientists, any results are better than no results, which is why so many members of the Weinberg lab descended thirstily on the *neu* gene.

Nobody in the lab knew exactly what the *neu* gene was. Nobody knew whether it was important or whether it had anything vital to say about the meaning of cancer. All anyone knew was that the gene existed, and that there were a minimum number of scientific papers that could be squeezed from it. The *neu* gene was guaranteed. As a result, the pace and direction of the project were dictated not by Weinberg's models or serendipity or the blast of a starting gun, but by the need for scientific survival.

• • •

The person who set the frenzy in motion was Alan Schechter — or Big Al, as Weinberg called him during his intermittent attempts at fraternizing. Big, tough, opinionated, bossy, shy, big-hearted Al; I never would have dreamed of calling him Al. Alan was difficult to get to know, and I didn't get very far. He'd say things like "If you knew me better," or "You can't say that, because you don't really know me," until at last I cried, "You're right, Alan; you're a complete man of mystery!"

Yet what little I did learn about him I liked. He acted tough, and then denied that he was acting tough. He removed all the seat belts from his four cars, and he couldn't understand why I found that so funny. He'd tell rollicking tales of his derring-do, of being held at gunpoint by the police in Spain, of crossing the border to East Berlin at two in the morning, of almost falling several thousand feet while rock climbing; but when I stupidly suggested (everything I said to Alan was stupid) that, so, he does court the tough life after all, he was insulted. "These things have nothing to do with being tough," he said. "I don't seek them out. They just happen to me more than they do to most people. That's just a fact."

Alan has something biting to say about almost everybody, but he is also subject to fits of pure kindness. He told me that any time I had a problem understanding a subject, I should ask him, and he'd try to help. Although he looks as though he subsists on a diet of steak and pork chops, Alan is a strict vegetarian — not out of concern for his health, but because he doesn't like the way livestock is slaughtered.

"Alan puts on this tough, macho act," said Snezna, puffing out her chest and lifting her arms away from her sides like a body builder, "but really he's such a sweet guy."

"I don't care what you say," I kidded Alan at one point, "I think you're a very nice guy."

"What do you mean, you don't care what I say?" he snapped. "I *know* I'm a nice guy."

I also liked Alan Schechter because he was raised in a penumbral ghetto in the west Bronx, just a few blocks from where I went to elementary school. He spent his youth hanging out in the streets with a group of boys much like himself.

"What did you do?" I asked. "Play stickball? Steal hubcaps?"

"No, I wasn't a *hood*," he said. "We weren't a bunch of juvenile delinquents. We just liked hanging out."

I asked him whether he ever went back to visit the Bronx.

"Why should I want to do that?" he replied. "My parents live in Jersey now."

I told him that most of the buildings in the old neighborhood were abandoned shells, if they were standing at all.

"Yeah, that's really weird when you think about it," he said, softening. "Most people can go back to the place where they were kids, and chances are their houses will still be there. But where we come from, there's nothing left. It's as if they've taken away a piece of your childhood."

Alan may not be able to do it in the Bronx, but he still enjoys hanging out. He prefers hanging out to doing benchwork, and he doesn't care if people think he's slothful. I have a snapshot of Alan talking on the lab telephone, the cord twisted around his waist and up across his chest, and the snapshot seems appropriate; I often saw him talking on the phone, and from his facial expression I'd guess that he wasn't discussing science. Weinberg occasionally groused that he wished he could figure out how to "put some fire under [Alan's] butt," but Alan's hindquarters seemed sheathed in asbestos. "I don't work that hard," he said. "If you knew me better, you'd know that my quality of life is more important to me than my job." Several of his colleagues said that he could get away with indolence because he was imaginative and extremely shrewd. "Alan may sometimes be lazy," said Rudi Grosschedl, the former Whitehead postdoc who was a friend of Randy Chipperfield's and who now runs a lab at the University of California, San Francisco. "But he thinks about science, and he always has ideas. In my opinion, he's one of the smartest people in Weinberg's lab."

Alan was shrewd enough to take on the *neu* project when it was considered an unreclaimable mess and a continuing lab joke. He was shrewder still to step aside right before the laughter died.

When Alan arrived at MIT, in 1982, there was no *neu* gene. There was the gene that Chiaho Shih had taken from rat neuroblastomas — brain tumors — and transfected into cultured mouse skin cells in 1980. The gene had generated firm, pulsing clusters in the recipient NIH 3T3 cells, but it hadn't yet been cloned or named. There was also an isolated protein lying around that a researcher had precipitated out from the focal cells. The protein weighed 185,000 daltons and was referred to as p185, an abbreviation of its molecular weight. The protein may have been the rat protein synthesized by the neuroblastoma oncogene, or it may have had nothing to do with Chiaho's oncogene. Few scientists in the Weinberg lab wanted to touch p185 or ponder about what it might be.

"The project was a shambles," said Alan. "Any time there was a p185 talk at group meetings, people would roll their eyes and say, Shut up already. They didn't believe the protein data were significant, and they didn't want to hear about it. I couldn't understand that, because I thought the neuroblastoma stuff was interesting — a lot more compelling than the *ras* work that was going on."

Bob Weinberg shared Alan's conviction that p185 was important, and he encouraged his new postdoc to tinker with the orphan protein. Alan soon determined the location of the protein within the cell, and he discovered that p185 was some sort of receptor: part of its head poked above the surface of the membrane, as though primed to catch a hormone or similar growth factor floating in the blood stream.

Now Weinberg was really excited. If the neuroblastoma oncogene turned out to be a receptor for a growth factor, the lab would have scored another first. At that point, scientists knew little about what any of the oncogenic proteins did in the cell. They couldn't relate the cancer proteins discovered thus far to familiar cell proteins that scientists had studied in the past. But scientists knew about receptors; they knew that cells are studded with dozens of receptors, which permit the cells to receive signals from the outside world. There are cell receptors that link to insulin, receptors that latch on to thyroid hormone, receptors for cholesterol molecules, iron molecules, and many other vital compounds swimming through the blood. By proving that the neuroblastoma protein was a perverted version of a membrane receptor, the Weinberg lab would have a clue about how the cancer protein worked.

Alan warned Weinberg that he wasn't sure p185 was a membrane receptor, and he certainly didn't know what it was a receptor for. Besides, said Alan, I still can't say if p185 is the cancer protein produced by the neuroblastoma oncogene. p185 may have nothing to do with tumors. It may be a piece of garbage that we've picked up during our precipitations.

Well, then, said Weinberg, it looks as though you're going to have to clone the neuroblastoma oncogene to resolve the question.

Alan was aghast. He wasn't *that* interested in brain cancer. Cloning the gene is out of the question, he said. How do you clone a rat gene from a mouse cell? Rats have repetitive DNA sequences, just as humans have *Alu* sequences, but the rat repetitive sequences look too much like the mouse repetitive sequences. There'd be no good probe to pull the gene out.

Weinberg replied, Why not isolate the healthy equivalent of p185 from normal rat tissue and show that it's different from the cancerous p185 that we've got in our 3T3 cells? Then we'd know we had a tumor protein. You've worked with proteins. You can do that precipitation pretty easily, can't you?

Bob, that's an even worse idea, said Alan. The differences between the normal rat protein and the cancerous rat protein are likely to be too subtle to distinguish experimentally. I don't want to spend five years on one oncogene and one protein.

He later told me, "Everything I could think of doing was too hard, and I was getting fed up with the whole thing. I was also working with Cliff at that point, trying to find the function of *ras*. Bob was pushing me along on that one, too — it was the usual lovely situation. I thought I would wait a little while with p185 until I could figure out some easier way to get at the gene."

With Alan's attentions diverted to *ras*, the neuroblastoma project idled for months, but a series of events in 1983 and 1984 dramatically resuscitated it. David Stern arrived at the lab as a postdoc and decided to try what Alan had turned down: finding the normal species of p185 in healthy rat cells. After months of struggle, he managed to precipitate a protein that he was pretty sure was a noncancerous p185. The accomplishment was significant. When he compared the two proteins, David felt confident that the tumorigenic protein in the lab's mouse cells was of rat origin — thus, that the protein was the product of a new cancer gene. Chiaho Shih had indeed transfected a novel oncogene into recipient cells years earlier, and that novel oncogene synthesized a membrane receptor. The Weinberg lab had something worth studying: a cancerous receptor.

Presented with David's results, Weinberg told Alan it was high time that the neuroblastoma gene was cloned. Alan retorted that he was still unsure about the whole project and that he had no intention of trying to clone the gene without a reasonable probe.

Weinberg may have wished that Alan had enough fire in his heart — or elsewhere around his person — to do a little hard work, but Alan didn't want fire. He wanted divine inspiration — or a reasonable facsimile from overseas.

• • •

In 1983 and 1984, two linked discoveries rocked the oncogene community, again boosting hopes that the scattered data could be sewn together into a tidy understanding of cell growth and overgrowth. Michael Waterfield of the Imperial Cancer Research Fund Laboratories in London and Russell Doolittle of the University of California, San Diego, in La Jolla, found that the gene for platelet-derived growth factor was 80 percent identical with an oncogene called *sis*, which had originally been detected in a monkey tumor virus. An 80 percent homology meant that the genes were one and the same: the simian retrovirus had adopted and mutated a monkey sequence encoding an essential cellular growth factor, PDGF.

The retrovirus, however, was not the splendid part of the discovery. The splendid part was the firm association between a cancer gene and a gene already known to have a fundamental role in normal cells. PDGF is the protein molecule, carried by red blood cells, that helps to heal wounds in the body by forcing tissue around the injury to proliferate. Nothing could be more proper. If there is any cellular gene that seems capable of becoming tumorigenic — that is, a proto-oncogenic time bomb — it's a gene for a growth factor such as PDGF, a modulator of ordinary cell division. "Finding that homology," said Waterfield, "was probably the most exciting moment of my scientific career."

"It's very rare in science that there's such a clean answer that it can be summed up in three words," he observed. "'PDGF is *sis*.' In those few words, everybody in the field could know what it meant."

The next year, the Waterfield lab presented another pithy summation. Julian Downward, a twenty-three-year-old graduate student, discovered that *erb-b*, an oncogene originally identified in a chicken leukemia virus, was 80 percent identical with a membrane receptor for epidermal growth factor, or EGF. Like PDGF, EGF is a hormone-type protein that somehow initiates day-to-day tissue growth, and the EGF receptor is the membrane's catcher's mitt for the growth factor.

The EGF receptor is *erb-b:* not very euphonious, but the significance of the association was once again clear.

"It was a spectacular discovery, with far-reaching consequences," said Weinberg. "The cellular gene encoding a growth-factor receptor had become converted into an oncogene. The take-home lesson was that any gene encoding a growth-factor receptor is in principle a proto-oncogene, implying an intrinsic ability to become an oncogene consequent to alterations in the DNA of the gene."

Almost overnight, the portrait of tumorigenesis looked completely simple. On the one hand, a gene for a growth factor — a wound-healing protein — had been converted into an oncogene. On the other hand, a gene for a growth-factor *receptor* had been mutated. The twin discoveries made much sense. All the genes in the cell that had anything to do with growth proteins, or the membrane vises for those proteins, were candidate cancer genes. Molecular biologists everywhere were galvanized into action. They began looking for molecular associations between the thirty or so known oncogenes and growth-related cellular genes. And Alan Schechter had his idea for a shortcut.

Alan had dithered with the neuroblastoma project for almost two years. Although he didn't know whether he had a significant gene or a significant protein, and was reluctant to attempt cloning a gene without a probe to fish it out, he did think that p185 was a receptor.

The EGF receptor was equivalent to an oncogene, *erb-b*. If p185 was also an oncogenic protein, thought Alan, maybe it was related to *erb-b* as well. And if p185 was the neuroblastoma cancer protein, he could pull out the oncogene from the mouse cells by employing a copy of *erb-b* as a DNA probe.

Alan obtained a radioactive clone of the *erb-b* oncogene and briskly screened the cancerous NIH 3T3 cells that Chiaho had sown with the DNA of rat brain tumors. Alan was just checking to see whether his idea had any merit. When he emerged from the darkroom with his results, he sighed with relief. The *erb-b* segment had found its mate in the cancer cells. The neuroblastoma oncogene was related to the receptor for epidermal growth factor.

Alan's results had double significance. First, the smear on his Southern blot meant that Alan had found a reasonable probe to clone the new oncogene. Second, the consanguinity of the neuroblastoma receptor and the EGF receptor suggested that the vexatious Weinberg protein was a tyrosine kinase. In those days, talking about tyrosine kinases was like invoking the jinni of oncogenes. Scientists had discovered that several cancer proteins were kinases; indeed, the connection between tumor proteins and kinase activity was about the only insight researchers had into the biochemistry of cancer. The word *kinase* is from the Greek *kinein,* which means to move, and that's what kinases do. The proteins filch phosphate molecules from ATP, the cell's energy currency, and plunk the molecules onto target enzymes in the cytoplasm. Outfitted with phosphate molecules, the substrate enzymes burst alive and begin jostling other proteins around them. Eventually and mysteriously, all that enzymatic shimmying and bustling reaches the nucleus, and the cell divides, matures, or behaves in a similarly theatric manner. Under ordinary circumstances, kinases help keep the cell alive and active; but when the kinases are mutated, they can contribute to tumorigenesis. By analyzing tyrosine kinases and their target enzymes, scientists hoped to chart the signaling pathways of several oncogenic proteins. They believed that some cancer proteins could be conveying their subversive message to the DNA by shuffling around phosphate molecules too often and to the wrong enzymes.

When David Stern saw that p185 was related to the EGF receptor, he more or less appropriated the neuroblastoma protein as his private domain. David Stern was obsessed with tyrosine kinases. One of his mentors in graduate school, Tony Hunter of the Salk Institute, had discovered the first tyrosine kinase — the *src* protein — and David had been raised with the conviction that tyrosine kinase activity was a big part of the story of cancer. Alan Schechter didn't argue with David. You want p185? he said. Be my guest.

Performing a few preliminary experiments, David demonstrated that p185 was indeed a kinase. When confined to a test tube and supplied with phosphate molecules and likely substrate enzymes, the protein began pinning the phosphate units on projecting amino acid knobs of the target enzymes.

The results were in, so Alan went to talk with Weinberg. Here's what we've got, he said. We've got an oncogene in these cells that's related to *erb-b*. It could actually be a mutant version of the rat receptor for EGF. Or it could be another receptor that's only distantly related to the familiar receptor. We won't know until we clone the neuroblastoma gene, but at least we have a DNA probe now to yank out the gene once we've built our cloning library.

Excellent, said Weinberg. But let's do it quickly. If it's a novel receptor that's related to the EGF receptor, I want to get it before somebody else does. Who do you want to help you with the cloning?

Alan replied, Give me Mien-Chie.

Mien-Chie Hung, a native of Taiwan, has a big, round, laughing-Buddha face. He was a new postdoc at the lab and was eager to please. He was honored that Alan requested his help. Although Mien-Chie didn't know much about oncogenes — as a graduate student at Brandeis University near Boston he had studied fruit fly genetics — he was a skillful cloner. "I was really glad to be working on cancer," he said. "I was glad to be working on something that could help humanity. I like basic research, but I wanted to contribute something. And I thought it would be nice to be able to talk to my friends about my work. When I used to tell them I worked on flies, they looked at me as if I were some kind of monster. But to say that I'm working on cancer, on brain cancer, sounds much more exciting."

Around the same time, Cori Bargmann was sniffing about for a project. Her last effort, an attempt to revert cancer cells to normal cells by interfering with the RNA of oncogenes, had been a total flop, and Cori wanted something that would work. She offered to participate in cloning the neuroblastoma gene, but she had a strict notion of what she would do. Alan and Mien-Chie could clone the cellular, or genomic, version of the gene: the complete, ungainly segment that encompassed both exons and introns, and that Alan had estimated measured almost 40,000 bases in length. She, Cori, would clone the complementary-DNA, or cDNA, version of the gene: the much smaller segment composed of exons alone. The exons are the parts of the gene that eventually are translated from RNA into protein. They are the sequences that count.

First, however, Cori waited. She wanted one more piece of information before she took out her pipette and set up her blots. If the neuroblastoma gene *was* the EGF receptor, it wasn't worth determining where the exons of the gene were; other labs had already done that.

"She was very clever about it," said Alan. "She had the right approach. She knew that the genomic cloning was going to be a bitch because the gene was so big, but she also knew that she shouldn't bother doing the cDNA work for a gene that had been cloned and sequenced. So she waited to see what I came up with. She was *very* clever."

Alan and Mien-Chie soon proved definitively that the EGF receptor and the neuroblastoma gene were two distinct genes. They found the human sequence that corresponded to their rough, uncloned copy of the rat neuroblastoma gene. Collaborating with Axel Ullrich of Genentech and Uta Francke of Yale, they determined that the human neuroblastoma gene was located on chromosome 17. The human EGF receptor gene is far, far away, on chromosome 7. At last the Weinberg team realized that they had a new oncogene, and they christened it *neu*, after neuroblastoma. The gene undoubtedly was a growth-factor receptor, but what it was a receptor for, the lab had no idea. They had yet to find the ligand, the hormone that binds to the receptor's external claw.

"We approached this whole thing backward," said Weinberg.

"Normally people isolate a growth factor and then look for the receptor that the cell uses to recognize the presence of the known growth factor. We'd discovered a growth-factor receptor without its ligand."

Neu was an important oncogene. Weinberg believed in the transfection assay, and he believed that any gene revealed by the assay must be central to cancer. And apart from *erb-b* and *sis, neu* was the only oncogene with a clear function in the cell: it was a receptor. So little was known about cancer genes and the proteins they specify that any shaving of information would make a given oncogene more experimentally appealing than the next.

The Weinberg scientists gradually carved the project into private parcels. David Stern already had claimed the *neu* protein as his domain. He would look for the receptor's ligand and for any clues showing how the receptor delivered its signal to the cell nucleus. Alan began by working with Mien-Chie on the isolation of the complete genomic *neu* gene, but he decided that his help was superfluous. Instead, Alan decided to look for evidence of mutant *neu* genes in human tumors, and Mien-Chie assumed full responsibility for the genomic cloning. Cori, who'd been clever, was now more than happy to clone and sequence the midget cDNA copy of the gene. Even René Bernards, then a newcomer to the lab, was given a subdivision: he'd try engineering an antineuroblastoma vaccine. The *neu* gene had vaulted to center stage of the Weinberg laboratory.

"I did my best to make it clear who was responsible for what," said Weinberg.

"The whole crisis might have been prevented," David told me later, "if Bob had only paid closer attention to what was going on in his own lab."

Cornelia Bargmann is one of the few almost perfect people I have ever met. She reminds me of John Tenniel's Alice, or maybe of the fawn that Alice encounters in the wood of no names. Cori's long hair is the color of fawnskin, her eyes are big and gray and soft, and she has the slender, gangling limbs of a fawn. Although she is young, in her mid twenties, she wears conservative clothes: a straight-line denim or plaid skirt with a pin in it, or a brown corduroy jumper with a checkered blouse. She is from Athens, Georgia, and retains a trace of a Georgia twang. "When I first met her," says her husband, Mike Finney, "she was still saying, 'I have to run a jail'" — her pronunciation of the word *gel*. Talking to a person one on one, she's demure and ladylike. When she talks to men, she flirts lightly and girlishly.

But scientifically, Cori Bargmann is no girl, and she is certainly not demure. As soon as she mounts a stage, she hardly seems to be a graduate student. She lectures about science clearly and confidently, her voice reaching out to the darkest seat in the auditorium, her self-possession rivaling that of David Baltimore or Bob Weinberg. I once watched her at a scientific meeting that she attended with other people from the Weinberg lab. Her lab mates generally clung to the scientists they already knew. Cori Bargmann darted around and sampled the available company, like a honey bee flitting from rose to peony to daisy. By the week's end, she had introduced herself to at least half the scientists in attendance and learned everything possible about their unpublished results. "I'm tired of talking to strangers," she declared to René Bernards at one point. "It's a really alienating experience. From now on, I restrict my conversation to my friends." A moment later, she wandered off to approach another table of aliens. No, Cori is *not* demure.

Nor should she be. She's simply too smart. Just a month after turning twenty, and having been elected to Phi Beta Kappa, she was a valedictorian of her class of two thousand at the University of Georgia. She applied to five graduate schools — MIT, Harvard, Stanford, Wisconsin, and the University of California — and all were desperate to have her. Two of the MIT professors I spoke with said she was the smartest graduate student they had ever encountered. "Boy, is she smart," said one, shaking his head and just stopping short of a low whistle. "She is so *sharp*."

"I have to constantly control my enthusiasm for Cori," said Weinberg, "because if I let it show, the other people in the lab would feel terrible."

He wasn't a very effective dissembler. Any time Cori had something to say at a group meeting, Bob would shift to the edge of his seat and shush up interfering comments. Not once would his eyes get that Vaselined look of boredom, as they often did when less favored people had the floor. Nor would Cori's colleagues need Weinberg to tell them how bright she is. Cori equals smart was a lab truism that I almost grew weary of hearing. "Whatever success she has she deserves," said Mike Gilman. "Cori is the best scientist in the lab."

Cori isn't vain, but she knows she's good. When Alan complained to her about the stresses of being at a place like the Whitehead and how relieved he'd be to say goodbye, she replied that she loved the competition — the keener the better. "He was shocked," she said. "He couldn't understand why I'd always want to be in the middle of a battlefield." Before David Baltimore finally managed, in 1986, to persuade a woman scientist to join the senior staff of the Whitehead, he speculated on the reasons that his gender-specific recruiting efforts had been unsuccessful. Perhaps women, he said, are less willing to accept the pressures and personal sacrifices of being "second to none." I repeated the conversation to Cori, and she responded that if Baltimore would only wait until she had a little more experience, she'd gladly sacrifice herself to help him fill his quota.

At the time she began cloning the abridged version of the *neu* gene, Cori had been in the lab for two years, and she was confident that she had mastered most of the necessary techniques of molecular biology. She was a graduate student who felt like a postdoc, and she was determined to do her part right and rapidly. She didn't want to spend six or eight years on her Ph.D. She regarded the *neu* project as potential thesis material, and she approached it calmly and rationally. She worked until midnight or one in the morning, six or seven days a week, without feeling that her social life suffered much. She cultivated chatty friendships within the Whitehead. Her husband kept similar hours at his MIT lab, two blocks away; sometimes he'd stop by to massage her shoulders, stroke her hair affectionately, and help her stay relaxed. Cori didn't rush, she didn't get frazzled, and Weinberg respected her so much that he let her be. "Bob has a good sense of what sort of motivation people need," said Jay Morgenstern. "He knows that the best thing for someone like Luis is a kick in the pants, and he knows that Cori prefers to be left alone." Cori agreed. "I'm fairly self-motivated," she said. "I get nervous if I think anybody is looking over my shoulder."

Mien-Chie had a slight jump on her in this project: he cloned the genomic version of the cancerous *neu* gene first, in September 1984. The gene is huge, and isolating it required, as Weinberg put it, some "whiz-bang technology," but nothing that Mien-Chie couldn't manage. As a graduate student, he'd cloned mammoth genes from fly DNA. "Until guys like Mien-Chie and René showed up," said Luis, "there'd been a real hole in the lab. There were a lot of amateurs bumbling around, trying to clone and sequence genes, but these guys are pros." Mien-Chie pounded the *neu* gene into a cosmid, a cloning vessel specifically designed to encase large genes. He then picked out the proper *neu* cosmid from a solution of millions of cosmids by probing the pile with *neu*'s cousin gene, the EGF receptor. Two months later, Mien-Chie isolated the corresponding normal copy of the genomic neuroblastoma gene.

Cori didn't care that she lagged behind Mien-Chie. Cloning the cDNA of *neu* wasn't the compelling part of her project, anyway. She needed the cDNA only to determine the sequence of the gene. The Weinberg scientists had to have the nucleotide sequence if they were to discover the mutation that caused the brain cancer; all else was preliminary to pinning down the mutation. If they succeeded, it would be only the second time that researchers had located a cancerous lesion in DNA. Of the thirty or so oncogenes that had been discovered by them, only the *ras* gene had been so precisely characterized. Many oncogenes appeared to be riddled with multiple mutations, and scientists couldn't figure out what was what; even the *src* gene, the hoariest of the oncogenes, long eluded mutational analysis. But the Weinberg lab had reason to believe that the *neu* error could be understood. They knew approximately how the chemical carcinogen that had given the rats their brain tumors attacked the chromosomes; and restriction enzyme maps of the *neu* gene indicated that the lesion was very tiny, perhaps as small as a point mutation. Mien-Chie couldn't very well sequence the genomic copy of *neu* in quest of a single base-pair error. The whole gene measured about 33,000 bases in length, and spelling out every one of those *a*'s, *t*'s, *g*'s, and *c*'s could take years. It would be much easier for Cori to sequence the condensed cDNA, shorn of the meaningless introns that play no part in the synthesis of either the normal or the cancerous *neu* protein.

The Weinberg lab had a plan — a seemingly reasonable and well-wrought plan. Mien-Chie and Cori would collaborate, as Cliff Tabin and Ravi Dhar had done years earlier on the *ras* project. Mien-Chie would mix and match pieces of the normal *neu* gene that he'd cloned with pieces of the cancerous *neu* gene until he had narrowed down the region of the mutation to a manageable size. As Mien-Chie mixed and matched, Cori would sequence the nucleotides on the tumorigenic version of the squat *neu* cDNA. After Mien-Chie had located the approximate region of the mutation, Cori would sequence the corresponding area of the normal *neu* cDNA to identify the precise discrepancy between the cancerous and healthy genes.

The plan required a good-natured acceptance of interdependency. Cori would need Mien-Chie's help. Not only would he tell her where to seek out the mutation in the gene, but he would have to provide her with a tool to clone the normal cDNA of *neu*. Isolating the cancerous cDNA was easy: the fact that the gene was cancerous gave Cori a method for selecting the proper segment during her experiments.

To clone the normal cDNA would be harder. According to the recipe for cDNA manipulations, Cori would need to begin with a special type of cell that produced copious quantities of normal *neu* RNA. (A cDNA is made by copying the messenger RNA of a gene into DNA in a test tube. That's why a cDNA is so compact and manageable; it's the equivalent of the gene message, composed of the protein-making exons alone.) Mien-Chie would have to breed those cells for her. He'd have to pass his cloned genomic gene through a difficult series of transfections and selections until he had developed a line of rat cells that pumped out fifty to a hundred times more normal *neu* RNA than any living rat would ever bother to transcribe. Only with geysers of healthy *neu* messages could Cori hope to generate a working cDNA clone of the gene.

Sooner or later, Cori would need those cells, and Mien-Chie knew it. But neither of the scientists worried about that at first. There was no hurry. Cori wouldn't have to clone the normal cDNA until much farther along in the project. Everybody was happy. Everybody understood what was expected.

Toward the end of 1984, Cori isolated the cancerous cDNA of *neu* and began sequencing some of its 4800 nucleotides. As Ravi Dhar, or any molecular biologist who's sequenced a gene, knows too well, sequencing is a delicate and error-prone procedure. A guanine band can look like a cytosine, an adenine like a guanine. Cori moved deliberately and cautiously.

Mien-Chie embarked on his designated task, mixing and matching the full-size genes. He began by dicing his clones with restriction enzymes. He was looking for an enzyme that would shear the genomic genes neatly in half; that would permit him to tack one side of the normal gene to the counterpart from the transforming gene. He was doing exactly what Cliff Tabin had done before him. The protocol was simple and sure. Mien-Chie pinched an enzyme into a test tube and then checked to see how many gene segments emerged from the reaction.

There are more than seventy-five restriction enzymes available for snipping DNA. Mien-Chie tried one after another. He tried the commonest enzymes. He tried the less common enzymes. He tried the most precious and costly enzymes listed in the New England Biolabs catalogue. He pinched, stirred, and counted gene segments.

He realized that he was in deep trouble.

The genomic *neu* gene was too big. No restriction enzyme existed that was specific enough for his experiments. In fact, the most precise enzyme available slivered his gene in *seven* places when he wanted only one. Mien-Chie couldn't do the proper mix-and-matches with seven different pieces. It was impossible.

Mien-Chie began to panic. It dawned on him that it would be incomparably easier to mix pieces of the normal *neu* cDNA with pieces of the tumorigenic cDNA. The short cDNAs would not suffer from the same enzyme problem that his lengthy genomic genes did. But Cori was working on the cDNAs. Would Cori get all the treats? Would she end up doing the sequencing *and* the mixing and matching? After months of work, Mien-Chie could very well be stuck with two large, utterly worthless clones.

Once his initial fears abated, however, Mien-Chie decided that all was not hopeless. He'd simply have to be smarter. He sat down with pen and paper, drew diagrams, studied restriction enzyme specifications, and made calculations. He devised a scheme. By craftily and imaginatively manipulating the conditions of the enzyme reactions, he would force the right gene pieces to meet up with their appropriate mates. He would make the mix-and-match project work, even with seven separate fractions of DNA. No gene jockey had ever managed such an intricate recombinant feat in the past — the maximum number of pieces that Cliff had had to worry about with the *ras* gene was four — but Mien-Chie accepted the challenge. He was, after all, a "whiz-bang" molecular biologist. And he didn't have a choice.

For weeks, Mien-Chie agonized over his calculations, refining the figures and conditions of the experiment.

In late January of 1985, Cori asked Mien-Chie whether he was preparing that cell line with the extra copies of the normal *neu* gene. Mien-Chie replied that he hadn't done it yet. Cori suggested that he could give her his clone of the genomic gene; she'd make the cell line. Mien-Chie hesitated. Give Cori his clone? The gene had been hard to isolate, and it belonged to *him.*

"I began to see how crowded this *neu* project really was," he told me. "I wasn't used to competing with people in my lab, and to me it looked as though Cori and I were becoming competitors."

One way or another, the cell line has to be made, Cori said.

Mien-Chie promised to do it. He needed the cells for a side project he was planning in any case.

Still he procrastinated. His plan for knitting together patches of the genomic gene might or might not work. He felt scared and defensive, and his biological reagents seemed to be his only bargaining chips.

Finally, Mien-Chie and Cori came to a loose understanding. An understanding so loose that Weinberg knew nothing of it, and even the bargaining partners were unsure of its bounds. Cori would hold off cloning the normal cDNA. Instead, she would continue sequencing the cancerous cDNA. Mien-Chie would develop the special cell line. At the same time, he would try, with the best of his technical legerdemain, to mix and match fragments of the two huge, genomic copies of *neu*. Perhaps he'd get incredibly lucky and find the mutation within the sprawl of 33,000 bases. Should he fail, Cori would clone the normal cDNA — using Mien-Chie's super-*neu* cell line — and then . . . Well, they'd see. They'd see who would mix and match the squat and simple cDNA's.

Mien-Chie bred a cell line that was engorged with high levels of *neu* RNA, and then he geared up to attempt his florid mix-and-match project. Quite suddenly, however, he had to put his experiments aside. His impressive productivity squealed to a halt. People in the lab began to notice that Mien-Chie often wasn't at his desk. And when they did see him, they noticed that he looked depressed. His moon-round face was getting rounder. Was he gaining weight?

Mien-Chie told Weinberg that he was having "personal problems." But he was too embarrassed to confide the nature of his problems to anybody besides Alan Schechter and the one or two other close friends he had in the lab. He assumed that the Americans and Europeans would scoff if they knew.

Bob Weinberg said that both Chiaho and Mien-Chie deeply respected him, and for the same reason. "They're Chinese," said Weinberg, "and they're taught to revere authority." Mien-Chie confirmed that a reverence for elders isn't an insulting racial stereotype. It's part of his heritage, and he didn't expect anybody who wasn't Chinese to understand the extent to which that respect guided him — and shackled him. He was very attached to his wife, King-Lan Cheng, and he enjoyed nothing more than when his young son, Victor, came to visit him at the lab. "The second word Victor learned to say was *oncogene*," said Mien-Chie. "The first word was *McDonald's*." But neither wife nor child could distress him enough to jeopardize his career. Only an elder had that sort of power: his mother.

In late 1984, Mien-Chie's mother became ill. She had a debilitating form of diabetes, and she was confined to bed. His father had died years earlier, which left the children to look after her. As the oldest son, Mien-Chie was expected to shoulder the greatest part of the burden. He was still relatively new in the Weinberg lab, and he was reluctant to take time off, but he couldn't say no. For three or four months, he spent most of his days taking care of his mother. Even after she recovered, Mien-Chie was not set free. "She began to complain about my being a scientist," he said. "I was earning less money than my two younger brothers. One is in business and one is in real estate, and she wanted to know why I was thirty-five years old, had a family, and had no money. She complained all the time. I don't know; maybe it was because my father was dead and she wanted her sons to succeed, in respect to his memory."

We were talking in the Whitehead cafeteria during an off hour, and though we were alone in the room, Mien-Chie was practically whispering. His round face shone with sweat, and a cowlick of his straight black hair waggled whenever he shifted in his seat. He seemed ashamed, rueful, flustered. His mother was just worried, he said, that his family would suffer if he was too poor.

"But isn't she proud to have a son doing basic research?" I asked. "Business and real estate are fine, but doesn't she think being a molecular biologist is more prestigious?"

"Nobody in my family understands my work," he said. "They understand salary." He giggled his distinctive giggle, a noiseless heaving of his shoulders.

"So then my mother decided that it was time for me to buy a house. That was the least the oldest son should do. I didn't really want to buy a house. My wife works. She's a technician for a small biotechnology firm here in Cambridge, but her salary isn't so big. And you know what a postdoc earns. But my mother insisted. She said she'd help out. To help us decide if we should buy a house, my wife and I threw the Chinese coins. You know what those are? They're an old Chinese way of making decisions. The answer came up one head, one tail — yin-yang — which means yes. So for a while I was really busy looking for a house and then getting a mortgage. We couldn't afford anything in Cambridge. My mother said we should keep looking.

"Finally we found a nice house in Natick, near Framingham. We could afford this house, with my mother's help, but it meant that I'd be very far from work. It meant that I'd have to drive thirty, forty minutes each way. So even when I got back to working in the lab full time, I had to leave all this time for traveling. I'd drive in in the morning, drive back home for dinner, and then say, Well, maybe I should drive back to the lab. But I hated that idea. Once I got home,

I didn't want to leave to drive half an hour again. What this has meant is that I don't spend long hours in the lab. I feel bad about that, but I don't have much choice. Since my wife works, we both have to look after Victor." Mien-Chie's experiments were further delayed when his mother decided that she wanted to go on vacation to Florida. Mien-Chie took wife, son, parent, and himself to Fort Lauderdale for two weeks.

Nine months had elapsed since Mien-Chie cloned the normal *neu* gene. On his return from Florida, he was eager to make up for lost time. Among other tasks, he perfected his *neu* cell line and, in early summer of 1985, gave the cells to Cori. "I think now I'm ready to work hard again," he said to me at that time. "I think I'll start trying to come back to the lab at night." He told me about his plan for mixing and matching what amounted to 66,000 bases of DNA (the sum of the normal and transforming genes). Because he had seven big pieces of DNA, he'd divided his project into seven parts. Some of the steps could be done simultaneously; others had to be done sequentially. If one of the seven operations yielded a recombinant clone that turned normal cells cancerous, Mien-Chie would stop, knowing that he'd located the region of the mutation. Otherwise, he'd proceed to the next step.

He drew me a diagram of his planned experiment, marking the species of enzymes he would use, percentages, restriction sites. The diagram was a whirligig of arrows and numbers. It certainly didn't look like the simple models that Weinberg sketched on the blackboard during group meetings. Mien-Chie said that he had the first of the seven steps cooking right then, and that within eight days or so he should know whether his initial hybrid clone was cancerous. "So what do you think?" he asked me, returning his felt-tip pen to his shirt pocket. "Do you think it will work?"

I laughed, amazed that he would solicit my opinion. "I have no idea. I hope so. What does Bob think?"

"He thinks that it's worth a try. You know what he does when he wants something to work? Have you ever seen him do this?" Mien-Chie crossed his middle finger over his index finger, and shook the linked fingers in the air. "He says, 'Let's keep our fingers crossed.'" Mien-Chie raised his other hand and crossed those fingers, too. "Maybe if I do what Bob says, it will work." A moment later, he let his hands fall back in his lap, fingers at ease. "If not, I think I'd like to work on drosophila again."

• • •

Eight days later, I stopped Mien-Chie as he was walking down the corridor. Although it was unlikely that such a major event as the identification of the *neu* mutation could have passed unheralded, I asked him whether he had any good tidings.

"Well, the first results were negative," he said. "It doesn't look as if the mutation is in that part of the gene, the first part. So now I'm going to look farther down. I'm looking at a couple of hybrid clones simultaneously. I've transfected them both into 3T3 cells; I'm trying to move things along. I should have the answers to those in a couple of weeks." He raised his right hand and crossed his fingers; his shoulders heaved with his silent laughter.

Two weeks later, I stopped by Mien-Chie's desk and guessed, from his impassive expression, that the second and third stages of his experiment had come up negative. "Really, the approach should work," said Mien-Chie. "There's no reason why it shouldn't. I guess I've just had bad luck so far in looking at the wrong part of the gene. I'll keep going. Bob is very encouraging. He says it could work."

I asked David Stern what he thought of Mien-Chie's efforts to find the mutation.

"It's the kind of idea that looks pretty on paper," he said. "But I doubt that it'll work in practice. There are too many opportunities for things to go wrong. All his reactions have to work perfectly. The precentages have to be perfect; the right pieces have to find each other. But real life isn't like that."

The next stage of Mien-Chie's experiments provided yet another lesson in real life. Hybrid clone number four was not tumorigenic. The result could mean one of two things. Either the mutation was farther down the line of the 33,000 nucleotides and Mien-Chie would find it during round five, six, or seven. Or David was right. The method might not work in practice, in which case Mien-Chie could have skipped past the lesion without knowing it; the mutation could lie in a region of the gene he'd already tested during rounds one through four. He remained valiantly jolly, his crossed fingers still waving in the air. I rooted for him. I wanted him to find that mutation. I didn't know what would happen if he failed.

And then I'd watch Cori, the cool, competent sylph, who needed no sympathy and crossed no fingers. Her sequencing trotted along, day in, night out. She deciphered the cDNA in groups of thirty to fifty bases, treating each collection with enzymes or chemicals that reacted differently to *a*'s, *t*'s, *g*'s, and *c*'s, and pipetting the complexes onto agarose gels, which could be x-irradiated and analyzed. Admittedly, hers was a straightforward experiment. She didn't have to concoct her own recipes and equations. Sequencing protocols are included in the standard molecular biology textbooks. But somehow, Cori sequenced better than most people. Her sequencing gels were perfect. She'd do all the chemical and enzymatic steps so precisely that the ladder rungs of black radiation left behind on the films of her gels could have been stroked by Albrecht Dürer.

"Cori, you make the most beautiful sequencing gels I've ever seen," René said wistfully, holding one of her films up to the light. He handed the film to me and asked, "You've seen these things published, haven't you?" I nodded. "And have any of them looked as beautiful as this?" I shook my head. "Yeah, most of them are so smeared with background shit," he said, "that when the guy tells you it's *a, t, g,* you think, Oh, yeah? Says you, buddy."

Often I'd find Cori alone in a room, her autorads on a light box, her chin propped on her fists, her long tawny hair falling forward, as she struggled to decode the nucleotides indicated by the lines of radiation. "Sometimes I stare at these for so many hours that my eyes start to cross," she said. "When I go to bed, I see sequencing gels dancing around before me." Any time she encountered a section of the gene that did not read clearly enough to satisfy her, she went back and did all twenty or fifty or hundred bases over again. "I'm probably being paranoid," she said, "but I don't like to make mistakes."

I asked her whether she ever got bored with sequencing.

"No, I actually enjoy it," she said. "There's enough interpretation required to be intellectually stimulating, but not enough to drain me. Now, if you ask me about writing papers — that I do hate. That is strictly boring, though I write quickly."

The summer coasted along, and the fifth and sixth stages of Mien-Chie's mix-and-matches came up negative. Neither of the crazy-quilt clones that he wove together from patches of the normal and cancerous *neu* genes transformed cells when he transfected the clones into recipient tissue. Yet Mien-Chie had hopes that the final hybrid gene — step seven — would be the one he'd been praying for. He continued to believe in the perversity of fate, a fate that would bury the *neu* mutation in the last place Mien-Chie happened to look. One week he gave a floor talk, and he discussed his method for mixing and matching such a lengthy gene. His talk was surprisingly entertaining. He projected the effervescence of a scientist just beginning an experiment, rather than of one who's crawling toward its completion. He joked, he flashed slide after slide, he spoke so breathlessly that the audience never stopped tittering. It was the most generous talk I'd heard: he used the pronoun *we* instead of *I,* and he gave credit to any Weinberg scientist who had ever so much as mentioned the word *neuroblastoma.*

But a good rap wasn't enough. Alan Schechter, Mien-Chie's best friend in the lab, was sitting behind me in the auditorium, talking to a graduate student from Richard Mulligan's group. "It's a really stupid thing he's trying to do," Alan said. "I don't know why he's wasting his time. Once the normal cDNA is cloned, the mix-and-matching will be done in a month." I winced at the word *stupid,* but I suspected that Alan was right. Mien-Chie was trying to do something the hard way when simplicity lay just around the corner.

A couple of days after his floor talk, I invited Mien-Chie and his wife and son out for dinner. I took them to one of my favorite restaurants in Boston, but Mien-Chie didn't have much of an appetite. He ordered a cheeseburger, ate two or three bites of it, and left the rest until it cooled to inedibility. King-Lan Cheng picked at a spinach salad. Only Victor seemed unaware of the pall that hung over the table: he squashed his hamburger into a hash of beef and bread and then stuffed the pieces into his mouth with his fingers. Mien-Chie talked briefly about his mother, but this time it wasn't an elder who depressed him.

"I thought, when I first came to the lab, that we were supposed to cooperate, help each other," he said. "I thought we were all in this together. I gladly gave Cori the high-expressing *neu* cells for her experiments. I thought if we all helped each other, we'd all take credit and get the work done faster."

"You're too nice," King-Lan Cheng said. "Other people take advantage of you." She turned to me. "I always tell him not to be so nice."

"But I guess I've learned my lesson," said Mien-Chie, his voice catching. "We're not supposed to cooperate. We're supposed to compete. Okay, that's the way it is. So we should compete. I should compete with Cori.

"Cori is very concerned about keeping her project to herself. You know, once when she was out of town I said something to Bob and to Phil Sharp about the cDNA experiments; I don't remember what. I was recommending something just because I thought of it, not because I wanted to do it myself. When she came back, she was upset. She told me, 'How dare you talk about my project when I'm not around? Don't do that again.' I was really shocked. I thought we were all in this together. But, okay, I was wrong. So now I'll just do things for myself, not help anybody else. Okay, I was wrong. No more being a nice guy. So okay."

"You've got to be mean," his wife said. "You've got to be selfish."

I didn't really understand what Mien-Chie was talking about, and his "okays" were making me nervous; each one was laced with a little more hysteria than the last. "Are you giving up the genomic mix-and-match experiment?" I asked.

"Cori is cloning the normal cDNA now," said Mien-Chie.

"And then what happens?"

"I tried talking to her about that," he said. "I thought she'd let me do the mix-and-matches with the cDNA's. I thought that's what we'd agreed. I helped her. I thought she'd help me."

"But what did she say?"

"She said she couldn't do that on her own. She said we had to let Bob make the decision. So that's what I get for believing people."

"You had a firm agreement that she'd give you her cDNA's so that you could do the experiment?"

"That's what I thought she'd do. I gave her my cells so that she could clone the normal cDNA easily."

"Is that a fair trade?" I asked. "I mean, you needed those cells for yourself, too, didn't you?"

"I just thought that she would do the sequencing and I would do the mix-and-matches. Otherwise, what have I done? What's my contribution? And what will I do? I told Bob I could work on drosophila. I could study oncogenes in drosophila. But he's not interested. He only cares about mammalian cancer, human cancer."

The waiter approached to ask us whether we wanted dessert, but Mien-Chie shook his head. He was sweating. He no longer looked like a laughing Buddha; he looked like the tragedy half of the tragicomic face on old theaters. "Everybody in the lab is fighting over the *neu* gene," he said. "There's not enough room for everybody."

With a lack of professionalism, as well as a bad habit of endowing inanimate objects with preternatural powers, I began to see the *neu* gene as a source of evil. I hated the gene. Mien-Chie wasn't the only one who was buckling under its weight. René was unable to get his anti-*neu* vaccine to work. He was trying to insert part of the cancerous gene into the arms of a vaccinia virus, the pathogen used to immunize people against smallpox. He'd then inject mice or rats with the recombinant virus and hope that as the rodents raised antibodies against the vaccinia proteins, they'd also develop immunity against the stowaway *neu* protein. But for a long time the most dramatic thing that happened in the course of his experiments was René's being bitten on the hand by an angered rat. The rat had been inoculated with a vaccinia-*neu* virus of uncertain potency, and René worried briefly about coming down with either smallpox or cancer; instead, he was feverishly ill for a few days as a result of a nasty bacterial infection transmitted by the rat's fangs. "As soon as I get better," he told me over the phone, "I'm going to break that rat's neck."

David Stern fared little better in his analysis of the function of the *neu* protein. For almost two years he tried to determine what the protein was a receptor for, but he couldn't locate the ligand. *Neu* didn't react strongly to any growth factor, hormone, or peptide (a small protein) that he fed to it. He tried exposing the receptor to extracts from human placenta tissue, which is rich in all kinds of growth-promoting nutrients. But nothing seemed to be *neu*-specific. "I even considered going through a catalogue of neuropeptides and ordering everything in sight," he said. The idea made sense. The *neu* oncogene had been cloned from brain tumor tissue, so it was possible that the normal *neu* protein was a receptor on brain cells designed to pick up a signaling neuropeptide. "But then I realized that there were two or three hundred of those little suckers, and I wasn't about to test them all."

Compounding the technical problems the scientists encountered with *neu* was Weinberg's increasing indifference toward the project. He had once been the biggest publicist for the gene, encouraging Alan to pursue it when other people in the lab thought it was an artifact and a joke. But the gene wasn't giving him what he wanted. He wanted to know what p185 was a receptor for; David had trouble finding the growth factor that fit. He wanted to know the mutation; years had passed without an answer. The other projects in his lab were listing or sinking. There were no pending breakthroughs on the function of *ras, myc,* or *fos*. There was no metastasis gene. Weinberg tried to put on a sunny face. At group meetings he continued to lay out his notes from conferences and conversations, hoping that one of his scientists might be inspired by others' experiments. "No matter how bad I feel, I have to smile and slap people on the back and tell them to keep going, that next week everything will be better," he said to me. "I can't let the people in my lab think that I'm getting discouraged. That would be completely demoralizing to them." Yet his low-level depression was hard to disguise. "We're dotting *i*'s and crossing *t*'s," he said. "Something has got to give somewhere."

The one bright spot seemed to be Cori Bargmann, who was as bright as a bright blue galaxy. I imagined the sort of scientist she'd be in ten years. Women have had trouble breaking into the upper ranks of science; Cori Bargmann would not be held down. While others in the lab wavered, she grew stronger. She constructed a test tube collection of cDNAs from RNA messages in the cells that were rich with the normal *neu* gene and proceeded to screen the catalogue of cloning viruses for the one carrying the proper cDNA. She worked calmly, steadily, and quickly. Something was going to give somewhere.

I wandered over to René's bay and flopped down into a chair. "René, what's going to happen with the *neu* gene?" I asked.

"The *neu* gene?" he said, alilt with sarcasm. "Oh, it will all come to a head sooner or later."

"You really think so?"

"Well, how much longer can this go on, two people doing the same thing?" he replied.

"They are doing the same thing, aren't they?" I said. "How did that happen? Why did Bob put so many people on one project?"

"You can guess the answer," he said. "Bob wants to find the mutation, and he doesn't care how it gets done. Besides, he had nothing better in mind for people to work on. Nobody has anything better in mind. You don't hear Mien-Chie or Cori offering to give up *neu* to help Snezna study PI turnover, do you?"

A couple of days after my exchange with René, I asked Cori how long it would be before she had the normal cDNA. I was under the impression that Mien-Chie still hoped to get his elaborate mix-and-matches to work, and the question was innocent. Cori, however, gave me a funny sideward glance and flipped her long hair behind her. "I hate politics," she said quietly. "I've never really had to deal with lab politics before."

It took me a moment to figure out what she was talking about. "You mean the mix-and-matches? Are you going to give the cDNA's to Mien-Chie?"

"Well, I would," she said. "I don't care about getting a paper out of it. At this point in my career, the most important thing is not the number of first-author publications but what Bob thinks of me. To get a good postdoctoral fellowship, I need his support. And I think

I have it. I think I've earned his respect. At this point I can afford to be altruistic.

"I know how much Mien-Chie would like to do the mix-and-matches," she said, her voice so soft that I had to lean forward to hear her. "But it's not really my place to make the decision. It's up to Bob."

There was a subtext to our conversation that only later became clear to me. Cori knew what Bob had decided. For the past couple of weeks, Weinberg hadn't been in the lab much. He was working up at Rindge, racing to erect the frame of the annex to his house while the weather was warm. In addition, he spent a few days in Denver at a function sponsored by the American Academy of Achievement, where he was serving as a "role model" for three hundred outstanding high school students. But any time Weinberg was out of town, he called the lab every day. And one of the people he invariably asked to speak to was Cori.

Weinberg returned from his trip out west looking fit and vigorous. He told anecdotes about the other American Achievers he met: Elizabeth Taylor, Lionel Ritchie, Adolph Coors, Gerald and Betty Ford. "Liz Taylor made a lot of noise," he said, "and Adolph Coors was as awful as you might imagine." Weinberg had been most impressed with Lionel Ritchie. "I wasn't familiar with his music," said Weinberg. "But he seemed like a very pleasant and intelligent man. Did you know he was trained at Juilliard?" It was a wonderful change, said Bob, not to be surrounded only by scientists.

Weinberg didn't take long to catch up with the state of affairs at the lab. He strolled around the bays and talked to people. I saw him talking to Shelly, Cori, René. I didn't see him talking to Mien-Chie.

The next day he talked to Mien-Chie.

I was walking past Mien-Chie's desk, and I nearly failed to notice that he was sitting there. He was quiet, and he was staring straight ahead of him, motionless. I stopped and sat down next to him.

"Anything going on?" I asked chipperly. He turned to look at me. I saw that his face was wet, but not with perspiration. He was weeping. His cheeks were gray, and his glasses were speckled with tears.

"What's wrong?" I cried.

"Bob took me off the project," he said. "He gave my project to Cori."

"Oh, God," I said. "When did you find out?"

"Bob told me this morning. He said that Cori would do the mix-and-matches. So that's it. He took away my project."

I sat there lumpishly, not knowing what to say.

"Bob told me that because she cloned the cDNA's, she should be the one to do the mix-and-matches with them. He told me he appreciated what I'd tried to do with the genomic gene, but that it was only fair for Cori to get the rest of the project. So that's it. All my work for nothing. Is that fair to *me*? I don't think it's fair to me. I'm a postdoc, Cori's a graduate student, but okay, that's what Bob wants."

I hadn't realized before that Mien-Chie felt entitled to the choice project at least partly because he was senior to Cori. He was her elder.

"What are you going to do now?" I asked.

"I don't know," he said. "I have a couple of things to finish up with *neu*; small things that I'm doing. But after that, I don't know. Bob talked to me about setting up a transgenic mouse system. He said maybe we could study the effects of the *neu* gene on the development of the mouse. He said I should talk to Rudi Grosschedl about it."

"Transgenic mice?" I said. "That would be quite an ordeal, wouldn't it?" Transgenic mice were the latest fashion in molecular biology. A number of scientists had begun microinjecting foreign genes into the embryos of mice as a way of studying gene behavior; Grosschedl, Baltimore's star postdoc, was one of them. For Mien-Chie to set up a similar operation, however, would require a major investment — in equipment, time, patience. It's not a lighthearted what-the-hell, let's-give-it-a-try sort of idea.

"Does Bob really want you to go into transgenic mouse research?" I asked. "I mean, what could you learn about *neu* by studying transgenic mice?"

"I don't know," said Mien-Chie haplessly. "I don't know how serious Bob is about it. It was probably just a suggestion. I don't know if I'm interested. I guess I'll talk to Rudi about it. I don't know."

He didn't seem to be in the mood to consider new experimental horizons. Instead, he started clutching at past successes. He talked about fruit flies. "Why should I learn about transgenic mice?" he said. "Why is that system better than drosophila? I guess I'm in the wrong lab. Bob won't let me study drosophila here."

Fearing that Mien-Chie was about to burst into tears again, I groped around for consoling words. I told him that I thought Cori felt terrible about the way things had turned out, and I asked him whether he'd talked to her that day.

"No," he said. "I do think she feels bad. It's not her fault. I don't blame her. I think she'll come and talk to me, but not just yet.

"I don't know. Maybe Bob wanted me off the project because I haven't been working hard enough. Or maybe it's because I'm Oriental, and people think I'm weird. I don't express myself well in English. But Chiaho Shih would never have let this happen to him. We're both from Taiwan, and we both went to the same school, but we're very different personalities. He's much more aggressive."

Mien-Chie went to check on some cells in the incubator. A moment later, Alan Schechter came over to his bench, which was opposite Mien-Chie's. Alan wore white shorts and a long lab coat that once had been white but now was stained with black and rust splotches from collar to hem. Not saying a word to me, he began pawing noisily through a pile of papers.

"So what do you think?" I finally asked him.

"About what? Life?"

"About Mien-Chie being taken off the *neu* project."

He snapped around and looked at me sharply. "Did he tell you about that?"

"Just now. He seems pretty upset."

Alan turned back to his pile of papers. "Well, he should have known it was coming. He should have known that once Cori had both clones, Bob wasn't about to tell her to give them to him. The writing was on the wall."

"I guess he hoped that Bob would let him do the experiment," I said.

"He was dreaming," Alan said matter-of-factly. "I'm not trying to excuse Bob. Bob was so out of it that he didn't know what was going on. Maybe he was hoping it would all just disappear. But Mien-Chie took too much for granted. He set himself up."

"Did you warn him about it?"

"I told him I thought he should talk to Bob and straighten things out," said Alan. "But it was none of my business."

Mien-Chie returned to his desk. "I guess there's so little left to do that my technician can handle it himself," he said pathetically.

Alan tried to stay busy a few minutes longer, but I could see that he was concerned about his friend's unhappiness. Shoving aside his papers, he sat down and propped his sneakered feet on a stool. "So what are you going to do next?" he asked Mien-Chie.

Mien-Chie began tossing out ideas, rapidly, desperately. He said that he wanted to look for the genes that regulate oncogenes: the genes that switch oncogenes on or off. He said he thought there might be tissue-specific cancer genes that affect only one organ or related groups of organs. He proposed a few notions about how he might be able to clone the genes that regulate those tissue-specific oncogenes. Alan argued against most of Mien-Chie's logic. Mien-Chie rallied examples from drosophila to buttress his points. He was back to drosophila. He'd decided he wanted to study fruit flies again, he said.

"Fuck flies!" said Alan. "Flies are boring! Flies are *dead*." I figured he was trying to coax Mien-Chie out of his funk, but Mien-Chie was adamant about the importance of drosophila.

The next day at lunch, Alan talked spiritedly about a job interview he'd had with a biotechnology company. They were very interested in hiring him, he said, but he doubted that he could tolerate working in industry. "They all wear ties," he said. "Even the scientists. And they're all clones of each other. They came to the interview with clipboards and papers, and they started highlighting things on the papers with yellow Magic Markers. Yellow highlights! It was the most pathetic thing I'd ever seen."

Paolo Dotto joined the table. Like most of the experiments in the Weinberg lab, Paolo's project — an attempt to study the effects of oncogenes on the skin tissue of living mice — was doing poorly, and that day Paolo's small mouth was pursed with anger. "Did you hear what happened to Mien-Chie?" he asked the assembled company.

"We know, Paolo," said Alan.

"I think it's terrible," said Paolo. "I think it's terrible that Bob did that to him."

"It was to be expected," said Alan.

"I don't think so," said Paolo. "Bob should have been paying more attention to what was going on."

"There are things you don't know, Paolo," said Alan. "Bob wanted to take Mien-Chie off the project a couple of months ago, but I argued against it. I thought Mien-Chie wouldn't have been able to handle it. I don't think you realize how weird he was being at the time."

"Well, I think Mien-Chie is being screwed," said Paolo. "He gave Cori all his reagents. He gave her his cell line."

"Yeah, but you don't realize that Mien-Chie had considered *not* giving away his cell line," said Alan. "I had to convince him that it would be more damaging to his career if he behaved like an asshole and not give out his reagents. And Mien-Chie should have realized that any gene that was cut in at least seven places by a given restriction enzyme was not a good candidate for mix-and-matching."

Paolo has a delicately beautiful complexion, and when he's upset his cheeks turn rosy; at that point, he looked as though he were wearing enough rouge for a streetwalker. "Bob should have been very clear and careful from the beginning about who was doing what," he said heatedly. "If he'd been paying more attention, he could have redirected Mien-Chie toward a project that would work."

"How much can you control a postdoc?" Alan demanded. "If Mien-Chie were a graduate student, then it's true, it would have been Bob's fault. But postdocs have to take responsibility for themselves."

"There are postdocs and postdocs," said Paolo. "A strong leader would never have let this happen."

"Mien-Chie thought he'd get the cDNA clones from Cori, and that's where he made his mistake," said Alan. "He shouldn't have trusted the integrity of his project to a half-baked agreement."

"I guess this just goes to show that you can't trust anybody," said Paolo.

"I'm not going to argue with you there," said Alan. "Who's talking about trust?"

Everybody was upset, and everybody was in a catty mood. Alan and Shelly began to tease Paolo about being Italian. Alan said that Paolo could be a "guinea" pig for the first experiments with transgenic humans. Paolo had never heard the slur before, so Alan explained what it meant.

"I guess I didn't know the meaning of that word," said Paolo, "because that's not something my friends would say to me."

"The first mistake you made was in thinking you were among friends," said Alan.

"I figured that out a long time ago," said Paolo.

Changing the subject, Alan began discussing other job interviews that he had scheduled; he'd pretty much decided to stay in academic science.

A day or two later, I dropped by Weinberg's office to ask him about Mien-Chie. He said he was on his way down to the cafeteria to get a glass of apple juice, and he invited me along. We sat in a dark corner booth and began talking, as we usually did, about me. Bob likes interviewing people at least as much as he does being interviewed. He asked me whether I'd been up to Fitzwilliam recently. Fitzwilliam, a small New Hampshire town about seven miles from Rindge, is the ancestral home of my father's side of the family. As a devout genealogist, Weinberg loved discussing my relatives. I replied that in fact I had visited the old Angier cemetery just a couple of weeks earlier. He asked about the prognosis for *Discover*, a financially troubled science magazine then owned by Time Inc. I said that his guess was as good as mine, but that the magazine wasn't likely to survive for the long haul.

"But we're not here to talk about *Discover*," he said, drumming his fingers on the table.

"Mien-Chie doesn't seem to be very happy these days," I said.

"What do you mean?" he asked.

"Well, I guess he's depressed about the *neu* project," I said. "I guess he was hoping to do the mix-and-match experiment with the cDNA's."

"That was a complete misunderstanding on his part," said Bob firmly. "It was not as if he was misled by anyone besides himself. I certainly never gave him the impression that I was prepared to ask Cori to hand over her clones to him. He didn't come in and ask me such a thing. If he had, I would quickly have disabused him of the notion. Such a request would not have been fair to Cori. It makes no sense at all to me that, having struggled so hard to make the cDNA's, she should then give them away to him so that he could do the most interesting thing one could possibly do with them — find the mutation. To the extent that there are frustrations and disappointments, they would have been enormously lessened had he talked about his preconceptions earlier.

"Mien-Chie tried his hand at localizing the mutation with the genomic clones, and while I thought that it was unlikely to work, I didn't forbid the attempt. He seemed eager to do it, so I let him do it. But just because that aspect of his project didn't work out was no reason for him to assume that Cori should sacrifice her project to him. You get my point? It wouldn't be fair to her."

"So he never consulted with you?" I asked. "You didn't know that he was hoping to do the mix-and-matches with the cDNA's?"

"No, I had no idea," Weinberg replied. "Again, if he had discussed it with me, I would have set the record straight.

"For me, it was a managerially difficult problem. If one wants to move somewhere on a project, one doesn't want to have a person working on it alone. One needs cross-fertilization and synergy. Granting that, the moment one has people working on related areas of the *neu* gene, then one has the problem of occasional overlap, bumping into each other.

"The easiest way to avoid that is to have everybody in the lab working on their own little areas. One person can work on oncogenes, and the other can work on high-energy physics.

"My goal has always been to have people working on complementary aspects of a project. My great horror has been the fear of having people in my lab compete with one another. It's very corrosive. But this is not the first time I've had this kind of problem."

"I guess Mien-Chie's upset because he felt that, since he gave his cell line to Cori for her experiments, he should get something in return," I said.

"I don't think he's justified in being upset," Weinberg said, a little testily. "There's also a question of the amount of time he's spent on his project. He cloned the genomic copy when? Nine, ten months ago? And what has he done with it? One has to ask how much time we give before we expect results. Two years, three years? There has to be a limit, and there have to be certain expectations of progress. None of us is immortal. If a person fails to demonstrate efficiency, one has to question his or her commitment to a particular project."

"But what will happen to Mien-Chie?" I asked.

"Mien-Chie will be a co-author on Cori's paper," said Bob. "And he's looking now at the effects of high levels of expression of the normal *neu* gene on cells. Something interesting might come out of that. I don't think he's fared so badly. I don't think, realistically, he has any right to complain. I'm not deserting him by any means. I'm not throwing him to the lions."

"Are you going to give him something else to work on?"

Weinberg stared blankly at the wall of the booth and tapped his fingers on the table. "The problem is not limited to Mien-Chie," he said quietly. "Over the next few weeks I have to do some very hard thinking. I have to regroup. I'm going to look at every project in my lab and ask, Is this yielding anything worthwhile? Where are we headed? Is this person wasting time? How much longer are we willing to throw good money after bad? Mien-Chie is not the only one who will come under my scrutiny. I'm going to have to clean house. I may have to drag some people off their projects kicking and screaming, but if that's what I have to do, so be it. I suspect that three months from now many of my people will be working on new projects. I don't know yet what those projects may be, which is why I've got some serious thinking to do.

"For example, I know that Snezna's phospholipid project is the sort of thing that we should be doing. Understanding the function of *ras* is a natural problem for us to work on. But I can't expect her to impale herself on my convictions. So I must come up with a better approach to oncogene function. I must come up with better ideas for people to work on. The same goes for Mike Gilman's project, Paolo's, David's, René's. I must decide what direction my lab will take. I've thought about transgenic mice. Should we get involved with that system? Is it worth the investment? Transgenic mice are very fashionable right now, but I'm not sure what sorts of questions we could answer through studying transgenic mice. I'm just not sure at all of the next obvious step for my lab."

I had never seen Bob look so unhappy and so earnest and so lost. He looked like the scruffy, sad-sack Charlie Chaplin of *City Lights*. The cafeteria was empty save for us, yet I felt oppressed by Weinberg's evident dismay. "It will be all right," I said. "You'll come up with something. Won't you?"

"Perhaps," he said. "If I think hard and clearly, perhaps I will." As though shaking himself out of a trance, he slapped his palm on the table and began sliding out of the booth. "Well, enough of melancholia. Have I answered all your questions? Have you had enough for the day?"

"I guess so," I said and stood up. Weinberg returned to his office; I stepped outdoors to squint and sigh in the afternoon sun. I felt creakily tired. I'd had enough of the day, and, for the time being, I'd had enough of the Whitehead Institute. I was ready to visit another lab. I wanted to go to a place where people didn't squabble so much, where experiments occasionally worked, where there were patches of happiness speckling the severity of molecular biology. More than happiness, I wanted contrast: a control for the Weinberg lab. Many of the scientists I'd interviewed had talked about Mike Wigler. They'd praised his expansive intelligence, and they had spoken enthusiastically of his prospects. "If there's anybody who's in a position to find the function of *ras*," said Mariano Barbacid, "it's Mike Wigler." I'd seen Wigler present two seminars at Harvard University. He didn't strike me as an easygoing, happy sort of guy; in fact, he seemed churlish and impatient. But I imagined that if Wigler and his researchers were on the verge of discovering the function of the greatest of oncogenes, they had to have at least a nodding acquaintance with joy.

The story of the *neu* gene was far from over. It would return later, beautifully and magically. To the astonishment of all, and especially of Weinberg, the gene would prove to be one of the earliest answers to cancer.

10 The Alternate Pathway

MICHAEL WIGLER is a child prodigy who just happens to be an adult. Great scientists are always "bright," but Wigler has the balletic, flamboyant, tireless intelligence of a *Wunderkind*. He's more sheerly intelligent than many of his peers, and his peers will admit as much. "Wigler has the most brilliant mind of any of us," said Bob Weinberg. "He doesn't suffer fools."

"Everywhere I go, people come up to me and tell me how intelligent Mike is," said Scott Powers, a postdoc in Wigler's lab. "I don't know why, but they feel compelled to compliment me on his intelligence."

"Wigler would try to play dumb in deference to us," said Gerald Fink, the Whitehead yeast geneticist. "He'd come to our meetings and claim that he didn't know the first thing about yeast. But then he'd ask these questions that out-yeasted the oldest yeast person among us."

Wigler has been a child prodigy since childhood. In elementary school in Garden City, Long Island, Wigler would shout out the answers to a teacher's questions before his classmates even had time to look up from their spitballs; his teachers had to handicap him during quizzes just to give other students a chance. He scored a perfect 1600 on the Scholastic Aptitude Tests. Majoring in math at Princeton University, Wigler enrolled only in graduate-level courses from his sophomore year on. He decided to go to medical school after college, but he was so bored with the medical course work that he took up tournament chess just to keep his mind limber. When he switched from medicine to molecular biology, he immediately clambered to the summit of his profession; as a graduate student working closely with Richard Axel at Columbia, Wigler helped to devise the technique for inserting foreign genes into mammalian cells — the transfection protocol that revolutionized DNA research. He was offered an independent position at Cold Spring Harbor after only six months of postdoctoral work — a ridiculously short period of time — and he soon became known as the protégé of James D. Watson, the director of Cold Spring Harbor and an icon of science. During the great oncogene races of the early 1980s, Wigler kept apace of Weinberg and even nosed him out once or twice.

So when Wigler claimed to have developed a system to outfox a human tumor cell, the scientific community paid heed. In the winter of 1986, Mike Wigler seemed to be on the verge of solving the biggest puzzle in oncogene research. He had a fail-safe method for discovering the function of *ras*. It was a logical, precise approach that suited Wigler's logical, mathematical temperament. Wigler's lab recently had discovered an analogue of the *ras* gene in one of nature's simplest and most genetically rational organisms: yeast. The yeast version of *ras* was shockingly similar to the human cancer gene. Through more than a billion years of evolution, the DNA sequence of the gene had been conserved, suggesting that whatever role the gene played in human cells it performed in fungus as well. The discovery of that kinship conferred on Wigler a distinct experimental advantage. Human cells and human genes are a muddle; yeast genes are easy, and yeast cells are as pliable as Play-Doh. Wigler could manipulate the *ras* gene in yeast to a degree and with a precision impossible in mammalian cell biology. And after he'd analyzed the biochemistry of *ras* in yeast, Wigler could raise that insight up the phylogenetic stepladder to humans. Relying on the malleability of yeast, Wigler would discover the nature of the most savage of cancer genes. Biologists everywhere were waiting to learn what Wigler would divulge.

Having heard so much of Wigler's brilliance and prospects, I wanted to get to know him. When he agreed to let me stay at his lab for a few weeks, in that winter of 1986, I felt almost absurdly happy. Wigler is an eccentric, boyish, wry-witted man who did his utmost to make his visitor feel welcome. I was happy because he immediately introduced me to the people in his lab, who seemed openly affectionate toward, or at least chummy with, their boss.

Above all, I was happy because I thought that at last I might see some real progress. I knew that the race to find the function of *ras* was the most delirious race of the day, and there seemed no better place to stake out a grandstand seat than in Wigler's lab, where the researchers had more than models or mantras or a scientific sense of smell. They had the perfect experimental system.

"It will take hard work and mental discipline on our part," Wigler told me. "But step by step we'll get there. We'll figure out the function, and we have as good a chance as anybody of being first. No, a better chance. There's really no biological system to compare with yeast."

Wigler, who doesn't suffer fools, was determined not to be one. With yeast, he couldn't lose.

Cold Spring Harbor Laboratory is only thirty-five miles from Manhattan, but it may as well be three hundred miles, it has such quiet beauty. To get there you drive along the same strip that anxious single professionals clog during the summer for their weekend pilgrimage to the Hamptons — the Long Island Expressway, which has been called the world's biggest parking lot. But you turn north long before hitting any resorts, and when you reach the Cold Spring inlet you can't believe that the shorefront has remained so chaste. Gentle hills swell up from the water's edge, and in the autumn the trees blaze bronze and butter and rose. The forty acres of the central laboratory grounds fall within a state-run bird sanctuary. More than 140 species of birds flutter through the area each year, a figure that apparently makes bird watchers salivate. If you're patient and lucky you may see a great blue heron, red-breasted merganser, ibis, egret, snowy owl, and even a westering bald eagle. At the very least, you'll hear the surreal music of circling geese and gulls.

The laboratory complex, too, is lambently beautiful. Many of the twenty-eight buildings are museum pieces dating from the nineteenth century, when scientists were gentlemen and aesthetes who made their work places look like their homes; the labs have white clapboard siding, sunbonnet roofs, colonnaded porticoes, and fat white chimneys. During the summer months, the grounds are overrun by thirty-six hundred scientists who flock from around the world to attend the Cold Spring Harbor meetings, the most famous fêtes in biology. But from September to May, the lab is a monastery for the hundred or so senior and junior scientists who make up the Cold Spring Harbor staff. The lab has no formal affiliation with a university, although graduate students from the nearby Stonybrook branch of the State University of New York occasionally apply to study with Cold Spring Harbor researchers. The neighboring towns, precious tourist hamlets, offer few diversions beyond boutiques selling Caswell-Massey soaps, exotic teas, and antique spinning wheels. In sum, the outside world does not impinge. There's nothing to prevent Mike Wigler from living for his laboratory — and nothing does.

Mike Wigler is not a handsome man, but he's not unattractive, either. He's a little bit of a schlemiel: he's about thirty pounds overweight, his taupe hair is a halo of frizz, his corduroy trousers hang too low on his hips, and the tongue of his shirt tends to flap out at the back. When his wife is out of town, say his postdocs, Wigler's sloppiness becomes artful. Then he may lumber about with large stains across his shirt or visible smudges on his eyeglasses; sometimes his glasses get so dirty that while he's talking to you he'll turn them sideways on his nose just so that he has a clear part to look through. Yet his face has a cuteness to it — when he smiles, he looks like a troll doll — and he's too aware of what's going on around him to qualify as a nerd. During casual conversation, he is easily distracted, and he'll sometimes stop in midsentence to stare off dreamily into space. It's a mistake, however, to assume that he's absentminded. From day to day — even from moment to moment — he knows the status of every experiment in his lab.

Indeed, the first big difference that I noticed between the Weinberg and Wigler "styles" was Wigler's omniscience. Wigler's lab belongs to Wigler; it is his fiefdom. He is always there, and he is always aware. Weinberg works from 10:00 A.M. to 6:00 P.M. five days a week, and he rarely comes into the lab on weekends; Wigler works from 9:00 to 6:00, goes home for dinner, comes back to the lab at 8:00 to stay until midnight, and he's never *not* there on weekends. Nor does he take vacations.

"I miss the lab when I'm away," said Wigler. "It's hard to say exactly what it is that I miss. Not the people, really. People come and go; that's the nature of a research lab. But the lab remains; it doesn't change. I think it's that essential element of lab-ness that I miss when I'm not here. When I'm not in the lab, I think about it. And when I think about the lab, I want to be there. If I'm in the lab and I have a headache, I may consider going home. But it's more likely that if I'm at home and get a headache, I'll come into the lab."

Wigler had begun the conversation in a prone position, sprawled across the Haitian-cotton couch in his office, but as he talked of his lab passion, he got so excited that he sat straight up, threw his left arm over the top of his head, and hugged a couch pillow against him with his right. "Sometimes I wonder what would happen if I woke up one day and didn't feel like coming into the lab. What happens if I just get bored? Luckily, I haven't had to confront that yet. Right now, coming into the lab is the most fun thing I can think of doing. It's as natural as breathing."

Most scientists consider travel one of the perks of a career in academic science, but Wigler so dislikes being away from the lab that he rarely bothers attending the many conferences to which he is invited. "I find what we're doing in the lab much more interesting than hearing about what other people are doing," he said. "That's a big change from when I was younger. I used to be fascinated by other people's ideas: listening to them would give me a hundred new ideas for things I'd like to try myself. Now I hear other people's ideas and I think, That's pretty stupid. What is that going to prove? Besides, meetings are usually held in really boring places like Colorado."

The ten or so postdoctoral fellows who work for Wigler no more want to be thought of as his subordinates than Weinberg scientists want to be thought of as Bob's. "Whatever you do, don't perpetuate the myth," said Scott Powers, "that Wigler is some sort of genius and we're just the hands that do the work." Yet Wigler's omnipresence is the genius of the Wigler lab. Shuffling from room to room, sneakered feet dragging across the floor, thumbs hooked in pockets, Wigler confronts his people with his stock query, "Anything new?" or its variant, "What's new?" The person may be sitting in the central postdoc office — where six of them share desk space — doing nothing more than smoking a cigarette or leafing through a journal, and still Mike will ask, "Anything new?" Sometimes he'll ask the question of the same person two, four, six times a day. Usually the postdocs will reply (with extraordinary tolerance and affection, I thought), "No, Mike," but on occasion they have some problem or result to discuss. By keeping in touch with the daily progress of his lab, Wigler helps salvage stranded projects and point out where mistakes were made.

"Mike can be overbearing, and all those 'What's news?' can be stifling," said the postdoc Dan Broek. "But I have to admit that Mike has kept me from wandering down a lot of blind alleys."

In numberless ways Wigler sets the tone for his lab. He's not the only person who's always there; everybody is always there. Although Wigler doesn't demand constant attendance, he notices when people are absent for two or three consecutive Saturdays. "He'll casually drop some comment like 'So, how's your golf swing coming along?'" said Scott Cameron, Wigler's only graduate student. "Or, 'So, been working on your tan lately?'"

By and large, however, the scientists need no jibes: they prefer the lab and the lab grounds to anyplace else. Their outside obligations are few. Most of the postdocs live communally in one of the four large Victorian houses owned by the lab, and the rent includes thrice-weekly maid service. The lab offers free breakfasts, nearly free lunches, and free afternoon snacks. There's a tennis court on the lab grounds and lab-owned sailboats. At night the scientists can drink at the lab bar, where beers are seventy-five cents and cocktails $1.50. Between experiments, the researchers play cards with Wigler, who is a bridge fanatic. After staying at Cold Spring Harbor for a while, I noticed that two people who hadn't known how to play the game when I'd arrived were toting around bridge instruction manuals. Wigler, say his postdocs, likes to have people around him. He likes to have people around to distract him when he's bored or has an unpleasant task to perform. For example, Wigler frequently volunteers to review grant applications submitted by other scientists to the National Institutes of Health and the American Cancer Society, but when he actually has to read and assign a rating to the proposals, he is like a fidgety, grouchy child. He'll wander out of his office every ten minutes to grump about the poor quality of the applications and the stupidity of the proposed experiments. He'll develop a sudden urge to understand the General Theory of Relativity or to discuss, for the third time that evening, a postdoc's experiment. "Wigler always wants to be surrounded by the people in his lab who interest him," said Carmen Birchmeier, Wigler's postdoc from West Germany. "If for some reason he finds himself alone at the lab, like on an early Sunday morning, he'll get on the phone and start calling people one by one."

Wigler's lab is not so much his home as it is his playground, and the people who work for him are not his surrogate children but his buddies. Because he spends far more time with his postdocs than he does with his peers, Wigler seems younger than his chronological age. He cherishes his reputation as an *enfant terrible*. In 1984, at the age of thirty-seven, he was designated one of *Esquire*'s under-forty superachievers, an honor that delighted him as much for its emphasis on his youth as on his accomplishments. When I said something to him over lunch about our coming from different generations, he abruptly stopped eating and pointed out that we were, in fact, members of the same generation. "I'm only ten or eleven years older than you," he said. "A generation is twenty years." His scientists believe that he is on their side. He is on their side because he is one of them.

"Mike cares about us. He cares about our careers," said Birchmeier. "And believe me, that doesn't happen in every lab."

"It's been a real privilege to work with Mike," said Powers. "He's committed to doing excellent science, and as long as you share that commitment, he respects you. I knew scientists at Columbia who cared more about academic politics than they did about doing good science. So I appreciate working with a person I can respect and who respects me when I fulfill my end of the bargain by working hard and not letting myself get away with sloppy thinking."

Hard work is the operative phrase at Cold Spring Harbor. The real reason Wigler likes to live in the lab and be surrounded by his staff is that he wants the harvest of their labor. When projects are going very well, the scientists may work hundred-hour weeks. Wigler doesn't do experiments himself, but he is so up to date on bench technology that he can guide and goad a project as though he wielded the pipette. Wigler knows what must be done to extract a lusty result, and he knows when an experiment is reaching its climax. Many times a postdoc will emerge from the darkroom to find Wigler lingering by the door, awaiting his fix. He is a vampire for data.

His addiction to results is why Wigler turned to yeast. Like others in oncogene research, Wigler had slammed into a wall when he attempted the next obvious experiment with the human *ras* gene. He wanted to know what the gene did in normal cells and how a mutation in the twelfth codon could catalyze cancer. But none of his skills could make sense of nonsense, and the *ras* signaling pathway in human tissue was biochemical gibberish. He stopped getting results and he started going through data withdrawal. He had no models in mind, and those which he heard about from other scientists sounded arbitrary and far-fetched. He began to wonder whether he had made a terrible mistake by leaving theoretical mathematics.

And then, one morning, Scott Powers, a lanky, shy, six-foot-five scientist from Ohio who's a ringer for Christopher Reeve as Clark Kent, presented Wigler with a Southern blot that cashiered Mike's gloom. On the blot was evidence of a homologue of *ras* in yeast

DNA. Although he wasn't yet versed in the nuances of yeast, Wigler instantly grasped the profound implications of the finding. He knew that yeast geneticists had developed elegant techniques for popping genes in and out of yeast chromosomes. He knew that the peculiarities of yeast reproduction allowed yeast biologists to replace a resident yeast gene with a doctored copy of the gene — the precise sort of barter impossible in mammalian cells. When scientists transfect foreign genes into cultured animal tissue, they are restricted to adding new genes on top of the old; they can't erase the activity of the indigenous *ras* gene in an animal cell before they slip in a mutant version of *ras,* which means that any information gleaned about the mutant gene is sullied by the unquantifiable contributions of the native gene. But yeast geneticists can exchange an existing yeast gene for a mutant version of the gene, tit for tat, and then see how the mutation alone perturbs the cell. As David Baltimore said, "In biology, often the only way to know what's right with something is to see what happens when it goes wrong." And nothing can be made to go wrong so readily as yeast. Wigler realized that, through the use of designer yeast genes and yeast gene swapping, he could reckon the effects of a mutant *ras* gene on the life cycle of the fungus. And those effects would clue him in to the biochemistry of *ras.*

He also knew that the task of discriminating between the *ras* protein in yeast and all the other yeast enzymes would be easy. A human cell has about 30,000 tangled, prattling proteins; yeast, a mere 4000. Finally, Wigler knew how quickly yeast experiments can be performed. Mammalian cell transfection takes a month or two — when it works. The species of yeast that Scott Powers had blotted was *Saccharomyces cerevisiae,* or baker's yeast, the same yeast that brews beer, bubbles champagne, and leavens bagels. Yeast cells divide once every hour or so, making it possible for a researcher to complete many yeast experiments in a week and sometimes overnight.

"Mammalian cell biologists would sell their grandmothers to be able to do with their cells what we can do with yeast," said Gerald Fink.

Wigler wouldn't have to market any relatives. Within days of Scott's discovery, Wigler had seen his future, and it was the brightest, yeastiest shade of white. He could leave the black box of the animal cell behind him.

"I was sick and tired of working with mammalian cells," Wigler explained. "There are no tools to approach them. You have no idea what's going on inside a mammalian membrane or a mammalian nucleus. It's like trying to figure out if God exists." He hesitated. Mike is rarely satisfied with the similes and metaphors he comes up with. "Well, maybe I shouldn't compare it to theology, but you do need a certain amount of faith, because there isn't much else to go on. That's the reason why I never wanted to be a scientist. There's too much luck involved. Relying on luck and guesswork is anathema to me. I'd rather not get an answer in science than spend time laboring blindly. It's too depressing.

"But with yeast, you have a system. If you know that the gene you're interested in exists in yeast, it's stupid not to exploit the system. Yeast genetics is probably the most powerful system available to molecular biologists today. It's almost like doing math. You're *thinking,* and thinking is something I enjoy doing."

Once he decided to plunge ahead with the yeast project, however, Wigler didn't take time out to think. "All the tools were available, and we knew that the only rate-limiting step was our stamina," said Jim Broach, a yeast geneticist who had collaborated with Wigler before joining the faculty of Princeton. "And when you're working with Mike, lack of stamina is not much of an excuse." From 1984 to 1985, the lab was open twenty-four hours a day. Wigler picked the brains of every yeast expert at Cold Spring Harbor, and he soon could follow the intricate Mendelian lineages of yeast in his head, as though he were playing a game of mental chess. He hired the hardest-working scientists in the world: he imported biologists from Japan. Living up to their reputation, Tohru Kataoka and Takashi Toda practically sacrificed their lives to the project. Takashi could work for three consecutive days without eating or sleeping. While he was sequencing the yeast *ras* gene, Tohru almost never removed the goggles that protected his eyes from radioactivity; when he did, the deep red groove across his forehead made by the goggles lingered for days. "I work hard, but you can't compare me with the Japanese," said Dan Broek. "They're not even Earthlings."

Together, Scott, Tohru, Takashi, and Dan performed interplanetary miracles. They created a strain of yeast that lacked a working copy of *ras,* and they observed that the mutant yeast spores perished at a defined stage in their life cycle: just before cell division begins. They cloned the yeast *ras* gene and then mutagenized one of its nucleotides in a test tube, fashioning a yeast version of the cancerous *ras* gene. When they forced a healthy yeast cell to trade its native *ras* for the clone with the point mutation, the yeast cell responded by lapsing into a mold mockery of cancer. Normal yeast cells stop dividing when they're deprived of sugar; the mutant strain continued to grow even in the absence of sweets. Healthy yeast cells propagate orderly colonies of offspring that spread out in neat white strips, like lines of cocaine; the "cancerous" yeast cells proliferated into an ugly tumorous clump. The experiment graphically demonstrated that, even in the most primitive of organisms, a point mutation in the *ras* gene invites cell anarchy.

The scientists were flying. Next, they knocked out the *ras* gene in yeast and substituted a copy of the *human* gene. They didn't expect the cross-species strain to survive; regardless of similarities, yeast and human genes are not identical, and the notion that a human protein could serve as proxy for a yeast protein seemed far-fetched. But survive the yeast did, albeit feebly. "That was one of Mike's favorite moments," said Scott. "When he saw that yeast and human *ras* had functional homology as well as DNA homology, he was as excited as a little kid. It meant that *ras* must be doing the same thing in yeast as it does in humans."

The next obvious question was: What did the *ras* protein do in yeast? Wigler and Jim Broach discussed the best way to approach the problem. They began to argue about techniques. Wigler wanted to try one method, but Broach insisted it would never work. Wigler stormed off to the library to find papers that would support his premise and serendipitously found the answer to the function of *ras*. It was buried in an obscure yeast paper that had nothing to do with *ras*, but Wigler noticed that the yeast strains described by the geneticists — a team from Japan — were the same as the mutant *ras* strains under scrutiny in his lab. The similarities could not be a coincidence. "It was like getting hit over the head with a hammer," said Wigler. "I'd started out just looking for precedents on how to screen for revertants, and there I was, staring at the function of *ras*."

In a few months of laboring so hard that even the compulsive Wigler researchers said they would not be able to repeat the performance, they had produced the biochemical and genetic proof for Wigler's theory. They demonstrated that the *ras* protein controlled the activity of a familiar and potent cell enzyme, adenylate cyclase, which produces an equally familiar and potent molecule, cyclic AMP (adenosine monophosphate). The two compounds already were known to operate in cells throughout the human body. They help spur the heart to race, the thyroid to secrete thyroid hormone, the liver to break down sugar, the ovary to release progesterone. And now the Wigler lab had fitted *ras* into the all-purpose signaling pathway. According to the Wigler schema, the *ras* protein was one of the so-called G proteins, the membrane-based proteins that are among the first to respond when a human cell receives a signal from the outside world — from, say, a passing growth hormone. (They're named G proteins because, in their active state, the proteins grasp a cell molecule known as GTP, or guanosine triphosphate). Once motivated, *ras* tickles adenylate cyclase, which in turn catalyzes the synthesis of cyclic AMP from the stored energy molecules within the cell. A little ring-shaped Schwarzenegger of a molecule, cyclic AMP then strong-arms a string of enzymes and their protein targets into action — including targets known to loiter in the cell nucleus, where any decisions to divide must be made. How a point mutation in the *ras* gene could debase the adenylate cyclase pathway severely enough to cause cancer was not yet clear, but the Wigler results looked assured, for in yeast, *ras* dispatches its signal from the membrane to the nucleus through adenylate cyclase and cyclic AMP.

The problem of *ras* appeared to be solved. It felt so gratifying to be able to say that *ras* stimulates cyclase. Biologists had not yet described every last ancillary enzyme in the adenylate cyclase pathway, but they understood more about it than they did about almost any other complex signaling pathway in the cell. That *ras* could be wedged into such a pedestrian trail was a relief. The first human *ras* oncogene had been cloned three years earlier, and it was about time that its function was known.

In October of 1984, Wigler announced the adenylate cyclase results at a cancer conference in New York City, and the audience was thunderstruck. Scientists were impressed not only with the implications of his results, but with the quality, breadth, and meticulousness of the Wigler lab's approach. They talked about Wigler's not having been satisfied to make his case through one type of experiment alone; he had zeroed in on adenylate cyclase from every possible angle. His lab had mated mutant strains of yeast to show that a strain of yeast deficient in *ras* could be rescued by a strain that had an abnormally active type of cyclic AMP molecule — a genetic demonstration that the two genes communicated. His scientists had mixed together *ras* and adenylate cyclase in a test tube under strictly calibrated conditions to show biochemical activation. And for each step of the experiment, the researchers had used a welter of controls to rule out alternative explanations for their results. "Some people say that it's impossible to be too careful," Scott Powers told me, "but if anybody in science can be said to overdo it on the controls, it's Mike." Bob Weinberg, who was to speak immediately after Wigler, preceded his seminar by declaring that Wigler's work marked "one of the milestones of cancer research."

"We were dancing on air," said Wigler.

And then Carmen Birchmeier came to Cold Spring Harbor from the University of Zürich. It was Carmen's task to kick the adenylate cyclase results up the evolutionary ladder from simple yeast to complex animals. It was Carmen's task to prove that Wigler had found the *true* meaning of a cancer gene.

Six months after Carmen arrived in the United States, Wigler joked sadly, "I was ready to ship her back."

A handsome, horse-faced woman with a throaty laugh and an astringent approach to work, Carmen was an expert on the *Xenopus* oocyte, another popular biological "system." This oocyte is the unfertilized egg of a strapping, spotted, eight-inch-long species of toad native to Africa. Like any egg, an oocyte (pronounced "oh-oh-site") is a single cell, but because its volume is about a thousand times greater than that of the average animal body cell, the *Xenopus* oocyte is useful for studying protein activity within the cytoplasm. Purified proteins can be injected into the eggs through ultrathin needles, and the eggs are easily observed under a microscope. As a graduate student in Zürich, Carmen spent almost four years microinjecting molecules into thousands of golden, glimmering *Xenopus* eggs. She was tired of toads; she wanted to study yeast. But Wigler required her needleworking skills. Amphibian cells, he reasoned, are close enough to human cells to broaden the scope of his lab's discovery. He asked her to see whether *ras* stimulated adenylate cyclase in oocytes as it did in fungus.

Carmen injected purified *ras* protein into a batch of oocytes and allowed them to incubate overnight. To her horror, the eggs responded by breaking out into a rash of white spots. Carmen knew that white spots were the sign of meiosis: the eggs were beginning to mature, as though anticipating fertilization by toad sperm. She also knew that meiosis was associated with a decline in adenylate cyclase activity, not with a surge. On a working vacation back in West Germany, Carmen examined the *ras*-treated eggs with a microscope designed for oocyte work and saw that they were indeed edging toward meiosis. Oh, Jesus, she thought. Mike is not going to like these results.

Even less likable were the results she got, back in the lab, when she measured the flux of cyclic AMP in oocytes injected with *ras*. They showed no measurable drop in cyclic AMP levels — as she had expected after seeing the meiotic dots that pocked the membranes of the eggs — but they showed no rise, either. Regardless of how much *ras* she pumped into the oocytes, they refused to synthesize abnormal quantities of cyclic AMP; the two compounds simply snubbed each other. Cyclic AMP assays are tricky — and dangerous. In the course of her measurements, Carmen had to inject huge amounts of radioactive substances, and she was never one to take inordinate safety precautions. "I got a very bad reputation," she said. "The Japanese were especially upset by my experiments, because, you know, they are really scared of radiation."

For her part, Carmen was scared of Mike's reaction to her dismal tidings. Yet she was confident of the validity of the results. In toad cells, *ras* did not stimulate adenylate cyclase. Recapping Carmen's work in rodent fibroblasts, other Wigler scientists could detect no significant relationship between *ras* and adenylate cyclase. Mike had to resign himself. What was true for yeast was true for yeast alone. Nature was acting like the worst sort of coquette. It had conserved the DNA sequence of the *ras* gene, but at some point in the evolution of higher organisms it had decided that the protein produced by *ras* should be reassigned to some other office in the crabbed bureaucracy of the cell, an office that had yet to be penetrated.

By 1985, the Wigler lab people had mothballed their dancing shoes. "Mike was disappointed. Everybody was disappointed. *I* was disappointed," said Carmen. "Mike could see that I'd done the experiments the right way, and there was no other interpretation for the results. He didn't take it out on me. But it wasn't one of the high points of my scientific career."

Wigler didn't take out his disappointment on Carmen, of course, and he didn't abandon hope. He could, at that moment, have given up on yeast. He could have decided that yeast was not such a "powerful" system after all, if results obtained from its witless cells couldn't be applied to humans. Ed Scolnick — the scientist who'd discovered the viral *ras* gene a decade earlier — also had toyed with the yeast version of *ras*, hoping to learn about human cancer. But once the oocyte and similar results came to light, he decided that yeast was offal. "Scolnick concluded that it was only worth going on with yeast if you're interested in yeast genetics rather than human cancer," said Jim Broach. "He wasn't interested enough in yeast to stick with it. He thought it would get him off track."

Wigler was not so easily defeated. He believed that yeast still could work. He believed that the *ras* gene in yeast had not one, but two functions. Yes, the gene modulated adenylate cyclase — and that probably was the primary task of *ras* in yeast. Many cellular proteins, however, are known to speak several languages, communicating with different target enzymes at different moments in the cell life cycle. A protein like *ras* is so important, thought Wigler, that even in simple yeast it was likely to make its presence known in many ways. And if the adenylate cyclase connection had not been conserved through evolution, maybe another of *ras*'s signaling pathways had been.

The idea of the Alternate Pathway was born. The idea sprang from Wigler's mind like Athena bursting from the skull of Zeus: a majestic abstraction and a virginal promise. The *ras* gene had to have another function in yeast. It *had* to. "I have little doubt the alternate pathway exists," said Wigler. "And I believe that some aspect of the pathway has been conserved in mammalian cells."

There were sound scientific reasons to believe in the alternate pathway. For one thing, although yeast seemed capable of surviving when its adenylate cyclase gene was mutant, it could not survive without its *ras* gene. To Wigler, the primacy of *ras* over cyclase in yeast suggested that in the absence of a target adenylate cyclase enzyme, *ras* shouted to the nucleus through another biochemical line.

For another thing, the similarity between the yeast and mammalian *ras* genes was too great to be irrelevant. Tohru Kataoka had already proved that the human *ras* protein could stand in for the yeast protein — the experiment described by Scott as "one of Mike's favorite moments." There had to be some reason that the two proteins were so close in shape as to be essentially interchangeable. There had to be some sort of functional kinship as well. And if there was a functional kinship, however subtle, that meant there could be an alternate pathway in yeast. When *ras* wasn't talking to adenylate cyclase in yeast, it must be whispering to a protein that resembles the mysterious confidant of *ras* in human cells. No other explanation made sense for the homology.

Wigler believed in the alternate pathway because he wanted to believe. He wanted yeast to work. He had his pride to consider. It was all well and good for Ed Scolnick to confide in Broach, Wigler's old collaborator, that he thought yeast was a waste of time. Scolnick wasn't the one who had proved the adenylate cyclase concordance. It was Wigler who was yeast's most famous champion. He couldn't give up after one little setback. More than an experiment was at stake. Wigler was defending a way of life, a vision of how biology should be. Yeast was biology's ticket out of the realms of soft science and onto the terra firma of math and physics. In paper after paper, Wigler had written of "the power of yeast genetics."

Besides, for Wigler, yeast was fun. He enjoyed thinking about tactics for getting at the alternate pathway. There too he could apply the elegant and logical rules of genetics; there too he was not reduced to fumbling in the dark. Under no circumstances did Wigler want to return to mammalian cells and trust his fate to luck.

During my stay at Cold Spring Harbor, the alternate pathway was the lab mission. I watched six of the postdoctoral fellows trying to find the alternate pathway. Nine of the nine — plus Wigler — believed that the pathway existed. That conviction lent the lab coherence and something resembling a harmony of purpose. Wigler's lab existed to find the function of *ras*. The oncogene was stanza, refrain, and coda.

"I can't imagine getting sick of the *ras* gene," said Scott Powers.

"If you're tired of the *ras* gene," said Dan, "you're tired of life."

The scientists approached the problem of the alternate pathway from divergent perspectives, bleeding yeast for everything it's got. They tried biochemistry, genetics, reverse genetics, gene swapping, suppressor screening. Many of the young researchers smoked cigarettes, and between experiments they'd gather in the central postdoc office to smoke and talk about yeast and *ras,* and to offer each other advice.

The lab had a large quota of Japanese scientists — a full third of the postdoctoral staff. By working hard, the three Japanese scientists set the standard to which the non-Japanese Earthlings aspired. Wigler had had to go out of his way to hire his Far Eastern wonders. The Japanese government doesn't provide scholarships to its young scientists for overseas study (although, paradoxically, the leading Japanese universities expect their professors to have had some experience abroad), and Japanese citizens are rarely eligible for American fellowships. Thus, Wigler had to pay the salaries of his Japanese researchers out of his own grants. It was money he considered well spent.

"I like the Japanese," he said. "They're intelligent, and something about their culture encourages them to think all the time. They don't act without thinking first. And I like them because they never complain. I sometimes wonder if it would be possible to *make* them complain. I mean, you tell them to do something, and they do it. I wonder if I were to keep piling on the projects, if eventually they'd start to scream."

Wigler didn't succeed in making the Japanese scream or even talk very much. Rather than talk about their experiments, the Japanese just did them, from earlier in the morning than anybody else to at least as late at night. Tohru Kataoka was such a technical wizard that Mike nicknamed him "the Amazing Tohru." As a middle-class Japanese gentleman, Tohru was generally polite — but not so polite as to brook distractions. When he was walking along the hall and somebody approached him from the opposite direction, he would lower his eyes, fire out a rat-a-tat-tat of clipped, impatient coughs, and twist his whole body sideways to allow the person the widest possible berth for the fastest possible passage. He hated wasting time on nonprofessional chores, such as choosing a wardrobe. I noticed that for two weeks in a row he wore exactly the same outfit to work: the same brown knit pants, the same blue-and-white-striped shirt, and the same snowflake-patterned sweater. The clothes were always immaculately clean, but they were always the same. Finally, on Day 15, I saw that he was wearing a different shirt.

Takashi Toda also avoided extralab dissipations. When the time came for him to be married, he allowed his family to search for a wife. And once a Japanese woman had been designated as suitable, Takashi met with his affianced a mere three or four times before the wedding. The Wigler lab had hardly realized that Takashi was engaged; then one day he showed up at the lab with his bride, Sachiko, in tow. Junichi Nikawa, the third Japanese scientist at Cold Spring Harbor and a funny, quiet little gnome of a man, had a girl friend back in Japan, but he couldn't manage to fly home for a vacation, and because her family was trying to discourage the relationship, she was not allowed to visit him in the United States. "I miss her very much, but I don't think we will see each other for one year and a half," he told me. "Not until I return to Japan for good."

The Wigler lab had lofty morale, a common purpose, and a mighty work ethic. But most of all, it had Wigler, and Wigler's faith in yeast.

"In the beginning, I pooh-poohed the whole yeast thing," said Wigler. "It wasn't my idea — it was Scott Powers who said, 'I want to look for *ras* in yeast.' And he kept at it until he found it. Then other people came to the lab and decided that yeast was a good thing to work on. And now that we've got the system going, we're not about to give it up. My guess is that the scientific community is reserving judgment on us at this point.

"It may take us a couple of years, but we'll get there. Of course, I realize that we're in a race. The search for the function of *ras* is definitely a race. Everybody and their uncles are looking for the function. It's possible that one of them will get there first. But the desire to be first is not a strong enough incentive for me to go backward. I'm committed to yeast. We're a yeast lab now."

The Weinberg scientists sometimes griped about Bob's indifference to their particular fate. Only one or two of the experiments in his lab had to pan out, they said, to keep his reputation afloat, and it was on the successful projects, they said, that Weinberg lavished his ministrations. The Wigler scientists, by contrast, felt that Mike was concerned for all of them. And he was; he had to be. So much of his lab was working on yeast that Wigler and his crew would swim or sink together. True, there were two or three other mammalian cell projects under way. One postdoc, for example, recently had cloned an intriguing new oncogene that seemed to be related to a retina cell gene which allows us to see in color. The novel cancer gene was unlike any oncogene found before, and Wigler hoped that the project might expand and become a separate protectorate able to support the efforts of several scientists. Nonetheless, his lab was a yeast lab. It *smelled* like a yeast lab. When yeast cells are in the incubator, proliferating into little white pillows atop their substrate of agar and sugar, they exude the scent of baking bread (more Wonder bread, alas, then *pâtisserie* bread). It *looked* like a yeast lab. There wasn't a surface to be found that wasn't stacked to the dormers with dishes of yeast colonies. The most important piece of equipment in the lab was the French press, a device that resembles a medieval thumbscrew and is designed to pulverize the thick walls of yeast cells.

And it *acted* like a yeast lab. Quickly. As quickly as a packet of Fleischmann's dried yeast can turn a squeeze of dough into something the size of a loaf, so a yeast lab can generate results. By the second week of my stay, I was watching results rising from the bench of every baker in the lab.

The first big result came from Scott Powers. Scott had been screening his yeast colonies for so-called suppressor mutations. He had begun with yeast cells that, containing the cancerous *ras* clone (the yeast gene with the lab-designed point mutation in it), grew like cancerous yeast cells. He had then looked for the rare yeast cell that spontaneously stopped being malignant — that returned to normal.

By the reasoning of yeast geneticists, such a revertant cell would have acquired a second mutation in a gene that interacted with the *ras* gene, and that second mutation would have canceled out the effects of the original *ras* mistake. The ability to screen for anti-*ras* mutations is another powerful technique of yeast genetics, and Scott was the Wigler lab's foremost practitioner of suppression. By screening for mutant genes that suppressed the effects of the yeast *ras* gene with the point lesion, Scott was selecting for genes located along the *ras* signaling pathway. Most of the genes he found would be members of the adenylate cyclase enzyme pathway, but if he was really lucky, he might sally right into a gene lying along the alternate pathway of *ras*.

Scott didn't find one suppressor mutation; he found dozens. Most of them were rubbish. In the freezer he kept vials of yeast colonies that he'd probably never bother taking out of storage. One day, though, he stopped me in the corridor and, with his characteristic shyness, preceded his announcement by asking me whether he could tell me something. He thought he had found a suppressor mutation, he said, that wasn't on the adenylate cyclase pathway. The gene was called sup-H, for suppressor-H.

"So you may have a gene on the alternate pathway?" I asked.

"Maybe," he said. "Or maybe it's off in left field. It could have nothing to do with signaling pathways. We've sent the strain off to Ira Herskowitz, at UC San Francisco, to have him take a look at it." Herskowitz was one of the most respected yeast geneticists in the world. Scott explained that his yeast cells with the anti-*ras* suppressor mutation displayed the strange side effect of being unable to mate — or fuse — with other yeast cells, as robust yeast cells will happily do. Wigler thought that Herskowitz would be able to tell them what was going on.

"But you think it might be on the alternate pathway?" I asked again. "Is that why it's called sup-H? For hope?"

"No, it's called H as in A, B, C, D," said Scott. "It's the eighth strain in a screen for suppressors. I guess we'll know by the end of the week. That's when Herskowitz is supposed to call us back."

The postdoc who was doing the most for the alternate pathway was Takashi Toda. "He's the workhorse of the project," said Scott. He turned to Takashi. "Don't you think so, Takashi? Don't you consider yourself to be the workhorse of the alternate pathway?" Takashi looked up from his work, stared at Scott for a moment, grunted, and returned to his scrutiny of his yeast plates. "He is," said

Scott. "Mike's depending on Takashi to get us our first lead."

Like Scott, Takashi was taking a genetics approach to the second pathway. The lab had already discovered that yeast could survive with a mutant adenylate cyclase gene but not without a working copy of *ras*. The mutant cyclase strain, however, was an infirm runt that grew at a fraction of the pace of zesty yeast. This strain did not lend itself to experimental manipulation. Takashi wanted to breed a hardier version of the strain to allow him to track the *ras* signal along its putative alternate pathway. If Takashi succeeded, the Wigler lab would be able to pick apart the components of the second *ras* pathway within a matter of months. It would be the best-case scenario.

"Takashi is the finest geneticist in the lab, apart from Mike," said Scott. "If anybody can get at the alternate pathway, he can."

As much as I enjoyed staying in the Wigler lab, I found the work grueling. The science was more difficult than anything I'd encountered before. Yeast strains are given code names that are almost impossible to memorize, and I was always forgetting which strains "rescued" and "complemented" which. I carried around a little index card in my pocket with all the pertinent mutations listed and cross-referenced, but even with the cheat sheet in front of me I had trouble understanding the scientific discussions, particularly when Mike was one of the participants. He was always thinking several steps ahead of everybody else.

I also found the hours difficult; the Wigler researchers worked too hard. Mike is the father of two. I thought crossly, Doesn't he ever go home?

Not when there's something worth waiting for. One morning I came to the lab earlier than usual. Although I didn't expect to find anybody but the Japanese at work, Mike already was pacing around outside his office. His shirt was barely tucked into his trousers, and his hair was tufted up in all directions, like meringue topping. He didn't look as though he'd slept very soundly.

"Hi," he said. "How late were you here last night?"

I was taken aback by the question. Was I too supposed to account for my schedule? "Not very late," I said. "I guess I left about nine."

"So you haven't heard about Takashi's result yet?" Mike said.

"No. What happened?"

"He's got a *cyr1-1* rescued by *ras*," Mike said excitedly.

I fumbled in my pocket for my little note of mutants, but I'd left it at home. "What does that mean?" I asked, helplessly. "What does that mean?"

"Essentially, it's the first hard proof we've gotten that there's an alternate pathway," Mike explained. "Go talk to Takashi about it. It's a really great result." But before I could leave, Mike had grabbed a passing postdoc named Dallan Young. "Did you hear about Takashi's knockout?" Mike asked him. Dallan shook his head; he wasn't one of the yeast people in the lab, and he didn't always follow the yeast work in detail. "He's got a *cyr1-1* rescued by *ras*," said Mike.

Dallan thought for a moment and then broke into a smile. "That's the cyclase-minus strain? Rescued by *ras*?" he said. "Wow! Is he sure? When did this happen?"

I still wasn't following. "You can't tell me what this means?" I interrupted pleadingly. "Just briefly?"

Now Wigler was responding to both of us at once; we were Larry, Moe, and Curly. "Last night. He has to do a Southern to make sure the cyclase is really gone. But it looks good. Go talk to Takashi. He'll explain it. He's never had any problems with his transformations. Carmen! Did you hear about Takashi's knockout?"

I ran down the hall to Takashi's room, still confused but knowing enough to gather that "Takashi's knockout" must really be one. I couldn't believe my great fortune. The real function of *ras* had been found — the function that would be conserved through evolution from yeast to us — and just at the time of my tenure in the lab.

Takashi was at his bench, hunched over a microscope. Takashi has the calm inscrutability of a Zen master, and that day he was no less calm and inscrutable for having made a big discovery. When I asked him to explain what he had done, he sat up straight on his stool and placed his hands on his knees, the portrait of a *bonzon*. The *cyr1-1* strain, he said, was the strain that he had been trying to cultivate for months. He had finally managed to shut down the adenylate cyclase gene in yeast and restore the feeble cells to brio through the introduction of a high-powered *ras* clone. That *ras* gene presumably was able to rescue the cyclase-minus cells by signaling along the alternate pathway with a little extra oomph.

"And so now you have something to work with?" I asked. "Now you have the strain that you can use to study the alternate pathway?"

"Mmmmm, maybe," said Takashi. "First I must make sure that the cyclase gene is really knocked out. I must do a Southern blot to make sure it worked. I think the chances are okay that the knockout was true. But first I must be sure."

"Have you ever done this type of thing and not had the knockout work?" I asked.

"No, yeast is very faithful in homologous recombination," he said. "One hundred percent of the time, I have had success in the past with these kinds of transformations, these recombinations. But still, I must do the blot to be sure that the cyclase gene is missing."

Takashi returned to his work, and I returned to find Wigler prancing in giddy glee. He had told everybody of Takashi's knockout, and he was ready to tell everybody all over again. At last he had data to play with. At last his belief in the alternate pathway had been vindicated. The cyclase-minus strain was the one the lab had most ardently sought. He called a mini–group meeting of Scott Powers and Dan Broek. He talked about who would do what next. Did Dan want to help Scott on sup-H, or should he work with Takashi on the knockout? He talked about the sorts of matings and crossing that Takashi could do to start tracking down the genes in the alternate pathway. Dan suggested that Takashi try seeing whether the human *ras* gene could work in Takashi's new strain as well as the yeast *ras* gene had. Dan reasoned that if the alternate pathway of *ras* had been conserved from yeast to *Homo sapiens*, then the human *ras* should work well in Takashi's new strain, which seemed to be unusually reliant on its alternate pathway. Hearing Dan's suggestion, Wigler almost bolted up from his couch.

"Dan, that's a *really* great idea," he said. "You can do that experiment. No, Takashi will have to do it. You don't know much about transformations. You can help him. Or maybe Scott can do it. No, Scott is working on sup-H. Okay, Dan, forget about helping Scott. You can help Takashi. I want that experiment done as soon as possible."

"Takashi's result is much more interesting than mine," Scott said, almost plaintively.

"I think you'll have a chance to learn some biochemistry now, on sup-H," Mike said to Scott. "Dan can show you some biochemistry." Wigler turned to me. "Isn't this fun? I think we're going to have a lot of fun with this one."

I nodded wildly, grateful to be there, at the trail head of the alternate pathway.

Scott Powers soon heard back from Ira Herskowitz on the sup-H strain of yeast. Herskowitz said that the sup-H gene was identical to a gene his California lab already had found, and together Scott and Herskowitz figured out what sup-H might be. The gene produces an enzyme that is a kind of protein haberdasher: it outfits certain proteins in hats of fatty acid; thus garbed, the proteins are able to migrate to their proper position in the membrane. The sup-H protein places a fatty acid cap on the *ras* protein and on some other yeast proteins — such as the proteins that yeast cells must secrete before mating with other yeast cells.

A mutation in the fatty acid modification gene explained the multiple eccentricities of the sup-H strain. The mutant *ras* gene can't turn the cell cancerous, Scott explained to me, because not enough of the mutant *ras* protein is being topped by the fatty acid that *ras* protein needs so that it can move to the membrane. And the strain can't mate, because it's not getting its yeast aphrodisiac proteins up to and through the cell membrane. In fact, said Scott, any yeast proteins that need fatty acid additions are likely to be out of whack in this mutant strain; none of the proteins is being properly processed.

I thought about the significance of sup-H. I thought about a protein that bustled around the cytoplasm, putting on fatty acids wherever they're required. I thought and thought, but I couldn't come up with any conceptual link between sup-H and the alternate pathway. "So sup-H is a processing enzyme?" I asked Scott. "It just modifies proteins? That's not really something on the alternate pathway, is it? It doesn't sound like it. It sounds more general than that."

"Yeah, it's more like a housekeeping enzyme," said Scott. "I guess I'm pretty disappointed. I'm not sure what sup-H is going to be good for. I don't know what it's going to tell us about signaling pathways. But Mike seems excited about it. He sees this as a way of finding other G proteins, since all G proteins are believed to need modification before they can migrate to the membrane." (G proteins are the membrane-bound proteins that are among the first cell components to react to a growth signal from the outside.) "People are always looking for new ways to find G proteins," said Scott, "so maybe this fatty acid gene is it. I don't know. Maybe Mike is just trying to keep me from being depressed. I don't think sup-H is such a big deal after all."

Sup-H did not look as though it would lead to revelations about the alternate pathway. But there was still Takashi's knockout, and there was still reason to hope. His knockout technique was the cleanest way to get at the second meaning of *ras* in yeast.

But disappointment is often contagious, and in a yeast lab — since single spores bloom into entire fungal armies by the hour — disappointment can be pandemic. A couple of days after Scott had learned that sup-H was in the millinery trade, Takashi's knockout was kayoed. Clobbered, flattened, and killed. And with the death of the great result, the innocence of the Wigler lab died too.

When he explained his original result to me, Takashi had been cautious. He needed to do a control, he said, to prove that the adenylate cyclase gene had been deleted, as he trusted it had. Only then could he know that *ras* must be working through some other pathway than the adenylate cyclase one. On completing his Southern blot control, however, Takashi discovered that the experiment hadn't worked after all. For the first time in his long experience with yeast genes, a yeast cell had failed to recombine as it should. The strain that Takashi thought had lost its adenylate cyclase gene had, in fact, kept it. Takashi hadn't rescued a mutant strain of yeast with *ras*. The "mutant" strain hadn't been mutant to begin with. Even Takashi lost his Zen-like calm when he saw the result. "This has never happened to me before," he told me. "I have never had this kind of nonhomologous recombination ever before." When he saw Mike coming, Takashi hid his DNA film behind his back. Wigler could not be hornswoggled. He demanded to see the blot. Takashi handed it over. There, in the lane where the adenylate cyclase probe had been applied to his DNA sample — a lane that should have been empty — was a thick black blemish.

Oh, well. Mike shrugged. Easy come, easy go. He told Takashi not to feel bad about the lousy result. Who knew how it had happened? Who cared? Takashi pledged to Mike that he would try again; he pledged to find the alternate pathway or he would "go back to Japan and take up farming."

Mike said, Fine, try again. But he sensed, at that moment, that the knockout approach wouldn't work. It hadn't worked before, which was why Takashi's sudden apparent success had been such a thrill. There was no reason to believe it would work in the future. With the letdown, Wigler became fatalistic. The knockout technique was the last conceptually easy way to get at the real, human function of *ras*, and it probably couldn't be done.

"Didn't Mike say this was supposed to be fun? Is this his idea of a party?" Dan Broek said. Dan is a blond, blue-eyed, fresh-scrubbed cherub, and even when he's smirking he looks angelic. "The alternate pathway," he added, "is a rocky road again."

Now the lab would need to apply "greater mental discipline," said Mike. The scientists would have to work harder. And they would have to face up to the unsavory truth that yeast was not so simple and elegant as it once had seemed. Wigler would have to be smarter. In yeast, the link between *ras* and adenylate cyclase was strong. Mike would have to come up with a clever idea, some way of circumventing adenylate cyclase, rather than trying fruitlessly to shut it off. He'd have to think about what else *ras* might be doing in yeast. He needed a model.

Several days after Takashi lost the bout, Carmen and Dan invited me out for a beer at a bar in nearby Huntington. They too were feeling fatalistic; they told me they would be happy to get drunk. Carmen said to Dan that Mike was upset because he wasn't getting any good data. "We have to give him some good data soon," she said, "or he's going to get really depressed." I asked them whether they thought that the *ras*-yeast project was in trouble.

"No, not trouble exactly," said Dan. "Mike is getting a little impatient with it, but I wouldn't call it trouble. Let's call it questionable."

"Are you worried?" I asked them. "Do *you* believe there's an alternate pathway?"

"Yes. No. Maybe," said Dan. "Do I have to bet any money on it?"

"There *has* to be an alternate pathway," said Carmen, banging a loose fist on the table. "The conservation is too good. Evolution isn't arbitrary. There has to be a reason for the gene to have been so highly conserved."

"So what do you think the alternate function is?" I asked.

"I don't know. Maybe regulation of the cyclic GMP pathway?" Carmen said. "Cyclic GMP [guanine monophosphate] is molecularly similar to cyclic AMP. It could be cyclic GMP. Ach, it could be anything."

"How long do you think Mike will stick with it?" I asked. "How long will he stick with yeast if he gets the feeling that there may not be an alternate pathway?"

"Oh, I think he'll stick with it for quite some time," said Carmen. "He's not working in yeast just because of *ras*. He likes yeast. He likes the system intellectually. He doesn't want to go back to mammalian cells."

"Are you kidding?" said Dan to Carmen. "Mike doesn't care beans about yeast. He's only working in yeast because he wants to get at the function of *ras*. If he could think of some way of doing the same experiment in mammalian cells, he'd be in the lab tomorrow, spraying down the rooms with fungicide."

As the days passed, I began to doubt that the Wigler lab would discover the function of the human *ras* gene any time soon. The entire field of oncogenes seemed to be paralyzed, and Wigler, like the rest, was confined to a wheelchair. His scientists worried about his moods. They'd ask each other whether Mike was depressed that day. They'd try to keep out of his way when he was in a surly temper, but with little success. Wigler looked to his postdocs to distract him from his unhappiness. Was anything new? he'd ask. Please, isn't anything *new*?

"Sometimes I can't stand it," said Dan. "Sometimes I can't stand the way he hounds you and nags you and watches everything you do. The other day I was coming out of the darkroom with some results, and they were a mess. I'd totally screwed up the experiment. I knew I'd screwed up the experiment, and I just wanted to be able to go back and quietly repeat it. But Mike came over and asked to look at the autorad. I said, 'No, Mike, the experiment didn't work and I want to repeat it.' But he kept pestering me until I had to show him the results. When he saw the gel, he got this look of total disgust on his face. His face was one big scowl, as though I'd just handed him a dead fish. He said, 'These results are awful. You really fucked up.' I was burning. I was furious. I said, 'I *know* the results are awful, Mike. I told you they were.' That's really the last thing I want to hear when I've made a mistake. He could have been supportive. Usually he is supportive when your experiments aren't going well. But then he gets into these overbearing, arrogant moods and he makes you feel as though you're a complete idiot."

In the absence of data, Wigler became volatile and unpredictable. One night he'd be festive and affable, entertaining his staff with anecdotes about his past. He'd talk about his parents, who'd been members of the Communist Party before becoming disillusioned by Stalin, or about the premarital romps of his old mentor at Columbia, or about his disastrous experiences in medical school at Rutgers. He'd been expelled from medical school because he was so bad at treating patients; he hated to see people die, so he'd run away from a patient's bedside when he was afraid the big moment was approaching. Mike's subordinates loved to hear his stories; they'd hang on his word, laugh in unison, and beg for more. At those moments, I could see how much they wanted to be his friend. He was the most popular boy in high school, able to command simultaneously a deep fondness and a slight sense of awe.

But then he might come in the next morning, glowering and snappish, and his researchers would be afraid of him. In the winter of 1986, Tohru Kataoka was particularly intimidated by Mike. Tohru decided to accept an offer of an independent position at Harvard

Medical School, and Wigler panicked. Although he knew that staff turnover was built into the basic research system, Wigler disliked it when postdocs had to leave his lab — just as he'd hated to see patients die, he said — and Tohru's pending departure came at an especially bad time. The Amazing Tohru was Wigler's best scientist, and with the yeast project skidding, Wigler needed Tohru's technical talents more than ever. Wigler had hoped that Tohru would stay on another year or two, but Tohru felt that the Harvard offer was too good to decline. Wigler doesn't hide his feelings, and he didn't hide his anger with Tohru. Tohru gave Mike a copy of his NIH grant proposal, in which he had outlined the projects he planned to pursue at Harvard. Tohru needed Mike's critique and approval of the application. Wigler, who was under pressure to review NIH grants, did not read it immediately. As the submission deadline approached, Tohru grew frantic. One day he stood outside Wigler's office door, preparing to confront him once Wigler got off the phone. Tohru paced back and forth, alternately ejaculating a rat-a-tat-tat of coughs and taking in sharp little sucks of breath. He ran his fingers through his black hair until it stood up every which way, punk style. At last Wigler emerged from behind his wood door, and Tohru asked him whether he'd read the proposal. Mike irritably replied that he'd read part of it and that much of it would have to be changed. He particularly objected to the middle section. He said that it made Tohru sound naïve and ignorant. "It will undermine your scientific credibility," he said. "You're a good scientist, but you wouldn't know it from reading your proposal." Tohru nodded repeatedly and sucked in his breath. He said that Scott Powers also had read the application and had suggested changes in the final section, which Tohru planned to incorporate. "You're going to change the final section?" Mike said. "So I don't have to bother reading it now, right?" Wigler went back into his office, and Tohru began throwing together papers from his desk and files. "Now I have to go home to rewrite everything," he said. "Now I must make all these changes, and I have only two days left."

"Poor Tohru," said Scott Powers, once Tohru had left the central postdoctoral quarters. "Mike knows that he's behaving irrationally, but he can't seem to help himself. He can't stand to see Tohru go."

Nothing upset Mike more, however, than to think his people were losing heart. As the yeast project continued to drag, Carmen and Dan decided to collaborate with another group at Cold Spring Harbor. James Feramisco and his postdoctoral fellow, Dafne Bar-Sagi, who worked with the mammalian version of *ras,* thought that *they* had discovered the true pathway of the oncogene. Their theory was that *ras* regulated the synthesis of a molecular precursor to prostaglandins, a group of fatty acids believed to mediate everything from blood clotting to labor pains. (Scientists were ever hopeful that cancer genes worked through familiar biochemical mechanisms.) Dafne and Feramisco wanted to strengthen their initial observations by testing their model in an array of organisms, so they asked Carmen to measure the effects of *ras* on fatty acid turnover in toad eggs, and Dan to look in yeast cells. Wigler agreed to the collaboration, providing that the experiments could be done rapidly and on the side.

Experiments are almost never simple, though, and Dan and Carmen ended up spending weeks, not just a few spare evenings, on the Feramisco project. Wigler grew angrier with each passing day, but he resisted playing the tyrant. "I'm skeptical about the whole thing, but if Dan and Carmen think there may be something there, I'm not going to stop them," he told me. Yet the investment of his lab's capital in somebody else's notion — even that of Feramisco, who was Wigler's friend — obviously rankled. One night Carmen invited me to watch her perform an oocyte injection for the Feramisco project. It was one of the most extraordinary experiments I'd ever seen. Viewed through a microscope, each oocyte was a perfect, glowing sphere, colored half yellow and half brown. As I watched, she pierced through the delicate skin of the egg with the glass microinjection needle. She extracted the needle, aimed it at another quivering orb, and plunged it in again. "I have to be careful doing this," she said, "or I end up with scrambled eggs." My eyes were welded to the microscope; I couldn't stop looking and gasping. The oocytes were like tiny planets in partial eclipse, the needle a crystal lance of the lord.

When I finally looked up, I saw that Mike had entered the room. His thumbs were hooked in his pockets, and he shuffled back and forth, watching us. "Are you working on the arachidonic acid stuff?" he asked Carmen, referring to the Feramisco experiment.

"I'm just going to try one more measurement," she mumbled.

He asked me if it was the first oocyte injection I'd seen. I replied that it was.

"It's pretty spectacular, isn't it?" he asked.

"You don't mean that, Mike," said Carmen. "You think it's rather boring."

"No, I *don't,*" he replied sternly. "I thought it was spectacular the first time I saw it. I still do."

Once he had left the room, she said, "He thinks the whole arachidonic acid thing is bullshit. I think I'm going to have to drop it pretty soon. Maybe this will be the last experiment."

"*Is* it bullshit?" I asked.

"Oh, maybe," she said. "It's just another model out of a hundred possible models. Mike doesn't want us wasting time on somebody else's crazy ideas, and I can't say I blame him."

The next day in group meeting, Wigler began by turning to Dan and Carmen and saying, "So it's agreed? We're going to drop the arachidonic acid work?"

Dan nodded. "I think, at best, I'm seeing a one-point-five to twofold stimulation in yeast. And that's with *ras* valine," he said. "It's just not a very convincing result."

Wigler looked at Carmen. "We're going to drop it, right?" he said.

Carmen winced. "Actually, I thought I'd repeat the measurements just one more time," she said. "I was up until four this morning doing injections, but some of them didn't come out right, and I'd like to try again."

"But why should *you* do it?" Mike demanded. "Why can't Dafne do the injections herself? I really don't think it's worth pursuing this wild goose chase any longer. I mean, you haven't seen anything significant so far, have you?"

"Okay, okay, you're right," said Carmen. "I won't do it."

"So it's agreed? You're going to drop it?" Mike said.

"I'm dropping it," she said. And did.

Yet for every loss of temper, Wigler displayed a dozen examples of his generosity. He helped Tohru to write a better grant application, and he gave him time off to do so. Wigler likes his people to work hard, but not at the utter expense of their social happiness. He knew that it was easy to be lonely at Cold Spring Harbor, especially for those people who had moved to the monastery from foreign countries and left their families behind. By tradition, Friday nights were no-work nights at the lab. So one Friday night, when Mike found Carmen still slogging away at her bench long after everybody else had departed, he anxiously asked her why she was still there. Carmen replied that she had work to do. Can't it wait until tomorrow? he said. Aren't you going out for dinner or a movie? She said that she had no plans. "I tried to tell him that I didn't really mind working alone," said Carmen. "But I could see that he felt pity for me, and it sort of made me laugh."

Wigler also fiercely defended the interests of the people in his lab. During the preparation of the paper reporting that *ras* controlled adenylate cyclase activity in yeast, Mike risked his good name to ensure that Takashi Toda received top billing. In an attempt to hasten progress on the report, Wigler had earlier agreed to a collaboration with the Japanese scientists who had been studying adenylate cyclase in yeast. He'd told one of the collaborators, Isao Uno, that he would be first author on the paper, but when Wigler realized how much more work Takashi had contributed to the project, he called Japan and informed Uno's supervisor that he was very sorry but he'd changed his mind. Takashi would have to be cited first. As Wigler talked, he could hear Uno's cries of outrage in the background. Wigler felt guilty about reneging on a promise, but Takashi was his responsibility, and Takashi's needs were paramount.

"Mike really put himself on the line for Takashi," said Scott Powers. "It wouldn't have made a difference to his personal glory if Uno were first. Mike would be last author in any case. But there was no question about who deserved first authorship, and Wigler fought to make sure that Takashi was treated fairly."

"I feel a strong commitment to my postdocs' careers," said Wigler. "If I were to put my own concerns before theirs, I'd have trouble dealing with my conscience."

Through his intermittent arrogance and tantrums, his impatience and what's-new nosiness, Wigler's conscience shines. He's no altruist, and he's as competitive as any other molecular biologist. ("If he thinks that somebody is trying to compete with him on a problem that interests him," said Carmen, "he'll throw half the people in his lab onto a project to make sure he gets there first.") Yet Wigler talks freely about his feelings, and he often thinks about how his work might be put to some practical, medical good. "I don't think it's a drawback to view our research in the context of human cancer. It's not even a side issue," he said. "On the contrary, I believe that thinking about what may be going on in human cancer can help you think more clearly about the problem at hand, even when you're dealing with yeast.

"Anyway, that's why we're here. That's why we're doing what we're doing. We're working in cancer research, and I'm not afraid to admit it."

The more I learned of Wigler's motivations for doing his sort of science, the more I admired him. One chilly, leaden Friday afternoon, he invited me to hear him play the piano. He'd been taking lessons for about eighteen months, but he warned me that he recently had fallen behind in his practicing. "I'll probably sound terrible," he said as we hurried across the grounds to the Cold Spring Harbor music room. I didn't believe him. I figured that Wigler would master anything he applied himself to.

He wasn't a complete liar. Although he didn't sound terrible that day, Mike is not musically gifted. He banged at the piano keys as though he were swatting mosquitoes. He started, stopped, started again, pausing to pick out the notes at one moment and rattling wildly through a passage at the next. He played pieces by Mozart and Bach, but his timing was so erratic that it was difficult to tell which was which. "I think my biggest problem," he said, returning the scorebook to his briefcase, "is that I have no sense of rhythm."

As we walked back to the lab, he told me how much he loved playing piano. "I'll never be Arthur Rubinstein," he said, "but piano gives me a kind of distracting peace that I don't get from anything else.

"Too often, people give up on those things they realize they don't have a genius for. That's especially true of very bright people who start out wanting to be scientists. I think a lot of young people get discouraged when they find that they're not an Albert Einstein or a Laplace. So they may quit trying, give up on science entirely, rather than using the talent they do have to contribute something to society. They decide it's easier to do something like go into business, where at least they'll be able to earn a lot of money."

I asked Wigler whether he had left theoretical math because he realized he might not be a Laplace.

"That's part of the reason," he replied. "I might have stayed in math if I thought I was going to be one of the immortals. I would have been very good. I think I would have been a first-class academic mathematician, but I knew fairly early on that I wasn't going to revolutionize the field. That knowledge, though, was only a small part of the reason I left."

We returned to his office, and, as always, Wigler rearranged the couch pillows before sitting down. "The real reason I left math was that I thought math emphasized some of my most negative qualities — my tendency toward self-involvement and alienation. Math has very little social relevance. I suppose in the long run it's useful to society, but in the short run math is more autistic than poetry. And I decided in college that I wanted to do something with my life that might be socially useful.

"I'd been having a lot of trouble in college. Not with my course work, but with my head. I was always depressed, and sometimes I wouldn't leave my dorm for what seemed like days at a time. It wasn't quite a nervous breakdown, but I was really low. So I dropped out for a semester and went back to my parents' house on Long Island.

"One evening, I was coming home on the train, and the sun had already set. The sky was bluish-gray and soft, very beautiful. I made a pact with myself. I promised that as soon as I got the pieces of my life back together again, I would change direction. I'd apply to medical school and try to do something that would help people. And I kept that promise. When I got back to Princeton, I applied to Rutgers. For some strange reason, they accepted me.

"Even after medical school didn't pan out, I still wanted to do something useful. I answered an ad in the *New York Times* for a position as a technical assistant in a biochemistry lab at Columbia. One thing led to another, and I realized that I'd be very good at basic research. I read Jim Watson's *The Double Helix,* and it was an epiphany. I decided to become a molecular biologist. I figured that was a field where I might be able to make a strong intellectual contribution, and have an impact on society at the same time.

"The work I did as a graduate student, with Richard Axel, was important in its way. It was important to show that we could transfer a gene into mammalian cells, get it expressed, and pull it back out again with an identifiable marker. But it was more important for what it implied than for what it actually was. We helped to develop a tool, but tools are only good for doing things like building houses. I'm sorry to say that we weren't the first people to transfect oncogenes. But once I saw the incredible importance of dominant oncogenes to the understanding of tumorigenesis, there was no question about what my lab would work on."

Wigler became uncomfortable when he dwelled too long on himself, and any time he talked about his past, he eventually shifted to a gentle self-deprecation. His lab had been successful, he said; *he* had been successful; but how much resonance did his success really have? I once caught him and Scott Powers poring over some NIH grant applications that Wigler was reviewing. They were counting up the number of times the applicants had cited Wigler's oncogene work in their lists of journal references.

"Do you know Mark Twain's novel *Pudd'nhead Wilson?*" he asked me later that day. "Essentially it's about a Yankee detective who lives down south, where he solves a murder mystery. He uses a lot of newfangled technology to reach his conclusions, and he turns the town on its head. It seems as though he's made a deep imprint on everybody, but the net result of all the excitement is that nothing changes. The town picks up where it left off, as though he'd never done a thing.

"Sometimes that's how I feel about my place in science. Nothing would be different if I had never existed. The history of oncogenes would probably still be the same. I don't think you could write Weinberg out of the story so easily.

"Oh, well. Who knows what will happen. This is one story that's still being written."

It was, finally, the *ras* gene as an open-ended narrative that kept Wigler going with yeast, even as outside observers gradually gave up hope. In late 1985, Mariano Barbacid had told me that Wigler was in the best position to understand the function of *ras*; but in mid 1986, after hearing Scott Powers present a seminar on the sup-H processing gene, he amended his prognostication. "I'm very skeptical now that yeast will give us the answer," he told me. "Wigler keeps looking, but what does his lab come up with? They discover some sort of yeast pheromone gene! We may not know a lot about cancer, but we can be pretty sure that it has nothing to do with pheromones."

Barbacid, however, wasn't doing any better in his lab on the biochemistry of cancer. No lab was doing any better. Indeed, Wigler believed that, with or without an alternate pathway, his lab had learned more about *ras* by studying yeast than had anybody else by fumbling about with animal cells. The Wigler lab proved that *ras* is a G protein that must clasp on to a GTP molecule to be active; they discovered a complex feedback control loop that keeps the normal *ras* gene under control but has little effect on the mutant *ras* gene; they discovered that *ras* is absolutely dominant over its signaling pathway — which essentially means that the buck starts with *ras*, and that critical decisions to grow are initiated by *ras* activity. "For every one discovery that has come out of the study of *ras* in mammalian cells," said Wigler, "we've made four or five in yeast cells. I'm talking about universal observations. Our discoveries have had a direct impact on the general understanding of *ras*."

Yet the big question about *ras* remained. How does it work in human cells? While there was any chance that the question could be answered through yeast, Wigler would continue with yeast in some fashion. Until the Wigler lab researchers understood the *ras* pathway in yeast so thoroughly that they could say, without hesitation, that *ras* stimulates adenylate cyclase, and adenylate cyclase only, Mike refused to give up. "There are still a lot of unknowns about *ras* in yeast," he said in 1987. "We still don't know, for example, whether *ras* acts directly on adenylate cyclase, or whether there's an intermediary protein between the two. If there is an intermediary protein, that argues that there could be an alternate pathway. We just don't know yet. The biochemistry is difficult and ambiguous. There's an excellent chance that our biochemical understanding hasn't kept pace with our knowledge of genetics, of DNA sequences. And, of course, you can't have a final answer about *ras* until you have the biochemical answer. From what I can see, nobody has the biochemical vocabulary to frame the right questions.

"I'm never consistent. I go back and forth in my belief that yeast is going to get us somewhere. I guess that means I believe there's a fifty-fifty chance that the basic understanding of *ras* will finally come from yeast. It's not as though something else has come along that demands we start galloping off in another direction and divert the resources of the lab away from yeast.

"What it comes down to is this question: 'What the hell is Wigler trying to do?' What I'm trying to do is no different from what molecular biologists have always tried to do. We try to understand a system to the point where we can explain all the quirks, where there are no quirks left. You can't just sweep the quirks under a rug because they're unpleasant. As long as the quirks remain, that means your understanding of a problem is incomplete. Right now, our understanding of *ras* — in both yeast and mammalian cells — is incomplete. We're up to our ears in quirks. So while those quirks remain, we have a job to do. Until we can answer a basic question like 'Does *ras* directly stimulate cyclase or not?' the usefulness of yeast has not been spent. I'm not staying with yeast out of some blind devotion to the system or some reluctance to admit defeat. If I suspected that yeast was a waste of time, I'd give it up in an instant. I have no romance about it.

"But one thing I have learned is that yeast is not so simple after all. Your basic yeast cell is every bit as complicated as your basic mammalian cell. The basic decisions about whether to grow or not are going to be the same, and they're going to be complicated. Signaling pathways are complicated. But life always turns out to be more complicated than you first imagined.

"Having said all that about the complexity of yeast, we know that the tools we have to approach the organism are still light years ahead of what we have in mammalian cell biology, so it's still worth my time and the time of my lab."

A yeast colony on a dish looks like discarded pocket lint, fleecy and trivial. Only by suspending sense and ego can one feel genetic fraternity with a white blur of spores. Yet Wigler tried to pick apart the signals of a single gene in those seemingly inconsequential cells, and he encountered the truest continuum of biology: frustration. Where he thought there would be simplicity, there was ambiguity. He thought he would find a single answer, and instead he realized he hadn't known the question. In two years' time, his lab learned more about *ras* than scientists had gathered in the preceding six, and yet they knew almost nothing. One month Wigler was yeast's most eloquent flack; the next, he was doomy and bored.

"Sometimes I have to sit down with Wigler and give him a little pep talk," said Weinberg in 1987. "Sometimes he starts talking so pessimistically about what he's done and what he hopes to accomplish. I tell him not to despair. I think that he'll eventually come up with very important discoveries in yeast. I think the final word on *ras* will be his, and I tell him so."

Far more than Weinberg, Wigler had made the function of *ras* his great crusade, and thus had less distraction from the frustration of repeated failure. Yeast might work or it might not. The perfect system, the system that couldn't lose, became no better than a flip of a coin. Adding to his feelings of helplessness, his mother became gravely ill in 1986. A long-time smoker, she developed lung cancer, which metastasized to her brain. During the worst period of her illness, Wigler spent very little time at the lab; when he was there, he carried depression with him. "It's hard for me because I know so much about her particular kind of cancer, and I have no illusions about what can be done for her," he said. "There's no effective treatment in her case. As soon as I found out what was wrong with her, I became resigned to it, and I started saying my goodbyes.

"This experience has made vivid in my mind what a horrible disease cancer is. I knew it before, of course, but witnessing it first hand had been a real eye opener. It's a terrible way to die. Now I tell people that they should stop research on heart disease. That's how *I* want to go."

Has his personal experience in any way strengthened his resolve to help, I asked, to turn his lab work to something useful?

"My resolve didn't need strengthening," he said. "I believe that basic science is the best way to attack human cancer, and that's why I'm doing what I'm doing. That's how my talents can best be put to use, and I know that what we learn about oncogenes will lead to real changes in cancer treatment. That's not a maybe; that's a for sure."

Our conversation ended soon afterward. Mike's bridge partner was calling him from the other end of the lab: Would he please return to the game?

11

Pink Cadillac

ON THE LAST NIGHT of my stint at Cold Spring Harbor, Mike Wigler made me miserable. He walked into the little office where I was gathering together my notebooks and papers, and he asked me about my book. How would it be structured, he wanted to know, and what sort of style was I planning to adopt? He loomed over me, the challenging professor, his thumbs hooked in his trouser pockets, his arms curving down from his shoulders like parentheses. Caught off guard, I stammered out a few disorganized ideas about presenting the problem of cancer, the traits of a cancer cell, how molecular biologists approach cancer, how they design experiments. As I spoke, his expression shifted from impassiveness to disappointment to disgust.

"So you're going to write a textbook," he said.

"A textbook?" I repeated, stung. "No. That's not what I was saying." I frantically ticked off a few more proposals, each one sounding more inane than the last. He shrugged. "I was just curious," he said, and shuffled out of the room, leaving me hunched over the edge of the couch with my arms hanging between my knees. Scientists accuse journalists of many infractions, but writing textbooks isn't one of them.

A moment later, he shuffled back in again. "I think you should write it like a journal," he said. "I think you should write about what you saw. Otherwise, it will be a very boring book." He paused. "You'll have funny stuff in it, won't you?"

"Funny stuff?" I repeated. "You mean ha-ha jokes? You mean people coming to work dressed in costume?"

"Not ha-ha jokes, but the humorous things that go on in a lab," he said. "You're going to talk about the fun of doing science, aren't you? I mean, yes, there's the competition, but there's also the great beauty of it."

Leaning forward and clasping my hands in supplication, I swore to him that if there was one thing I'd learned to appreciate during my months in the lab, it was the beauty of science. If I weren't a writer, I'd be a scientist, I said. I love science, I told him, and I promised to try conveying that love. "Well, that's good," he said, doubtfully. It was obvious that he thought I'd missed the point completely.

Dan Broek stuck his head in the door. "Did Carmen tell you we were going out for a beer?" he asked me.

"She looks as if she needs one," said Mike.

On the train back to Manhattan that night, I considered whether I had missed the point. I must have missed the point, because at that moment I was having a hard time seeing the humor and the fun. I was having difficulty understanding the why of science. What, after all, had I seen? I saw experiments not work. I saw young scientists sparring over turf, authorship, errors, bench space. I saw people crying and getting sick. I saw senior scientists at their wits' end, confronting crises or fleeing crises. I had traveled from one famous lab, where big, promising projects fizzled and died, to another famous lab — the brash, upstart lab — where the solution to cancer turned out to be yet another red herring. I saw Mike Wigler train the full arsenal of his brilliance and the most sophisticated technology on primitive yeast cells, only to discover that yeast was cribbing from a different evolutionary script. I saw that great intelligence, dedication, long hours, tenacity, intuition, were meaningless if nature didn't obey a prefab model. I began to suspect that Mary Godwin Shelley and Boris Karloff were right. Scientists are mad. Stark, sputtering mad. Real discoveries are rare. Scientists spend most of their time carrying on arcane dialogues, their words twisting back on themselves like Möbius strips.

It was not my most precise revelation.

Two weeks after leaving Cold Spring Harbor, in early 1986, I returned to the Whitehead Institute. About half a year had elapsed since Mien-Chie was thrown off the *neu* project and the Weinberg lab hit what René called "the lowest low in its history." And I finally got the point. I saw the beauty and the humor of science. There was a pink Cadillac parked in the Whitehead lot, and the entire institute was getting ready to put on the dog.

Bob Weinberg is a legend, and nested within his large legend, like marioshkas, are many little legends, the people who not only made the important discoveries in the Weinberg lab, but lent themselves easily to narrative. Their traits, their idiosyncracies, can be summed up in a sentence or two, for these are part of them and survive their departure. Chiaho Shih became a legend for finding the first human oncogene and for being an antisocial thistle. Cliff Tabin was famous for the *ras* mutation and four-part ligations and for his mouth. Hucky Land mixed together two oncogenes to transform primary tissue, and he always dressed like an auto mechanic.

Before Cori Bargmann published a single first-author paper, she was famous among the MIT faculty for being *smart*. Stephen Friend did some great experiments as a postdoctoral fellow in the Weinberg lab, but he too gained a suprascientific reputation: he was the nicest guy ever to work at the Whitehead. Before I met Steve, Mike Gilman and René Bernards both talked to me about his niceness.

"You'll really like him," Mike said over the phone. "He's such a sweet guy. We were all worried before he came, because he was going to be using Randy Chipperfield's old bench. Nobody had used Randy's bench since he died. Then we heard that Steve was one of these M.D.–Ph.D. people, and we were really worried. Another doctor, we thought. But we were wrong. He's more of a scientist than a physician. And he's fun to have around."

"This guy is unbelievable," said René. "He didn't know any molecular biology until he came, but he's a really fast learner, and he's got a lot of energy. He runs around getting everybody excited. He always wants to try new things."

Hearing René praise another scientist, I had to pay attention. Who was this man, I marveled, who'd arrived at the lab in January of 1986 and by February had won such critical acclaim? Is he a miracle worker or an opportunist? Is he falsely sweet, like aspartame, or cleanly sweet, like a sound night's sleep? When I met Steve Friend, I discovered that he lives up to his surname. He has a way of making you feel, within a few minutes of introduction, that he is an old, trusted friend. He's not so much sweet as he is open. He gives people the benefit of the doubt, which encourages people to do the same with him.

I didn't, however, get much of a chance to talk with Steve on the first day of my visit. He was too busy. That night, the entire Whitehead staff was going on its first institute-wide field trip, to the Wonderland track in nearby Revere. They were going to the dog races.

"Dog races?" I said when René told me of the plan. I hadn't heard mention of dog racing since my high school days in Michigan. "Whose idea was that?"

"Steve's!" René said gleefully. "Isn't it amazing? The guy's only been here a month."

A consummate impresario, Steve was determined that his hundred new best friends would go to the races in style, and he had a thousand details to attend to. He had to arrange for the chauffeured limousines to be at Main Street by seven. He had to be sure that every limousine came equipped with a color TV; some even had private bars. He had to make reservations for the Wonderland banquet. Most important, he had to set up the prerace seminar in the Whitehead auditorium, at which scientists would learn the history of dog racing and the proper way to place a bet.

I'd never seen the Weinberg lab so crackly with energy. Scott Dessain, a technician, presented a talk at the race seminar about the invention of the motorized jack rabbit that the greyhounds chase around the track, and somehow Scott had managed to dig up old photographs and patent diagrams of the prototype. Mien-Chie, an old dog bettor from Taiwan, explained the fine art of betting. You can bet a Quiniela Box, he said, choosing three or more dogs; if two of those dogs finish first and second, you win. Or you can bet a Trifecta Box by picking three or more dogs, and should three of those dogs finish first, second, and third in any order, you win.

If you *really* want to hit the jackpot, he said, you can bet a Superfecta Key Entry. You must call the winning dog and the three dogs that will finish second, third, and fourth. If you choose the right Superfecta Key, said Mien-Chie, you might win $12,000, $15,000. You will go home a very happy gambler.

In the evening, the scientists decked themselves out in appropriate finery: gold lamé, silver lamé, strapless black gowns, tuxes and top hats. Bob Weinberg, who had been hospitalized briefly that afternoon for a stomach ailment, recovered sufficiently to show up wearing a Tom Wolfe–style white suit. At the seedy race hall other gamblers were dazzled by the display. "Are you people artists?" one woman asked Snezna, who looked gorgeous in a snug black cocktail dress. "Are you in the theater? You're all so elegant!" The Wonderland management honored the scientists by naming one of the twelve races the Whitehead Event. We sat at tables covered with white linen and ate chicken slathered in nondescript continental sauce. ("No rabbit?" asked Scott.) Most of us lost money but hardly noticed.

Steve fitted into the lab, yet he changed it. He has the good looks that are nearly prerequisite for a position with Bob Weinberg. His hair is crow-black, his eyes green, his features firmly organized, and his complexion as ruddy as a yachtsman's. He laughs often, goofily, and warmly. The Wonderland track was just the beginning, he said. Next on the agenda were mud-wrestling and tractor-pulling competitions. "I believe in experiencing everything at least once," he said. "And with tractor-pulling contests, once will probably be enough." He cajoled the most indolent scientists to start training for the Boston Marathon. He joined the local rowing team that David Stern, Bob Finkelstein, and Ruth Wilson (the DNA extracter) belonged to, and he joined the glee club that David and Cori sang in. When André Bernards purchased a 1976 pink Cadillac for $75, Steve became one of its most enthusiastic drivers and promoters; he supplied the pair of fuzzy dice that hung on its rearview mirror. "I found them in the street," he said. "Somehow their origin seemed appropriate." So celebrated did the pink Caddy become that David Baltimore boasted about it wherever he went.

Steve's acquisition of the M.D. had been followed by a residency in pediatric oncology; his Ph.D. was in protein chemistry. He knew nothing about manipulating DNA. "I literally had never held a pipette or used a restriction enzyme," he said. He needed help, and he got it from the people he jogged, rowed, and sang with. He got help from the gamblers and the Caddy lovers. He had most of the Weinberg lab eating out of his hand and helping him design his experiments. The one person who didn't take the bait was Weinberg. Bob doubted that Steve's project would yield anything big. "It's a potentially interesting project," said Bob in February, "but we'll have to wait and see what he comes up with." Steve Friend was studying a problem that was unrelated to anything Weinberg had worked on in the field of oncogenetics. A year or two earlier, Bob and Hucky Land had considered the same problem but had dropped it for lack of any clever ideas on how to tackle it. Bob might have been happy to leave it dropped had it not been for Steve's geniality. When Steve came to Bob and said, "I want to study retinoblastoma," Bob couldn't say no. But what he could say was: It's not likely to work any time soon. And what he could do was to leave Friend to monkey around as he pleased, which suited Steve fine. Steve needed guidance, but he felt he needed technical guidance more than he did intellectual aid. "I see Bob as a sort of film producer," said Steve. "He makes the work possible. He greases the wheels; he gets the money; he creates the right environment for getting things done."

As Weinberg produced, Steve wrote, starred, and directed. He solicited help, he learned quickly, and he had the energy of a thermonuclear device.

Retinoblastoma is a gruesome cancer of the eye. It exclusively strikes children under the age of eight. Sometimes infants are born with the milky-white tumors already clouding the retina, the innermost layer of cells at the rear of the eyeball that catches light and sends it along the optic nerve to the brain. Unless it is treated, the disease invariably ends in death, and the current therapies are harrowing. Retinoblastoma victims usually end up losing one or both eyes or must undergo whole-beam radiation, which can dangerously soften the skull and trigger the development of secondary cancers.

Fortunately, the disease is as rare as it is grisly. Only one in twenty thousand youngsters falls prey to the cancer, compared with an incidence rate of one in five thousand for leukemia, the most common childhood malignancy. What's more, when retinoblastoma is caught early enough, the survival rate is better than 90 percent. Retinoblastoma is not a major public menace among cancers.

From a scientific standpoint, however, retinoblastoma has long been a fascinating disorder, a cancer that is believed to hold clues to the genesis of many, if not all, human malignancies. It was retinoblastoma and tumors like it that kept Mike Wigler from bothering to attempt oncogene transfection when the notion first came to mind. It was retinoblastoma that made Weinberg's lovely success with the *ras* gene so surprising.

I first learned of the retinoblastoma gene from Harold Varmus in the summer of 1985. He asked me whether I'd ever heard of "recessive oncogenes," or "anti-oncogenes." I hadn't. "They're the wave of the future," he said. "They're the next problem to be cracked." He rustled through his files and extracted a review he'd written on oncogenes that had, toward the end, a section on anti-oncogenes.

"It's a great term, *anti-oncogene,*" I said. "It sounds like something a journalist might have made up."

Varmus admitted that the term was sensationalist, even presumptuous. Not much is known about these anti-oncogenes, he said. According to the simplest model, the cell has proto-oncogenes, such as *myc* and *ras,* which promote cell growth, and anti-oncogenes, which inhibit cell growth. You know about the Weinbergian *ras* theory of cancer, right? said Varmus. The twelfth codon of *ras* is changed from a glycine to a valine, and somehow that mutation steps up the gene's activity.

But it's just as easy to imagine a gene that exists to block cell division. And if that gene is accidentally deleted from the chromosomes, there would be nothing to stop cell proliferation. Scientists have reason to believe that gene loss, rather than gene mutation, is responsible for retinoblastoma, said Varmus. In a strikingly high percentage of retinoblastoma tumor cells, a portion of chromosome 13 (humans, remember, have twenty-three pairs of matched chromosomes) is missing. Not chemically altered, but missing: the chromosome is shorter than it should be. The loss isn't random. From patient to patient, the same region of the thirteenth chromosome is partly or wholly absent, a region called q14. The pattern suggests that there's a gene located within the q14 territory that bears a crucial relation to the serenity of eye tissue. Once the gene is deleted, the retinal cells replicate out of control. So retinoblastoma is very different from *ras*-induced cancers, said Varmus. You don't find a gene mutated into a hyperactive state. Instead, you don't find a gene where a gene should be.

Varmus added that there were other cancers known to behave like retinoblastoma. Wilm's tumor, a childhood kidney cancer, is thought to result when an anti-oncogene located on chromosome 11 is knocked out of a kidney cell. Childhood neuroblastoma probably arises when a gene is lost from chromosome 1 in a nerve cell. These genes have not yet been cloned, said Varmus, but once they are, they'll open a new chapter in oncogene research. The chances are excellent that the loss of recessive, growth-controlling genes is a key cause of many human cancers. The cloning of one will help focus the search for the rest. Varmus predicted that the retinoblastoma gene would be the first to be isolated. Eventually.

I started a new file, called "Recessive Oncogenes," and consigned the study of the subject to my futures list. There seemed plenty of time before retinoblastoma would be newsworthy.

Over the following months, however, the subject of anti-oncogenes kept cropping up. I heard retinoblastoma discussed at scientific meetings. I heard about Webster Cavenee, a molecular biologist then at the University of Cincinnati College of Medicine, who was said to have done more work than anybody else on understanding recessive oncogenes. Scientists mentioned recessive genes in passing, during interviews, to buttress their argument that cancer cannot be reduced to the *myc* and *ras* genes. Cancer is complicated, they said. "And as complicated as we think things are," said Tony Hunter of the Salk Institute, "they're sure to be more complicated than that."

The retinoblastoma gene was like a storm cloud, rolling closer, darkening, picking up moisture and mass as it moved. I learned of the challenge of cloning the retinoblastoma gene. Scientists were trying to isolate a gene that was not only missing from tumor cells but had a completely unknown function in normal cells. They couldn't find it by putting tumor DNA into normal tissue and screening for foci, as Chiaho Shih had done to collar the *ras* gene. There is no retinoblastoma gene in retinoblastoma tissue; it's like looking for a needle in a haystack that you know doesn't have any needles.

Nor could they pull out a chromosome 13 from a normal eye cell, crawl around the DNA of the chromosome, and expect to stumble on the gene. True, they knew roughly where the gene was located: in the q14 band. That knowledge narrowed the search to 10 percent of the chromosome. But the q14 band is still quite vast enough. It comprises five million base pairs of DNA. Five million. How would scientists know, while scratching through five million nucleotides, that they'd found the 20,000 or 50,000 bases of the retinoblastoma gene? What was their assay, their trail marker, their hope?

An impossible task, and an irresistible one. The retinoblastoma gene was a new kind of cancer gene, unlike any oncogene cloned to date. The lab that found it would set a paradigm, and scientists love to set paradigms.

Another derby was under way. That's what molecular biologists were saying: everyone wants to clone the retinoblastoma gene. But for once, Bob Weinberg was not in the race, so I postponed getting too excited. The contestants were in California, Toronto, Montreal, London, and I couldn't follow every last event in the oncogene track meet.

And then I visited Mike Wigler's lab, where one scientist, Graeme Bolger, was attempting to clone the retinoblastoma gene, and suddenly the race became riveting. Graeme is a physician from Montreal who had worked with retinoblastoma children. He described to me the pathology of retinoblastoma: the appearance of a tumorous retina that shines like frosted crystal beneath the glare of an ophthalmoscope; the tragedy of babies who'd lost their eyes to the disease. He explained that the retinoblastoma gene is thought to be responsible for a second type of malignancy as well: osteosarcoma, or bone cancer. In many osteosarcoma cells, the same q14 band of the thirteenth chromosome is completely or partly missing.

"You're talking about the cancer that Edward Kennedy's son had?" I asked.

"That's the one," he replied. "The full name is osteogenic sarcoma."

"So Ted Jr. had a deletion of the retinoblastoma gene in one of his bone cells? That's how he got the cancer?"

"Well, there's no way of knowing without analyzing his tumor cells, which I imagine are long since gone. They're certainly not available to me," said Graeme. "But if the model for retinoblastoma and osteogenic sarcoma is correct, it's a possibility."

Scientists were racing to clone a cancer gene that may have been behind one of the most famous cancer cases of the twentieth century. They were looking for a gene that led to cancer in the youngest of infants. The connection between basic research and real human cancer had never seemed more lucid or more poignant. I asked Graeme whether he thought he was close to cloning the gene. He said no. He'd made strides in improving his method for seeking the gene, but he was still at an elementary stage of the experiment.

Within the next six months? I begged. He replied that frankly he doubted it.

I soon learned that other retinoblastoma researchers were no farther along than Graeme. I was crestfallen. Was this another search for the function of *ras* or *fos* or *myc*? I knew the answer, of course. Scientists can't predict when a breakthrough is going to occur. I just wished that somebody would reach the peak after all these frustrating plateaus.

When I heard that Steve Friend was working on retinoblastoma, my optimism revived. If somebody is industrious as he was looking for the elusive anti-oncogene, its capture could not be far off. But Steve disappointed me, too. "I'm not trying to clone the gene," he said.

"Oh," I replied.

"Bob had been worried about that when I first came to the lab," said Steve. "He told me that Hucky Land had fooled around with the cloning for a little while, without success, and he wondered why

I'd try to do the same thing. I told him I wasn't interested in cloning the gene."

"So what are you doing?" I asked.

Steve said that he was trying to develop an "expression system," a method for understanding the function of the retinoblastoma gene. Some scientists, he explained, believe that the retinoblastoma gene works to shut off the signal of a type of *myc* gene in human cells called N-*myc*. The retinoblastoma gene, he said, could be an anti-N-*myc* gene — whence the term *anti-oncogene*. It could prevent eye tumors, not by sending a stop signal to the cell nucleus, but by damping N-*myc*'s potentially cancerous "go" signal. There was some evidence to support the theory. Several labs had discovered that the N-*myc* oncogene is expressed at unusually high levels in retinoblastoma tumor cells, as though N-*myc* had lost its muffler.

Steve hoped to determine whether the retinoblastoma gene acted as the muffler for N-*myc* or whether the overactivity of N-*myc* was merely coincidental in eye tumors. He didn't need a cloned copy of the retinoblastoma gene for his initial experiments, he said. He could study N-*myc* expression. For example, he could transfect extra copies of the N-*myc* gene into normal eye cells and see whether they turned cancerous. Conversely, he could try forcing retinoblastoma cells to stop dividing by adding a battery of chemicals known to help cells differentiate, take on the characteristics of adult eye tissue. Mature eye cells almost never divide. Assuming that N-*myc* is the division signal for retinal cells, he said, the application of differentiation chemicals should mock the retinoblastoma gene and shut down N-*myc* expression.

"There are lots of little fun things we can play around with," he said. He peered at the agarose gel he was preparing. "The color of this looks strange. Does it look strange to you?"

I replied that it looked like the same shade of purple as any other gel I'd seen. A few minutes later he commented on the beauty of the weather, and a few minutes after that he talked about running in a ten-kilometer race. Steve doesn't talk in a straight path; he weaves every which way, like a shopping cart with a wheel out of alignment. Steering him back to the subject of his project, I asked him what he'd do if it turned out that the retinoblastoma gene doesn't control N-*myc* expression — if it does something else altogether. Would his expression system reveal the alternatives?

"I think this gel is dripping," he said, scrutinizing the edges of the glass frame into which the gel had been poured. "No, maybe not.

Um, well, N-*myc* must be at least part of the story. Why else would you see high levels of the gene in the retinal cells of infants and in retinoblastoma but not in adult eye cells?

"Besides," he continued, "I figure that with so many people trying to clone the RB gene, somebody's going to get it pretty soon. And once I have this expression system up and cooking, I'll be in a good position to get a copy of the clone and run with it."

If Steve was cooking, it was nothing you'd find in Fannie Farmer's book. He had contacts at every hospital in the Boston area. The phone would ring, and it would be for Steve. A retinoblastoma patient had just been "enucleated" — a cumbersome avoidance word for removal of the eye. An osteosarcoma had just been carved from the limb of a teenager. Did Steve want to pick up some of the biopsied material for his experiments?

"He'd come back with these really disgusting blobs of things floating in a vial," said Mike Gilman. "You didn't want to know what they were, but he'd always tell us."

Steve plated out the biopsied material on petri dishes and allowed the tumor tissue to replicate until he had enough cells to pulverize for RNA analysis. René showed Steve how to do Northern blots: screening the tumor RNA with radioactive probes to gauge the arousal of the N-*myc* gene. René also provided Steve with the control he needed. For his model to have any meaning, Steve had to compare N-*myc* expression in tumor cells with N-*myc* levels in normal eye tissue. But normal human tissue of any sort is notoriously hard to culture. Healthy cells resent being removed from their organ of origin to the hostile environment of a plastic dish, and they usually die after a few days. For all his connections, Steve didn't have access to a limitless supply of wholesome human eyes. He needed eye cells that could be cultured; he needed a retinal line that would thrive indefinitely *in vitro*.

"That's why it's so great to be at the Whitehead," said Steve. "There's always somebody who can help you." Calling his old lab back in Amsterdam, René asked a technician to dig through the freezer and pull out a few samples of his Ad-12 retinal cell line, which had been sitting in storage, unused, for the past two years. As a graduate student, René had bred the cell line from the eyes of a nearly six-month-old fetus that had been aborted for medical reasons. He had prompted the eye tissue to survive *in vitro* by transfecting in the DNA of the adenovirus, an agent of respiratory diseases in humans. For unknown reasons, the viral DNA kept the retinal cells alive without grossly altering their shape or turning them into cancer cells.

"It was a big deal at the time," said René. "It was the first human retinal cell line. But I couldn't think of anything special to do with the cells at the time, so they were frozen away.

"The funny thing is that a year or a year and a half ago, Web Cavenee came to give a seminar at the University of Leiden. I heard him talk about retinoblastoma, and he mentioned this business about trying to clone the gene. After the seminar, I went up to tell him about my Ad-12 cell line. I thought it would be a good way for him to look for the gene. I thought it was very possible that the RB gene was expressed in these cells. After all, the fetus was twenty-three weeks old, and if the model is right and the RB gene turns on to help the eyes differentiate, then that would be the time you'd expect to see RB expression. At six months, the eyes are maturing rapidly. So these Ad-12 cells seemed like a good place to look. I said all this to Cavenee, and he agreed. He said he was interested. I offered to give him the cell line, and we arranged for him to come by my lab, which was right near the lecture hall." René pointed out the window to a parking lot across the street from the Whitehead. "It was as close to where he spoke as that lot is to this building. No great effort to get there, right? Well, he was supposed to come at four. I waited until eight or nine. He never showed up.

"Tough break for Web Cavenee," said René. "What a jerk. But I'm glad he never came. I'm glad that I saved the cells for Steve."

Packed on dry ice, the Ad-12 retinal cell line arrived at the Whitehead a week or so after the call overseas.

But Weinberg didn't immediately pay attention to the expression system under construction. Cori Bargmann, his best and brightest, had just given him a piece of news about the *neu* gene that exceeded his brightest hopes.

In the wake of the shakeout over the *neu* project, many of the Weinberg scientists assumed that nothing good could come of the gene. Too many people had worked on *neu* for too long, with little to show for the effort. The researchers had found the oncogene in the brain tumors of rats, and they'd figured out that *neu* was some sort of membrane receptor. They didn't yet know what growth hormone it was a receptor for, however, and they had no idea whether the oncogene was significant to human cancer. Weinberg had assigned Cori Bargmann the most compelling part of the project — finding the *neu* mutation — but even there the researchers doubted that she would reveal anything novel about tumorigenesis.

Having lost the mutation account to Cori, Mien-Chie fled from the Weinberg *neu* project with hardly a glance backward. He never bothered developing a transgenic mouse system to examine the effects of *neu* on mouse growth, nor did he look for the equivalent of the *neu* gene in fruit flies. Instead, he started looking for a job as an independent researcher. When he won a position at the M. D. Anderson Hospital in Houston, he sounded like a prisoner unexpectedly released on parole. He talked about how great life would be in Texas. "Did you know there are *two* Chinatowns in Houston?" he asked me. "There's an old Chinatown and a new Chinatown. The new Chinatown is really a Little Taiwan. Most of the people there are from Taiwan, like me. It's so beautiful, this new Chinatown. Very clean and modern. I think we'll feel very much at home there. We've been in Boston too long."

The *neu* gene project had taken on a farcical air, and the Weinberg scientists started placing bets on where in the gene Cori Bargmann would find the mutation. There were two likely possibilities.

"It's in the extracellular domain," said René. "And, hey, I'm a big guy. I'll bet a six-pack of Heineken on it." The extracellular domain of the *neu* protein is the claw that rises above the cell membrane to catch a wayfaring growth factor. If the mutation had damaged the shape of the extracellular claw, the *neu* protein could behave as though it always clutched its corresponding growth factor. The receptor would remain eternally excited, shooting a steady ejaculate of signals down to the cell nucleus.

"You'll lose, big guy," said David Stern. "It's going to be in the kinase domain." The kinase domain of the *neu* protein swings inside the cytoplasm and activates target enzymes by pinning phosphate molecules to them. A *neu* protein with a misshapen kinase tail might continue thrashing around and passing out phosphates when it should be dangling at ease. Cori Bargmann sided with David. "It's in the kinase domain," she said. But the scientists knew that neither alternative would be particularly earthquaking. There was a precedent for both types of mutations. When the receptor for epidermal growth factor is transformed into a cancer gene, the DNA of the receptor suffers insult to both its extracellular and kinase domains, resulting in a protein that is warped at each end. The Weinberg scientists had adopted enough of the Weinbergian philosophy to feel that providing additional support for previous results is not a scientific style to aspire to.

Luis Parada was so bored by the obvious options that he wagered his beer on a third part of the *neu* protein. "I'll bet on the transmembrane domain," he said. René replied that it was a good thing Luis was leaving the lab soon to do his postdoc in Paris. The section of *neu* that spans the membrane of the cell constitutes only 2 percent of the total receptor. It was unthinkable that the mutagen responsible for the rat brain cancer had zeroed in on such an insignificant region of the gene. No, the *neu* gene was not another *ras* gene. There weren't going to be any surprises.

Yet Cori didn't mind that she was assigned an obvious experiment. She thought the *neu* gene was "fascinating" and "worth the efforts of a whole lab in itself." Calmly and elegantly, she spliced together sections of the cancer cDNA and the opposing parts of the normal cDNA. I'd stop by her bench and ask whether she was getting close. She'd reply that she had the region of the mutation narrowed down to 1000 bases, then 600, then 300. When she had eliminated enough of the gene to make her work manageable, she stopped to sequence the area of the DNA where she thought the lesion must lie.

Toward the end of winter of 1986, I asked again whether she was closing in on the mutation.

"I think I know where it is," she said.

I waited for her to tell me, but she didn't say any more. She just looked at me. It was as though I had asked her whether she knew the time and she'd replied yes.

"So where is it?" I finally asked.

"It's in the transmembrane domain," she said.

Cori talks softly, and I thought I had misheard her. "It's in the kinase domain?" I asked.

"No, the transmembrane domain."

"But that was supposed to be a joke!" I cried. "Luis made the bet as a joke."

"I've always thought that Luis had a nose for science," said Cori. "Even when he's joking, he has a nose for science."

"But isn't this amazing?" I asked. "What sort of mutation is it? How can it be in the transmembrane domain? What does it do?"

She explained that a point mutation in the transmembrane domain had changed a thymine nucleotide to an adenine. As a result, the amino acid dictated by the triplet of DNA bases now read glutamic acid rather than valine. The biochemical difference was significant: valine has no electric charge, but glutamic acid is a negatively charged amino acid. Somehow, the addition of an electric spark to the little belt that cinched the waist of the protein and anchored it to the membrane drastically altered the behavior of the receptor.

"There are two possibilities to explain how the change from valine to glutamic acid may affect the receptor," she said. "One is the push-pull model. According to that idea, a receptor that's not bound to its growth factor is pushed slightly outward. Once it's bound, it pulls in. The negative charge could keep the receptor always pulled inward. For some reason, that's not a model I like.

"I prefer the clustering model. Now, according to that scheme, receptors don't work as individual units. Instead, what you have is a series of receptors on the membrane — in this case, a series of *neu* receptors — and they all have to bind with growth factor more or less simultaneously. At that point, they move toward each other in the membrane, forming a clump of receptor-factor complexes. Only when they're clustered are they able to send along their signal." The insertion of a glutamic acid into the transmembrane domain, she explained, could trick the *neu* protein into believing that it was surrounded by other *neu* receptors even when it stood alone.

"It's hard to believe," said René. "Cori took a boring oncogene and made it into something really interesting."

I asked him who got the beer, since Luis already had left the lab for Paris.

"We drank it in his honor," said René.

Cori didn't waste much time celebrating. So astonished was she by the presumed position of the mutation that she decided to confirm and reconfirm her results. She went back and sequenced the normal and cancerous *neu* genes all over again, using two different sequencing techniques. She locked herself up in a room and studied the sequencing gels until her vision blurred. "Bob wants me to hurry; he thinks I've been careful enough. But with something like this you can't afford to make a mistake," she said. "I can't get as wildly excited as he is, because then we'll feed each other on faith alone, and that's dangerous. If I'm going to talk about a new kind of mutation, I don't want to have to say later, Sorry, guys, I was only kidding."

To extend her results beyond a single clone of the *neu* gene, Cori collaborated with Mariano Barbacid's lab people. They examined neuroblastoma tumors from rats that had been injected with a number of noxious chemicals. In five different experiments, the researchers discovered that the *neu* gene had indeed been mutated in exactly the same spot of the transmembrane region: an adenine nucleotide sat where a thymine should be.

The Weinberg lab had found the first mutation in a cancer gene years earlier, and then other labs confirmed that the *ras* lesion is shockingly common: when they looked at *ras* in a tumor, its twelfth codon often read valine rather than glycine. The Weinberg lab identified the second specific mutation in a cancer gene and then confirmed that this mutation, too, is predictable. Only one node in the transmembrane domain is vulnerable to sabotage. Cancer was beginning to look not only simple, but chillingly precise — even rigid. Hit 32,999 bases of the *neu* gene, and nothing happens. Hit number 33,000 and place an adenine nucleotide where a guanine should be, and the gene erupts.

"It's the consistency of the result that makes it so exciting," said Weinberg. "When you find the same result in five independent cell lines, you know you're on to something biologically significant."

The biological community concurred. Cori presented her data at a huge oncogene meeting in Maryland. In preparation, she'd agonized over details: she'd made the slides herself rather than entrusting the task to an illustrator; she'd packed an iron in her suitcase so that her skirt and blouse would be perfectly pressed. She stood on the stage, a willowy, fearless girl-woman, her clothes crisp, her voice crisper, and she announced the discovery of this odd mutation in the transmembrane domain. Her talk was flanked fore and aft by half a dozen other talks, but when the crowd dispersed for a coffee break, all I heard from the scientists around me was babble of the *neu* mutation. They were astonished that a small electrochemical change in the tiny, sheltered section of the protein could fuel such perfervid activity in the cell.

The *neu* gene finally was news, and other labs began to take an interest in its tumorigenic powers. Most hoped to make the oncogene confirm or deny a favored model of membrane receptor activity. A team of biomedical researchers in California and Texas, though, suspected that the *neu* gene was not only biologically significant; it was medically significant as well. And in a sense, Dennis Slamon and his colleagues at UCLA and the University of Texas Health Science Center in San Antonio had the last laugh with the *neu* gene.

In the summer of 1985, when Mien-Chie was working on the genomic *neu* gene, Alan Schechter had begun to examine whether the gene was mutated in any human cancer cells. His effort was desultory at best. He did discover mutant *neu* genes in several cancers of the human bladder, skin, and blood cells, but he didn't pursue the implications of the discovery. At the time, Alan was more concerned with finding a job than he was with *neu* or the rabid turf wars, and Weinberg didn't think the human part of the project pressing enough to bequeath to somebody else.

But science progresses by accretion and synthesis. Spurred by the results of the Weinberg lab and others, Dennis Slamon followed up on the link between *neu* and human cancer. The *neu* oncogene was originally found in rat brain tumors, but Slamon had cause to suspect that the gene was less relevant to human brain tumors than it was to one of our three biggest cancer killers: breast cancer. Stuart Aaronson of NIH had reported that the *neu* oncogene was amplified in a cell line cultured from human breast tumor tissue. Healthy cells have only one copy of the *neu* gene, but these cells had *neu* in triplicate, quadruplicate, sextuplicate. When cells are maintained in dishes, however, many queer things happen to their chromosomes, things that turn out to be biologically meaningless. Slamon wondered whether the *neu* gene might be amplified in fresh breast tumors, too, and whether such abnormalities might contribute to the cancer. His interest in *neu* was further piqued by Cori Bargmann's results on the outlandish membrane mutation.

Collaborating with William McGuire of the University of Texas Health Science Center, Slamon began by studying the tumor cells of 189 women who had a relatively advanced form of breast cancer: the malignancy already had spread to at least one lymph node in the armpit. Most of the women had been patients of McGuire, and some had been dead for years, but McGuire had kept frozen biopsy samples and careful medical records for each. Thus, any data that Slamon divulged about anomalies in *neu* could be correlated with a patient's actual prognosis and eventual fate.

Slamon purified the DNA from the tumor cells and probed the material with a copy of the human *neu* gene (which he'd obtained from Axel Ullrich of Genentech, and which Ullrich referred to as the HER-2 gene, for human EGF receptor 2). Slamon's results revealed a spectacularly consistent pattern of *neu* activity. The more severe the woman's cancer — the more her lymph nodes had been infiltrated — the more wildly amplified was the *neu* gene in her tumor DNA. In women whose tumors had invaded three lymph glands, for example, the cells harbored as many as twenty copies of the *neu* gene. What's more, the degree of *neu* amplification corresponded bleakly with a patient's prospects. Those women who'd relapsed and died after the shortest interval following initial treatment turned out to be the patients with the greatest number of *neu* genes in their tumor cells. Women who had only one copy of the gene had an 80 percent chance of surviving two years without a recurrence; for women with five or more copies of *neu,* that two-year survival figure plunged to 35 percent. Somehow, the excess of *neu* genes lends a breast tumor cell a hideous sort of growth advantage.

Slamon has since expanded his analysis of the *neu* gene to 123 patients whose lymph nodes were altogether free of the disease. It is this second study that could have a profound impact on cancer care. By the accepted medical protocol, patients with advanced breast cancer automatically are placed on a course of chemotherapy and radiation, and knowing the status of *neu* amplification won't make much difference in their treatment. But in cases where the neoplasm is confined to the breast, doctors currently are more conservative, restricting intervention to mastectomy or lumpectomy alone. Sadly, however, 20 to 30 percent of the women with clean lymph nodes eventually relapse after either form of surgery. That figure is not high enough to warrant the use of aggressive and devastating chemotherapy in every case of early-stage cancer, but it's high enough for physicians to wish they had some way to identify who is at risk of recurrence. Doctors then could prescribe adjunct treatment immediately, while the scattering of tumor cells that escaped the surgeon's scalpel is meager enough to be destroyed by toxins and radiation. Slamon's question is whether the *neu* gene is a potential prognosticator for those early-stage women at risk of relapse. Is the gene already amplified in their cancer cells even while the breast tumor is small and confined? Slamon's results would not be complete until some time in 1988, but in the fall of 1987, he expressed great optimism. "I wouldn't have worked so hard and so fast on this project if I wasn't hopeful," he said. "But we think all the data point at *neu* amplification as a powerful prognostic indicator. If it's true, it will be a very important correlation — a *clinically* important correlation."

Should Slamon's equation between *neu* and breast cancer hold up, thousands of women may benefit. Approximately one in eleven or twelve women contract breast cancer. The sooner the cancer is treated, the better a patient's prospects for survival. If an aggressive form of the cancer can be identified at an early stage and treated with appropriate force, the thousands of women who otherwise would relapse after surgery may be able to walk away, cured.

Nor should his results require years of animal testing and controlled trials to be incorporated into clinical medicine, as would the discovery of a new drug. If oncologists are convinced that *neu* amplification is a consistent sign of a more dangerous type of breast cancer, they may begin screening the tumor cells of their patients with *neu* probes as early as 1989. "This could be very, very exciting," said Slamon. "The use of *neu* for clinical prognosis would be the first instance of using basic oncogene research against a major cancer killer. It's encouraging to me that we're finally getting some of our lab results over to the patient's bedside."

Slamon's work on *neu* amplification may eventually converge with the Weinberg lab's discovery of the membrane mutation. Slamon speculates that the reason the *neu* gene became amplified in breast tissue in the first place was through a point mutation in the patient's healthy proto-oncogenic copy of the *neu* gene. A carcinogen could have mutated the transmembrane section of the gene, and that mutation could have liberated the *neu* gene from its normal cellular constraints, culminating in the crazy photocopying of *neu*. In other words, a point mutation in the normal *neu* gene may have been the initiating event, and the lethal amplification of the mutant gene may have followed as a result of the primary lesion. By the fall of 1987, Slamon was sequencing one of the amplified *neu* genes from breast tumor cells to see whether it had the membrane mutation that Cori had discovered in the rat copy of the oncogene.

If the amplified *neu* genes are all mutated, and if they're all synthesizing mutant membrane receptors, Slamon believes there would be a way to attack those receptors with a drug designed to recognize mutant *neu* proteins on the surface of mammary cells. As he envisions it, the drug would lodge itself in the claws of the receptors and immobilize them, blocking the transmission of their cancerous signal to the nucleus. The drug would be a powerful weapon against breast tumors.

For the time being, however, the strength of Slamon's work lies in its potential utility for cancer prognosis. It is not THE Answer to Cancer, but it looks like AN answer — a magnificent early spin-off from basic research.

Weinberg told me how sorry he was that he had chosen not to pursue the studies of human *neu*. "I didn't think it was going to be conceptually interesting to look at human tumors, so the amount of time we spent on it was slight," he said. "I decided not to push it further. Stupidly, I didn't push it further. It looks as though *neu* will be a good diagnostic instrument for breast malignancy."

But Weinberg didn't sound tetchy or depressed as he talked about his lack of foresight. Instead, he was grinning, and his black-brown eyes were as bright as a city night. Weinberg felt vindicated. He'd known all along that the oncogene was important. His lab had discovered it, nurtured it, brawled and despaired over it, and thrust it into the public view. Bob Weinberg could not do everything on his own, and translating fundamental science into clinical tools had never been his bailiwick.

Besides, he no longer needed *neu*. By the time Slamon revealed that *neu* was amplified in mammary cancer, Weinberg had something he thought was much, much better. He had found the grail that half his peers had been vying to find.

12 Superfecta Key

THE LIGHT RAIN was mild and sweet, more of a mist than a drizzle, and even the dusky pink face of the Whitehead Institute seemed grateful for the teases of spring in April of 1986. It was the sort of day when nothing can possibly happen, so you may as well give yourself over to a lazy joy.

The first person I met, when I entered the lobby of the institute, was Snezna. It was only three weeks since I last saw her, at the dog races, but I was impressed anew by her lucid beauty. She hugged me warmly and told me that I looked great.

"I look great?" I said. "No, you look great. You must have found another strange new growth factor in milk."

"Vivek is finally back," she said, and grinned with cheery lewdness. Vivek had spent the better part of the past two months in Socorro, New Mexico, mapping galaxies at the Very Large Array of radio telescopes that is Socorro's greatest claim to fame. "I was climbing the walls for him," she said, and clawed the air in demonstration.

"Are you running off somewhere?" I asked.

"I'm already late for an appointment at Harvard," she said. "But I'll talk to you later. I'm glad you're here." She hesitated. "You'll be glad, too."

"Oh? Why's that?"

But she turned to hurry off, repeating that she'd talk to me later.

The second person I met was Mike Gilman, and he too looked as fresh as the spring rain. He told me that his latest experiments were going like "gangbusters." Several months earlier, Mike had come across a powerful new experimental technique called a DNA-protein-binding assay, which allowed scientists to identify cellular proteins that fastened on to the DNA molecule. Once again, Mike became the first Whitehead scientist to exploit the new technology. Ditching his *myc-fos* project, which he'd decided was futilely irreproducible, Mike employed the binding assay to return to his old love: gene regulation. He had found a complex of proteins that controlled the promoter of the *fos* gene during cell division, helping to transcribe *fos* DNA into RNA. Although he knew that the project was too narrow in scope to enthrall Bob Weinberg, Mike believed that his DNA-binding proteins would, in the long run, reveal more about the *fos* oncogene than his celestial model of signaling pathways ever did.

"I just met Snezna," I told him. "She looked happy. You look happy, too. Are you guys planning to go to the dog races again, or what?"

"Well, I don't know about her," he said, "But I'm happy because I'm finally getting something worth publishing. I'm starting to write up a paper — my first paper from the Weinberg lab, if you can believe it."

The third person I met was André Bernards, who was loitering around the Weinberg lab. He was happy because he'd had the foresight, before the rain began, to throw a tarpaulin over the pink Cadillac. The Cadillac was a "convertible" that could not be converted to a roofed state. The car hadn't actually begun life as a convertible, but at some point in its history its top had been ripped off in an accident, which is why André had been able to buy the car for $75. We stared out the third-floor window to the parking lot below and watched the tarpaulin billowing gently in the wind. I asked him whether he knew the song "Pink Cadillac," by Aretha Franklin. "Oh, we have a tape of it for the car," he said. "And Bruce Springsteen's song, too. This is no amateur operation."

We settled into chairs, and he brought me up to date on the *abl* oncogene, his project in David Baltimore's lab. Like the *myc* gene, *abl* seems to be involved in several kinds of human leukemia. André was trying to corner the entire gene and its accompanying regulatory sequences, but it seemed to be a longer gene than he'd originally anticipated; now he thought it might extend for hundreds of thousands of DNA bases, which would make it one of the larger human genes yet known to humans.

Steve Friend came by the bay. His face was flushed, and he looked as though he could barely keep his feet from sprinting off ahead of him, but he stopped to chat. André told Steve he'd be glad to know that the Cadillac was safely covered.

"Thank God, for the sake of the fuzzy dice," said Steve.

André excused himself to return to work. Hoping to discourage Steve from doing the same, I settled deeply into my chair. I was pleased to see that he wanted to talk.

"So what's new?" I asked him. "That's what Mike Wigler always says to his people. 'What's new? Anything new?'"

"Well, let's see," said Steve, rocking back and forth against the edge of the lab bench. "When was the last time you were here? Two or three weeks ago?"

I said that was about right.

"Well, um, in the time since you were here, I think I've cloned the retinoblastoma gene."

Before you even know you're reacting to a startling piece of news, the mind explodes with every squalling, strident, competing reaction possible.

I'd asked a silly, empty-calorie question, expecting the usual answer: Not much. Instead, Steve stood there, his arms crossed, his face ruddy with excitement, saying that he'd cloned the gene that everybody in cancer research wanted to get. In no particular order — because they had no order at the time — these thoughts ricocheted through my skull:

I thought that he was kidding.

I thought that he was deluding himself.

I thought that, once again, I'd misunderstood what had been said.

I thought — I'm being serious — that the reason his face was flushed was that he was drunk, although it was only 11:00 A.M.

And then I realized he was telling the truth.

"Oh, my God," I squeaked. "I can't believe this. Oh, my God. When did this happen? When did you do it?"

"I just got the full clone last week. Since then, we've been looking at expression levels of the clone in retinoblastoma and osteosarcoma tissue. We've also been looking at DNAs for internal deletions of the clone."

He went loping off on a technical explanation that I couldn't follow. I wasn't concentrating. I was still in shock. When Cori told me that the *neu* mutation was in the transmembrane domain, I was surprised largely because the possibility had struck me as a joke. Now I was overwhelmed because Steve had just told me that the Weinberg lab had performed a miracle.

Steve tugged me back to earth by his repeated warnings that he was not yet certain he had the right gene. There was much to be done before he'd feel secure. Recalling all the past occasions when fantastic results had vaporized into irreproducibility, I stopped gasping and exulting. Still, never before had any scientist said to me, so baldly, "I think I've done it." I couldn't help feeling blessed by this astounding turn of events.

"What does Bob say?" I asked. "What does he think about this? I haven't seen him yet today."

"He's being very cautious," said Steve. "And that's good. He doesn't want to get too excited at this point. He's trying to keep the whole thing in perspective."

René came by the bay. One look at me, and he knew that I knew. "So you've told her the news?" he said to Steve. "And did you tell her it was a SECRET?"

"René's right. We don't want to let any of this leak out," Steve said to me. "We have to get a lot more proof before we can talk about it in public."

"Who would I tell?" I said to René.

"Well, for starters, Mike Wigler's lab," he replied.

I promised them that I wouldn't utter a word to the Wigler scientists, *Time* magazine, or anybody else. And then I asked Steve to please explain what he'd done and what additional proof he needed before he'd be sure he had the retinoblastoma gene. Together Steve and René outlined the history of the cloning, lobbing ideas back and forth to each other on how next to proceed. They talked in whorls, tangles, and arpeggios: the success had burst upon them so suddenly that they still didn't know exactly what they were seeing.

My notes from that first session were largely an incomprehensible jumble. Believing that the retinoblastoma gene was a distant prospect, which in any case would be reached by some other lab, I'd avoided getting serious about it. Now I felt like a college student who'd procrastinated studying until the night before finals. There was still time to catch up, but I had to cram. I had to understand how Steve had arrived at this epiphany. I discovered that the tale began in the nineteenth century and reached a climax with Thaddeus Dryja, one of the few scientists around who can rival Steve Friend in niceness.

Most human tumors afflict internal organs, and thus are invisible to all but the surgeons who enter, scalpels in hand, to have a look. Retinoblastoma is different. If left untreated, the tumor will cause the eye to bulge from its socket; if left further untreated, the tumor will spread outward until it obscures the face. I saw a photograph of a retinoblastoma child from Papua New Guinea, where there's little access to medical care. The left half of the boy's face and head was nothing but a grotesque mass of tumor, the color and apparent consistency of cottage cheese. The boy was holding his father's hand. The father was grinning for the camera, and what remained of the child's mouth was grinning, too.

As an obscenely visible cancer, retinoblastoma fascinated early artists and physicians. In the National Museum of Guatemala there is a spectacular Mayan stone bust of a retinoblastoma victim, depicting with almost medical accuracy the engorged, protruding eye characteristic of advanced retinal cancer. Petrus Pawius, a Dutch anatomist, produced what is probably the first clinical record of retinoblastoma in 1597. "I opened the cranium of a three-year-old infant," he wrote. "It had been suffering for some months with an enormous tumor of the left eye, such that the whole ocular globe with all the muscles made an outward projection, and this mass had increased in such proportions that the protuberance had the size of two fists." Despite surgical intervention, Pawius's young charge died two weeks later.

From the sixteenth until the mid nineteenth century, stymied doctors speculated wildly on the cause of the disease, variously attributing it to fungal infection, trauma to the eye, or an excess of candlelight. Some authorities suggested removal of the eye as a possible treatment, but few doctors followed the advice. For one thing, it seemed that by the time the tumor was big enough to deform the eye, the child already was doomed. For another, anaesthesia had not yet been invented, and the prospect of operating on a sentient, screaming infant was too repellent. Retinoblastoma children invariably died before the age of five.

The middle of the nineteenth century saw the introduction of two priceless medical tools: the ophthalmoscope and ether. Physicians now could peer into the dark recesses of the eye and spot a tumor when it was only a millimeter or so in diameter. The child could then be put under, and the diseased eye or eyes plucked out. Partly or wholly blind though they were, many retinoblastoma children survived to adulthood and had children of their own. And as they begat, a startling pattern emerged. The offspring of retinoblastoma parents sometimes developed the eye cancer themselves. In 1886, H. de Gouvea, a Brazilian physician, recorded the first documented case of inherited retinoblastoma: a man in Rio de Janeiro, who as a boy had had his tumor-filled right eye removed, fathered three children who became afflicted with the malignancy. Other examples of retinoblastoma families were reported and marveled at. Never before had a cancer seemed so closely linked to bloodlines.

The cancer wasn't always inherited, however. In many instances no familial link could be found to explain a case of retinoblastoma. By a similar token, some retinoblastoma survivors produced perfectly healthy children. To complicate the matter further, there were retinoblastoma victims whose parents were disease-free, yet who then went on to spawn retinoblastoma children; they were, in other words, the progenitors of a new retinoblastoma bloodline. The tumors of the three classes of cancer sufferers were identical under the microscope, indicating that it wasn't the disorder that varied, but the genetic patterns by which it was transmitted.

Retinoblastoma was a true conundrum. Some patients contracted the disease early in infancy and had three, five, or more tumors distributed across both eyes. Other children didn't develop the malignancy until two or three years of age and had only one tumor in one eye. Whatever the pattern of the cancer, it defied glib analysis, and clinicians were too concerned with treating the disorder to worry about defining it scientifically.

In the late 1960s, Alfred Knudson, then at the University of Texas, synthesized the disparate bits of information about retinoblastoma into a grand intellectual framework. He considered the important variables: age of onset of the cancer and total number of tumors in a single patient. He plotted the information on a graph and compared the curves to studies of leukemia incidence in mice that had been exposed to radiation.

Knudson's calculations led him to a bold conclusion. Retinoblastoma, he said, was a two-hit disease. A single retinal cell had to be battered by two mutations to its DNA before it would turn cancerous. Knudson proposed that those children who had multiple tumors early in life were born with a built-in defect — their first "hit" was part of their genetic legacy. They carried the retinoblastoma trait, and thus their eye cells had to undergo only one more mutational hit before cancer developed. The children who fell into the category of retinoblastoma carriers were of two types: either they inherited the trait from a retinoblastoma parent, or the chromosomes of the mother or father had suffered mutational damage during the development of the egg or sperm cell. In either instance, the offspring was bestowed with the retinoblastoma trait at the moment of conception and thus was extremely susceptible to the disease.

By contrast, the older children who had only one tumor suffered from so-called sporadic retinoblastoma. They began with perfectly normal chromosomes, but at some point during the growth of the eye a lone retinal cell had the great misfortune to be struck twice. Two times in the course of ocular development, the DNA of one cell was wounded — perhaps by a mutagen or cosmic ray, perhaps by the random errors in chromosomal replication. That doubly damaged cell then proliferated into a tumor.

Knudson's two-hit hypothesis straightened out many of the muddles about retinoblastoma. The children with hereditary, or germinal, retinoblastoma developed multiple tumors because they bore the genetic defect in every cell of their bodies, including *all* their eye cells. There are about a hundred million retinal cells, which means there are many targets for a second mutational hit. And each time a genetically fragile eye cell receives a blow, it replicates into a separate tumor. The model also explained the extreme youth of germinal retinoblastoma victims. Their retinal cells were predisposed to cancer, and the likeliest time for the second mutational events was when the eyes were growing most rapidly — during the last two months of gestation and the first few weeks after birth.

Extending the two-hit theory to victims of sporadic retinoblastoma, Knudson explained why such children had only *one* eye tumor and were older than the germinal patients when the cancer began. Sporadic cases have no inborn genetic defects — either in their eye cells or in any other body cells — and thus must undergo two separate hits to a single retinal cell before the malignancy can grow. The odds of more than one normal retinal cell being hit twice are close to zero. No child who was chromosomally fit — even a child of Job — was so unfortunate as to endure more than the freak, two-pronged accident that hammered his one eye cell into cancer. And that eye cell needed a little time to be attacked twice: hence the older age of spontaneous retinoblastoma patients.

Knudson didn't attempt to explain the nature of the hits that spawned retinoblastoma. The mutations might have damaged the same genes on the matched pairs of chromosomes, or they might have crippled two entirely different genes. "At that point there was no chromosomal data to go on," said Knudson, "and you can only speculate so far without running into the realm of fantasy."

Reality began to intervene in the mid 1970s, when cytologists localized the retinoblastoma trait to the thirteenth chromosome. They noticed that in about 10 percent of the patients, one of the two copies of chromosome 13 was truncated. If the patient had the germinal form of the disorder, the thirteenth chromosome was abnormally short in all body cells; if the patient suffered from sporadic retinoblastoma, only the tumor cells possessed midget copies of the chromosome. With the improvement of chromosomal staining and banding techniques in the late 1970s, retinoblastoma researchers further elucidated the chromosomal aberrations. They determined that when a portion of the thirteenth chromosome was missing, it was always in the same q14 region, three quarters of the way down the chromosome. But researchers were perplexed. If there was a link between chromosomal deletions and retinoblastoma, why did they find such deletions in only one out of ten victims? Why couldn't retinoblastoma be like, say, Down's syndrome, in which all the afflicted have one extra copy of chromosome 23? Why should the chromosomes of 90 percent of retinoblastoma victims look healthy?

In 1979, the guilt of the thirteenth chromosome took on a new twist. At New York Hospital in Manhattan, affiliated with Cornell Medical School, Robert Ellsworth, David Kitchin, and David Abramson, who collectively treat more retinoblastoma children than anybody else in the world, observed a tragic phenomenon. Almost 20 percent of their germinal-retinoblastoma patients, once cured of eye cancer, developed bone cancer when they were teenagers. Osteosarcoma in the general population is very rare, so the New York doctors assumed that there was a genetic connection between the retinoblastoma trait and bone cancer. To prove the link, David Kitchin examined osteosarcoma tissue from patients who had never had eye cancer; like retinoblastoma, osteosarcoma can arise spontaneously. (Edward Kennedy, Jr., suffered a spontaneous osteosarcoma.) Astonished but triumphant, Kitchin discovered that in about 10 percent of the spontaneous osteosarcoma tissue, part of the q14 band of the thirteenth chromosome was gone.

No quibble about it, the q14 band was significant to both eye and bone cancer. But at the time, few people noticed: Kitchin's paper was published in a modest genetics journal of small circulation. Weinberg, Wigler, and the rest dominated the prominent journals and the media with news of the *ras* gene and point mutations. The smattering of researchers who studied retinoblastoma knew that the *ras* model of cancer could not explain their specialty. DNA from retinoblastoma tissue, when transfected into normal cells, didn't produce foci. Besides, there were those deletions of q14. Deletions implied not genes turned on but genes departed.

The idea that retinoblastoma is a recessive cancer — a cancer of two missing genes rather than of genes gone awry — was another one of those ideas which occurred to several people simultaneously. Al Knudson thought of it; so did Ray White and Webster Cavenee of the University of Utah, and Brenda Lee Gallie of the University of Toronto. Buried within the q14 region of chromosome 13, they decided, was a gene necessary to stanch the growth of eye and bone cells. And when part of the band is deleted, the gene is lost, along with whatever genes and junk DNA surround it. Al Knudson's two-hit model proved uncannily accurate. Before a cell could lunge toward cancer, it must lose both copies of its retinoblastoma gene, one on the father's chromosome 13, the other on the mother's chromosome 13.

All that remained to clinch the recessive model was to account for the 90 percent of retinoblastoma and osteosarcoma tumor cells that had no chromosomal deletions. Web Cavenee, who at the time was a postdoc in Ray White's lab, took up the challenge. Isolating normal versions of chromosome 13, he picked off pieces of the chromosome at random and crafted the segments into about two dozen DNA probes. He then applied those probes to the chromosomes from fifty samples of eye and bone cancer tissue. He was seeking submicroscopic abnormalities, signs that chromosomes with no large-scale deletions of q14 were nevertheless damaged.

Cavenee had a grand slam. *In every single case,* his DNA probes revealed severe biochemical trouble on chromosome 13. "Usually, a scientist expects to see something five percent of the time, or fifteen percent of the time," said Cavenee. "But after you see it again and again and again, you think, Hey, there's something going on here." The majority of tumors had become what geneticists call homozygous for chromosome 13. One copy of chromosome 13 had been thrown out, and the remaining copy had been reduplicated in its stead. Thus, the tumor cell had two copies of, say, the mother's chromosome 13 and no copies of the paternal chromosome.

"Homozygosity is a red flag for geneticists," Cavenee explained. "When genetic diversity is lost, that usually means that recessive, 'bad' genes are allowed to come to the forefront. In the case of retinoblastoma, what we can assume is that there was a nonfunctioning copy of the retinoblastoma gene on one chromosome and that it was reduplicated when the good chromosome was lost. The cell was left with two nonfunctioning copies of the gene."

The evidence that retinoblastoma is a recessive cancer was now air tight. An eye cell had to lose both copies of its retinoblastoma gene before it turned cancerous. The cell might lose a whole copy of chromosome 13, or it might lose only a tiny, pointlike piece of the q14 territory. In any case, the retinoblastoma genes had to be killed.

Cavenee's paper on the fifty tumor samples was published as a major, eight-page article in *Nature* in 1983. The decision by the *Nature* editors to highlight the report signaled a sea change. Retinoblastoma no longer was a subject of limited concern. It was the "wave of the future," the flip story of the *ras* gene. If scientists were stymied by dominant oncogenes, perhaps recessive oncogenes would help reinvigorate the study of human tumorigenesis. Perhaps scientists needed to adopt a parallel tack, pursuing the function of dominant and recessive cancer genes simultaneosly. Perhaps there could be no breakthrough in understanding genes like *ras* and *myc* without an understanding of the genes that conspire against *ras* and *myc*.

Scientists had to decipher the retinoblastoma gene. And to manage that, they had to clone the gene. Web Cavenee's *Nature* paper set off the retinoblastoma race.

"The people who work in retinoblastoma have a long history of cooperating with each other," Cavenee told me. "We've shared our data, our reagents, our tumor samples, our probes. We talk to each other on the phone all the time. I don't mean to sound Mary Poppinsish about it, and we're not all best friends by any means. But it's not like Ford and GM or the people in the AIDS field. There's no cutthroat competition about it. It's really been a pleasant enclave in biology."

On the night that Steve took the Whitehead scientists to the dog track, some of us trundled down to the stables to examine the greyhounds that would be running in the Whitehead Event. The animals were anorectically thin, but they yipped, wagged their tails, and stuck out their tongues just like any other frisky, friendly dogs. They looked so cute that I wanted to pet one, but when I asked a dog owner for permission, he laughed and yanked on his dog's leash to make sure I wouldn't try.

"Right before a race," he said, "these fellows are so hungry to get out on that track, they'll take your thumb off before you know what hit you."

I have to be gravely ill before I can be persuaded to visit a doctor, but that's because there are so few doctors like Ted Dryja, an ophthalmologist at Massachusetts Eye and Ear Hospital in Boston.

He's a dream doctor, a genuine humanitarian. Dryja treats many retinoblastoma children, and he is forced to subject them to unfathomable pain, but he is so gentle that I could imagine him soothing a panicked child with a soft touch and a murmur of empathy. He looks almost like a child himself. He has huge blue eyes, baby-fine hair, and the poreless, cherry-tinted skin of a boy.

Dryja is a pure M.D., with no formal education in molecular biology, but after completing his training in ophthalmology, he decided that it might be worthwhile to learn something about DNA. By then, he had already enucleated more retinoblastoma children than he cared to recall, and he wanted to apply any basic research skills he might acquire to trouncing the disease he despised. Working with Marc Lalande and Samuel Latt at Harvard Medical School, he began studying chromosome 13. He wasn't doing anything too fancy. He was just cloning little segments of the chromosome, each piece measuring about 4000 bases. Chromosome 13 has about sixty million bases of DNA, so each sequence that Dryja isolated represented a tiny fraction of the whole. "It was a way of getting my hands dirty," he said. "I was just learning the basics of recombinant DNA technology." After he had cloned the pieces into separate viral vessels, he would go back and figure out approximately where on the chromosome the fragment had come from. That's the retro order in which chromosomes must be studied: first, sequences are tugged off at random; then their position on the chromosome is mapped.

As his skills improved, Dryja began collaborating with Web Cavenee. Dryja cloned several of the DNA probes that Cavenee used to demonstrate the loss of genetic diversity in retinoblastoma cells. The collaboration was short-lived. After the publication of the *Nature* paper — on which Dryja was the second of many co-authors — Cavenee became quite famous, but Dryja remained relatively obscure. He felt that Cavenee wasn't giving him enough credit. "He talks about work that's a year or two old as though it were something brand-new," said Dryja, "and he never bothers to mention the people who helped him with that work."

Dryja decided to strike out on his own as a basic researcher. He wanted to find the retinoblastoma gene. For his experimental approach, he continued doing more or less what he'd done before, picking off pieces of chromosome 13 at random and then identifying their place of origin on the chromosome. He also attempted a bit of "chromosome walking." After localizing a probe on the chromosome, he'd clone additional segments situated to the left and right of the initial probe. Dryja was hoping either to pull off a piece of retinoblastoma gene on the first round or to "walk" next door to the retinoblastoma gene once he'd found the chromosomal address of the initial segment. But he would need the luck of a leprechaun to succeed. The odds of his happening onto the retinoblastoma gene were about one in ten thousand.

In early 1985, when Weinberg was mulling over the problem of retinoblastoma, Weinberg and Hucky Land paid a visit to Ted Dryja at the Eye and Ear Hospital. On hearing of Dryja's approach, Weinberg practically snorted with laughter. "I said to Hucky, 'Hucky, this is a method that has virtually no chance of succeeding,'" Weinberg recalled. "Dryja was taking a strictly structural approach, jumping onto the chromosome and walking along, inching along, hoping that he would stumble on the gene. But the chromosome is vast, and it was unclear to me how Dryja would know if he got the gene anyway. He'd have no functional assay that would tell him he was in the right gene. I thought it would take him ten years at the minimum."

Dryja told Weinberg that he believed he'd know when he got there. He reasoned that if he did isolate a piece of the gene, he should be able to find a retinoblastoma patient who was missing that piece.

"Admittedly, the odds against that were as bad as the odds of cloning the right probe in the first place," said Dryja. "What I needed was a patient who had homozygous deletions — who was missing the segment on *both* copies of chromosome thirteen. But most patients don't have double deletions. Usually what you see, in cases where there are deletions, is a deletion on one chromosome and the other chromosome of normal size. Presumably the second copy of their retinoblastoma gene has been knocked out by a submicroscopic mutation. I wasn't even sure that there was such a thing as a double deletion, but that's what I'd need for my assay to work. If I isolated the DNA from tumor cells, and those cells had any version of my probe DNA remaining at all, I wouldn't be able to tell that I had the right probe. But I thought there was no harm in trying."

Even for Bob Weinberg, Dryja's scheme was too far-fetched. Bidding Dryja the best of luck, Weinberg and Land strolled back across the Longfellow Bridge to the Whitehead.

Although there was no easy or obvious tactic for cloning the gene, Dryja's competitors were somewhat saner than he — that is, they relied less on the whims of fate and more on brute force. In 1983, Robert Sparkes of the University of California, Los Angeles, found a chromosomal marker that was within shooting range of the retinoblastoma gene: the gene for an enzyme called esterase-D. He didn't know what the enzyme did in the body, and he didn't much care. Sparkes was interested in the esterase-D gene only because it, too, seemed to be located in the q14 band of chromosome 13. And Sparkes knew that, with the isolated esterase enzyme in test tube, it would be a straightforward job to work backward from protein to the gene that decrees its production, a maneuver that could not be performed for the mysterious retinoblastoma gene and its equally mysterious protein. Sparkes and his colleagues, William Benedict and Wen-Hwa Lee, cloned the esterase-D gene in mid 1983, but they kept their coup quiet for a few months. Wen-Hwa Lee wanted a head start in exploiting esterase-D as a guide to the retinoblastoma gene. He began walking along the chromosome from the position of the enzyme gene. He walked like a cartoon man crawling on all fours across the desert. The q14 band is five million bases long. Lee didn't know which direction the retinoblastoma gene was in, whether it was 10,000 bases away or a million. But always he had the esterase-D gene as his reference point, his beacon, which is more than Dryja had for his chromosomal creepings. "Our technique was very methodical, very careful," said Lee. "Dryja needed luck. We just needed time, and we knew we'd get there. Step by step, we'd get there."

In 1984, before the Sparkes report on esterase-D was published Brenda Gallie and Robert Phillips of the University of Toronto independently cloned the gene for the same enzyme. But they opted against anything so daunting as walking along the chromosome from esterase-D. Instead, they scrounged about for extra markers in q14. They hoped that if they found a scattering of genes across the band, the markers collectively would point them toward their quarry. Or, rather, they flailed around in the lab while discussing the possibility that such an approach might work. "Quite honestly, we didn't have a clear idea of how, exactly, we could get the gene," said Gallie. "We knew it was a race, but it seemed to us that everybody was in the same position. We were all in the dark."

And then, in 1986, Ted Dryja called Gallie and said that he needed her help. Immediately. He thought he was on to something, but his lab was too small to prove it. Could Gallie please analyze one of his DNA probes and tell him whether what he thought was true really was true? Could she see whether he had a piece of the retinoblastoma gene?

• • •

Ted Dryja's lab is worse than small. It's a closet. He's lucky to have the lab — Mass Eye and Ear is more a hospital than a major research center — but good fortune doesn't make the space any less claustrophobic. When I interviewed him in his office, a little cubbyhole at the rear of the lab, I felt that if I made any sudden moves, I'd bruise myself on a filing cabinet or fall out the window.

"There are three others besides me in the lab," he said. "As you can see, we're not the Whitehead Institute. We're not equipped to do RNA analysis or anything too elaborate. We can do some cloning, but there's a limit. So when I got my probe, I thought I'd be better off collaborating with somebody who had the space and resources to do Northern blots or whatever else was necessary. I wasn't going to take no for an answer."

Indeed, Dryja had been battling the no's for eighteen months, including some no's of his own. He'd pinch off a piece of chromosome 13, walk briefly to the left and right of the probe and pinch off a few more pieces, apply the probes to tumor DNA, and throw the whole slop away: the tumors had the probe sequences. Then he'd leap onto the chromosome again, scoop out another fragment, walk a little, test the clones — and throw them away.

Ted began to suspect that he was wasting his time and should stick to treating patients. He wasn't cloning the retinoblastoma gene. He was cloning trash. He was only in his early thirties, but he began to imagine himself after three decades as a retiree in West Palm Beach, still picking off pieces of chromosome 13 and tossing them away.

Dryja's daymares about his future evaporated virtually overnight. One morning he came to the lab, glanced at the exposure of a DNA gel he had set up the evening before, and knew, with the rare, pure intuition that slays commonplace caution, that he had the right probe. The gel had thirty lanes' worth of data on it, the DNA from thirty retinoblastoma patients. The probe that he'd applied to the tumor DNA was designated H3-8. And two of the thirty lanes were blank. The probe had not found its match in the cancer cells: the thirteenth chromosomes of those two patients were doubly deleted, and among the items missing was the sequence corresponding to Dryja's H3-8 segment. "I couldn't sleep for a week after I saw that gel," he said. "I mean that. I didn't sleep."

Dryja realized that the H3-8 probe might not be a piece of the retinoblastoma gene; it could be part of a neighboring gene that had been deleted along with the cancer gene, or it could be something else altogether — coincidental chromosomal loss entirely outside the q14 band. But he didn't believe the skeptics' arguments. He believed he had the gene by the sleeve. His probe could be used to clone the whole retinoblastoma gene. If only his lab were more amply equipped, he could take the project to its glorious conclusion on its own. But it wasn't, and Dryja didn't crave the glory, anyway. He wanted the gene, and he was willing to share the credit with anybody who could help him.

Dryja called his friend Gallie in Toronto and sent her and Phillips the probe. "That's the most amazing thing about Ted," said Brenda Gallie. "He doesn't care about himself and winning races. He just wants the science to move ahead as quickly as possible."

A few weeks later, Gallie reported back to Dryja that she was very sorry, but she didn't think he had the right gene. She had screened both tumor cells and fetal retinal cells for signs of probe expression. If H3-8 were a part of the retinoblastoma gene, it should be turned into RNA in normal fetal cells and not be expressed in eye cancer cells. Gallie told Dryja that she didn't see expression in either tissue type, and she thought he'd cloned an inconsequential bit of the chromosome.

"I didn't believe her," Dryja said. "I knew the probe was significant."

Dryja wasn't sure whom to turn to next. He discussed his H3-8 probe with a number of his colleagues. He talked excitedly about the double deletions he'd seen in two separate tumor samples. And one of the people who shared his enthusiasm was Steve Friend. Dryja knew Friend from the days of Steve's internship, when Dryja had been the instructor of the tumor boards at Children's Hospital. Dryja was aware that Steve had begun studying retinoblastoma at the Whitehead, and the two men kept in touch. Steve's experiments weren't going anywhere yet, but he told Ted that he was getting the swing of the technology. He was still investigating the role of N-*myc* expression in retinoblastoma, and he was churning out one RNA blot after another. He had this terrific cell line of normal fetal retinal tissue, he said, that René Bernards had ordered from Holland. Steve had made all sorts of blots with the RNA of those cells.

In fact, he said to Dryja, it would be a trivial thing to throw your probe onto my blots and see if there's any hint of expression in the fetal retinal cells. The blots are just sitting there, waiting to be used. Why not let me give it a try?

Dryja wavered. Collaborating with Gallie had been one thing; she was his peer, no more nor less celebrated than he. If Dryja were to sacrifice his probe to the lab at the Whitehead, however, and if Steve were to unearth anything important with it, the event could be interpreted as another victory for the great Weinberg lab. Little Ted Dryja would be engulfed and forgotten.

Then again, if there was any group capable of cloning the retinoblastoma gene, it was the Weinberg team.

Before handing over his prize to Steve, Dryja met with Weinberg three times to forge the terms of the liaison. "I assured him that I would never try to steal his thunder," said Weinberg. "He'd done the great bulk of the work in getting the probe, and anything that came of it would be credited to him. I've already had my share of glory."

A week or two later, in late February of 1986, there was glory enough for all. Steve gave Ted the answer he wanted to hear: yes.

When Steve smeared the DNA of Dryja's probe onto the RNA of the adenovirus-treated fetal cells, he saw a smoky streak, meaning matrimony. When he introduced the probe to the RNA of retinoblastoma cells, he drew a blank. The probe was expressed in normal fetal eye cells but was silent in tumor tissue, just as the retinoblastoma gene should be.

"I think that from the moment we saw the expression," said Weinberg, "we knew we had the right gene."

They had the probe; now they had to clone the whole gene. They had to clone the gene before Wen-Hwa Lee marched into it from the esterase-D location. Lee had since left Bill Benedict's lab to hang his shingle at the University of California at San Diego, and Dryja heard that he was getting awfully close to the gene.

They had to clone the gene before Benedict, who was barreling ahead without the help of Lee. Benedict had twenty or thirty people working on the retinoblastoma project, and though he refused to be specific about his approach, he let it be known that he was very cunning.

They had to clone the gene before Brenda Gallie realized she'd been wrong and attempted to isolate the gene herself. She still had a copy of the H3-8 probe. Dryja's initial collaborator was now Steve Friend's competitor.

They had to clone the gene before Cavenee. Nobody knew what Cavenee was up to, but he certainly was talking a lot to the press.

It was all on Steve's square shoulders now. Dryja was busy enough back at Mass Eye and Ear. Steve had never cloned a gene before, but he had his wooing ways. This time, though, he had to suffer several rejections. René refused to be wooed; he was trying to do something that would get him a first-author paper. David had little more cloning expertise than Steve. Mike Gilman cared only for protein-binding assays. Cori was swept up in the *neu* gene. She did, however, have a suggestion: ask Snezna.

"Snezna knew as much or more about cDNA libraries than anybody else around," said Cori. "And I thought it would be good for her, too. I thought it would be good for her ego to be involved in the most important experiment in the lab."

In late winter of 1986, Steve asked Snezna, and Snezna, den mother and Mother Teresa of the Weinberg lab, dropped everything she was doing to devote herself to Steve. Goggle-eyed, tireless, as though wired on crack, they worked fifteen-hour days. Vivek didn't mind Snezna's schedule; he was a scientist and knew all about the narcosis of a pending discovery. Diane, Steve's wife, didn't mind, because her career as a classical horn player finally was taking off, and because she knew her husband. "For as long as I can remember," she said, "he's worked fifteen-hour days."

Snezna didn't exactly teach Steve how to clone a gene; Steve served as Snezna's hands. She'd tell him, now do this and this, and he'd do this and this. Once, when Snezna had to leave the lab for a few hours, she left a note for Steve with a little plastic tube taped to it. In the tube were a drop of water and a drop of oil. "Steve," said the note, "I can't be with you while you're doing the reaction, but it should look like this when you're through." Her tutorial tone was in order. Cloning the gene required building a cDNA library of all the messages in the fetal retinal cells, and with Canada and California snapping at their heels, Steve and Snezna couldn't afford to err. Snezna wanted the library to be perfect the first time around, so she fluttered over Steve nervously. Said she, "It was like trying to teach a hemophiliac to ride a bicycle."

By early spring, they had a working clone of the retinoblastoma gene, or at least what they thought was the retinoblastoma gene. They still lacked proof. Steve began hybridizing the clone to the DNA of retinoblastoma and osteosarcoma tumor cells. He was screening for tiny deletions of part of his gene. Ted Dryja, when first exploring the possibility that his probe was deleted from tumors, had simply sought large-scale double deletions that encompassed the probe DNA. Steve needed submicroscopic excisions of portions of his gene alone. He located three or four examples of chromosomal deletions that carried off part of the clone along with surrounding DNA. But the deletions disturbed him. Although each took away a fragment of the gene, they were all unidirectional: they deleted only the left half of the putative gene and continued leftward through neighboring chromosomal DNA. There was no logical reason that a random chromosomal deletion shouldn't affect the right half of the retinoblastoma gene and rightward DNA. In fact, Steve's data suggested that the true retinoblastoma gene just might lie to the left of his clone. Still, he was optimistic, since each time he did find a submicroscopic deletion in cancer tissue, it included a chunk of his clone.

This was the status of the project when Steve first shared the big secret with me. He and René agreed on the sort of proof that was needed: a deletion that swept into the gene from the right. If he had a right-sided erasure that had resulted in eye or bone cancer to complement the left-lying deletions, he could confidently claim in print that however the gene is disrupted, it's his clone that must be deleted.

That afternoon I had lunch with Weinberg in the cafeteria. "Steve told me about the retinoblastoma gene," I said. "I think this is the greatest thing I've ever heard."

"Yes, we're extremely fortunate," he said. "The whole thing just came to us from out of the blue. Did Steve tell you the story behind it?" I replied that he had.

"There's just one thing I'm worried about," he said, slowly sipping his soup and assuming his most solemn expression. "When this discovery comes out in public, I want to be sure that Ted Dryja gets all the credit. He's done all the hard work, finding the probe and walking along the chromosome. He's the one who's been slaving away at it for three years. Steve Friend, while he has done an admirable job, has only been working on this for three months. I'm very concerned that Dryja receive all the credit due him. Do you get my drift?"

I nodded.

"It would be very easy for people to think that somehow I'm responsible for this and say, 'Well, Weinberg, you've done it again.' But I'm not responsible. I've been a bit player, an innocent bystander, you might say. It's Ted Dryja who should be showered with glory. Yes, I'm very concerned about how this might be interpreted."

Weinberg looked so distressed that I began to feel we were talking about Dryja as though he had some sort of fatal disease. I timidly if presumptuously offered the opinion that scientists, at least, are generally good about dispensing proper credit to their peers.

"Perhaps," said Bob. "I'll certainly do my best to emphasize his role and de-emphasize my own. And I'll do the same in any discussion I may have with the press."

I told him that I would write that Bob Weinberg had nothing to do with the cloning of the retinoblastoma gene, at which point he changed the subject and asked me whether I'd been up to Fitzwilliam lately.

The following week I went out to Cold Spring Harbor. One of the postdocs was leaving the Wigler lab, and I wanted to say goodbye. Everybody demanded to know if I'd been back to the Whitehead. I said that I had, and that things were going very well there.

"What do you mean, 'very well?'" asked Scott Powers.

"Just that," I said, recalling the admonition I'd received to keep everything secret.

"Have they found the function of *ras* yet?" he pressed.

"No, nothing new on the *ras* gene."

"Are you talking about the retinoblastoma-associated sequence?" he said.

I stared at him and then burst into laughter.

"Mike told us that one of the Weinberg postdocs has cloned a sequence that seems to be associated with retinoblastoma deletions," said Scott.

"Who told Mike that?" I asked.

"I don't know. Probably Weinberg. They talk a lot on the phone, you know. So what more can you tell us about it? Should Graeme be worried?"

I insisted that the whole thing confused me and that I didn't understand it. But what I really didn't understand was scientific secrets. I hadn't yet heard a secret that lived up to its definition.

By late April, most people in the oncogene field seemed to know about the retinoblastoma gene. Retinoblastoma specialists knew about it in intimate detail. Steve Friend had met Bill Benedict at a conference in Colorado that February, and now Benedict felt justified in calling Steve once or twice a week to "see how he was doing." Steve would be in the middle of an experiment, dashing from one bay to another and balancing petri dishes in either hand, when the phone would ring: Bill Benedict for Steve Friend. Steve was polite but cagey. Benedict asked Steve, just out of curiosity, what Brenda Gallie was doing with the H3-8 probe. More politely still, Steve replied that he hadn't the faintest idea. "Benedict can be a pain in the neck sometimes," Steve muttered.

"I met him a couple of months ago," said Cori. "You know what sort of guy he is? He's the kind who wears string ties and a cowboy hat and walks around slapping everybody on the back."

"Is he jealous?" I asked Steve. "Is he close to cloning the gene, too?"

"If he were," said Steve, "he sure wouldn't tell me about it."

By May, Steve obtained a gorgeous piece of evidence that he had the right gene. It was a retinoblastoma cell line that had a tiny deletion, not creeping in from the starboard side of chromosome 13, but smack in the middle of his cloned gene: an internal deletion. The only thing wrong with the patient's cells was an absence of a sliver of Steve's clone; the rest of the chromosome appeared to be normal. Steve tried to temper his excitement. It was only one example from one patient. Any number of chromosomal anomalies or experimental artifacts could explain the deletion, none of which would concern retinoblastoma.

Steve needed more tumors. Most of the bone and eye cancer cells he tested displayed no deletions of his clone, internal or otherwise. Such results didn't surprise him. He assumed that undetectable point mutations, not deletions, were responsible for knocking out the majority of retinoblastoma genes in patients. But point mutations and theories couldn't prove his case for the clone. Because retinoblastoma and osteosarcoma are rare, his pickings were slender. So when he heard in June that a preteen boy at New York Hospital had just had an osteosarcomic lump removed from his leg, Steve didn't want to trust potential evidence to Express Mail. Scott Dessain, the technician, flew down to New York to fetch a sample of the tumor.

The moment Scott returned, Steve isolated the bone cell DNA and blotted it with his clone. The autoradiograph of the reaction was ready two days later, and the news flashed along the Weinberg corridor as flames spring up through a dry forest floor: Steve had found another internal deletion. The boy from New York was chromosomally normal save for a small gap in the middle of the DNA corresponding to Steve Friend's clone.

"Some people have all the luck," said René. "Internal deletions are probably the least common way to kill the retinoblastoma gene, and Steve's got two of them."

Steve presented his results at that week's group meeting, marking the first time Weinberg had heard all the latest news. Weinberg asked many pointed questions. Some of Steve's data bothered him. Why, for example, did Steve never find expression of his gene in any retinoblastoma cells? If point mutations often disable the gene, the DNA should still be transcribed into RNA, even if the messenger RNA couldn't be turned into a working protein. But Steve hadn't seen a single case where there were abnormal transcripts of the gene; instead, he'd seen no RNA at all. Steve suggested that the gene was so sensitive to mutations that even the smallest error prevented its initial transcription. Weinberg wasn't convinced. True, it was better to see the cloned retinoblastoma gene consistently turned off in eye and bone cancer than to see it always turned on, but the expression studies weren't entirely satisfactory. "It's very perplexing," he said. "I'd be happier if we occasionally found abnormal transcripts."

More confounding, Steve had seen expression of the gene where one might have predicted there would be nothing: in a variety of unrelated control tumors from the lung, bladder, and kidney. According to the model, the retinoblastoma gene should be tissue-specific, switching on in normal eye and bone cells to damp the growth of those organs alone. Yet Steve found gene activity in malignancies of many different tissues.

"The model would predict that the retinoblastoma gene is specific to retinal and osteo cells," said Weinberg. "Why should the gene be expressed in all these tumors?"

Maybe the model is wrong, Steve argued. "Why shouldn't it be expressed all over the place?" he said. "It could be that expression in these cases is meaningless. It just happens. It could just be random transcription. In any case, there is *no* transcription in retinoblastoma and osteosarcoma, and that's what matters to us."

Weinberg emitted a few humphing noises. "It's all very confusing," he said. "Obviously we don't know everything that's going on here."

Steve's DNA results were far happier. "The internal deletions are very persuasive," Weinberg concluded. "I think they're our best evidence in favor of the clone."

René said that he thought the RNA work simply needed some fine-tuning, and Weinberg agreed. He ended the meeting with a plea for secrecy. "I know a lot of this has begun to leak out," he said. "But I think we should get ourselves onto firmer ground before we say anything more about it. I entreat everybody to keep this information restricted to those in this room. Please." Steve seconded the call for silence.

Afterward, I asked Weinberg how he felt about the retinoblastoma gene. "Before today, I was skeptical," he said. "Now, I'm seventy-five to eighty percent sure we have the right gene. The internal deletions are very, very good." He nodded his slow nod of conviction. "I think recessive oncogenes are going to open a new chapter in oncogene research. A major new chapter. I predict that the story of recessive oncogenes will be as important as the original dominant oncogenes were." Coming from the man who helped found the field of dominant oncogenes, that was high praise indeed.

Over lunch, Shelly Bernstein said to Steve that there was no doubt about the gene left in anybody's mind, and everyone at the table agreed. A current seemed to course through the Weinberg scientists. They weren't jealous of Steve's success; they were delighted with it. "I couldn't wish it on a nicer guy," said David Stern. Their communal joy reminded me of a comment Weinberg had made several months earlier: "When a person is well liked in the lab, that person's triumph is a crown for all to wear." (Conversely, he added, when the person isn't so popular, "it may be a crown of thorns.")

Only one person felt left out. I stopped by Snezna's bench after lunch to ask her why she hadn't had anything to eat yet. She'd looked gloomy during the group meeting; now she looked as dark as a nimbus cloud. "I'm not hungry," she said.

"Is there anything wrong?" I asked.

"Oh, it's nothing. It's just Steve."

"You don't think he has the right gene?"

"No, no, I think he has the gene. It's just that . . . I mean, I'm very happy for his success, it couldn't have happened to a sweeter guy, but I wish he would slow down sometimes. I wish he would stop and take the time to thank people for what they do for him. He's always rushing around in a mad heat. And he talks about this thing as though he'd done it all. He could never have done this himself. He needed help from a lot of people, and it's time he admitted that." She stopped to wipe an angry tear from her eye.

"This will sound really petty, but I was so upset at group meeting that I almost walked out of there," she continued. "He was just talking on and on, as though he were the big man who had cloned this gene all on his own. And he didn't even make any sense. Could you follow what he was saying?"

I told her she was asking the wrong person.

"Well, he didn't make any sense to me. And then when he asked everybody not to talk about this outside the lab! He's the one who's been telling everyone he knows. He'd tell a stranger he met on the street about this gene.

"I'm worried about Steve. He's working so hard that he's running himself into the ground, and he's also starting to alienate people. Believe me, I'm not the only one to feel this way. Other people are beginning to complain about him, too. He expects that it's natural for people to neglect their own work to help him, but it's not, and it won't last."

Steve was busy and crazy and running himself into the ground like a repair crew's jackhammer. He was too busy to take time out for jogs around the Charles River or weekend excursions to mud-wrestling contests. He was too busy to worry about people's feelings. His wife was pregnant with their first child, but Steve was too busy even to think about fatherhood. He was a first-year postdoc who had just made what could be the biggest discovery of his scientific career. That was all he thought about. He wanted to tighten his retinoblastoma data and get out the paper. Later, he would discuss with Diane the decoration of the baby's bedroom. Later, he would make up for his neglect of Snezna in the best way he knew how. But for now, he said, "I just want to get this paper out."

As Steve cloistered himself in the lab and prepared the retinoblastoma report, news of the SECRET continued to spread. That summer, when I attended the annual oncogene meeting in Frederick, Maryland, I overheard a discussion between Scott Powers and Mitch Goldfarb, now an associate professor at Columbia. Mitch was flipping through the conference proceedings in obvious irritation. "I thought there was supposed to be some announcement about the retinoblastoma gene," he said to Scott. "But I don't see anything about it listed in the program."

René and Cori had come to the meeting to discuss their own experiments. They were under strict instructions from Weinberg not to say anything about the retinoblastoma gene other than "We have isolated a fragment of chromosome thirteen that seems to be associated with the retinoblastoma locus, and we are now in the process of trying to clone that fragment." They were called on to repeat the line many times throughout the meeting, as scientists cornered them and tried to extract information. Inder Verma, a principal investigator at the Salk Institute who had heard extensive rumors about the gene, asked René very detailed questions, but all René would tell him was "We have isolated a fragment of chromosome thirteen . . ." Not easily rebuffed, Inder Verma buttonholed Cori. Said Cori to Verma, "We have isolated a fragment of chromosome thirteen . . ." Mariano Barbacid was so desperate that he asked *me* about the retinoblastoma gene.

"I didn't even know Bob was interested in retinoblastoma," said Barbacid, almost peevishly. Earlier in the meeting, Barbacid had become somewhat defensive when a speaker credited Bob Weinberg alone with discovering the point mutation in the *ras* gene. Barbacid likes to remind people that his report on the point mutation was published in *Nature* along with Weinberg's. Now Bob Weinberg was striking a mother lode again — and in an entirely new area of oncogene research. It wasn't fair.

On the last day of the conference, I talked to Tony Hunter, the meeting chairman, a gentle-voiced Briton with a wild beard that reached almost to his waist. I was aware that Tony knew of the retinoblastoma gene, but I wasn't sure how much he knew. "Have you heard everything about it?" I asked. "All their latest results?"

"Oh, yes, I think I'm fairly up to date on the details," he said. "I've also heard that somebody else has cloned the gene."

I gawked at him. "You're kidding."

"No, it's a guy out in San Diego," said Hunter. "Have you heard of him? His name is Wen-Hwa Lee."

13

Through the Looking-Glass, and What May Be Found There

THE NEXT WEEK I went back to the Whitehead Institute. Hunter hadn't known how close Wen-Hwa Lee was to proving that he had the retinoblastoma gene. All he had learned was that Lee had cloned the gene by walking along the chromosome for 200,000 bases from the esterase-D gene. I was wondering whether the Weinberg people would be upset or nervous about this gale wind from the west. They weren't: Wen-Hwa Lee didn't have the chance of a Dove Bar in a microwave. "You're just in time for another great moment in science!" René called down the corridor, pointing his finger at me.

"What is it?" I said, running down to his bay.

"Steve is getting ready to send off his paper to *Nature,*" said René. "The delivery guy is coming at five-thirty to pick it up, and it's flying to England tonight."

A blur of color rushed by, which I made out to be Steve Friend. "I have to go see about the figures," he jabbered as he flew. "They're not up from the art room yet." The graphs, autoradiographs, diagrams, and restriction maps that accompany a scientific paper are usually the last items to be completed.

"Bob has been on the phone to Peter Newmark, the editor of *Nature,* three times the last couple of days," said René, "trying to set this thing up."

"What do you mean, 'set this thing up'?" I asked.

"I guess he wants to make sure that Newmark realizes how important the paper is," said René.

"So what did Newmark say?"

"He didn't make any commitment," said René. "That guy's got guts. Imagine wanting to see the paper before he accepts it."

That day, Bob Weinberg was out of town, as he often is during the summer months, but he called the lab several times from his wife's family's home in Vermont to make sure everything was proceeding smoothly. Ted Dryja had come over from Mass Eye and Ear to help with the finishing touches. He and Steve retired to the word-processing room to compose the abstract — the little summary paragraph that appears at the head of the paper. The abstract was startlingly bold. Throughout the body of the paper, including the lengthy title, the authors had scrupulously avoided saying that the gene they had cloned was THE retinoblastoma gene. They knew they didn't have the absolute proof that would be demanded to make the claim; such proof would require their demonstrating that the protein from their gene could revert retinoblastoma cells to normal eye cells, and Steve was a long way from having attempted that sort of complicated experiment. All he had was a piece of DNA that was expressed in fetal eye cells, wasn't expressed in eye cancer cells, and was occasionally deleted from retinoblastoma and osteosarcoma chromosomes. Though it looked like the right gene, the authors had to be linguistically ginger. But the first sentence of the abstract declared: "Based on its chromosomal assignment, a cDNA segment has been isolated which is homologous to the transcript of the retinoblastoma gene." In plain English, "We have cloned the retinoblastoma gene."

"What's *this?*" cried René, as he read a typeout of the abstract. René was the second author on the paper, and therefore had a personal stake in its accuracy. "We went through all this trouble not to call it the retinoblastoma gene, and then BOOM, here it is. Bob's going to be really upset when he sees this. *I'm* really upset. Steve!"

Steve came over to René's desk. His face was flustered-pink, the shade of André's Cadillac. "What's wrong?" he asked.

"Does Bob know about this abstract?" René demanded. "Does he know we call it the retinoblastoma gene?"

"Of course he knows. He helped us write it over the phone," said Steve. "Don't worry. The editors will probably flag us on it anyway."

Steve mimeographed copies of the complete paper and handed them out to several people in the lab for last-minute comments and corrections. Cori Bargmann and I went into Bob Weinberg's empty office to read it. She sat at Bob's desk, hunched intently over the paper, her pencil in hand. She circled references, made proofreading corrections, compared one paragraph with another. As she was finished reading a page, she passed it on to me. Although I'm embarrassed to admit it, I have rarely felt so excited. It may not have been like reading the Magna Carta from King John's hand with the ink still wet, or Watson and Crick's original paper on the structure of the double helix, but it probably was as close as I was going to get to a bit of history off the presses.

Cori hurried out to find Steve. "Do we still have time to make changes?" she asked him. "I think there are some things in the figure legends that need to be corrected." Figure legends are the explanatory lines below the diagrams, autoradiographs, and so forth. Scientists often don't bother reading the legends, preferring to make their own interpretations of the data, but Cori, being close to perfect, was a perfectionist.

"Sure, sure, we have plenty of time," said Steve. "The delivery guy won't be here for forty minutes." He and Cori hurried into the computer room.

When they were gone, René told me about the cover letter that Weinberg had written to Peter Newmark to accompany the retinoblastoma paper. Bob had made two stipulations. The first was that *Nature* publish the report as a major article, in the front of the journal, rather than as a shorter Letter to *Nature,* as run-of-the-mill papers are presented. He also said he wanted the paper to be published on its own and not with any related papers about retinoblastoma.

"Bob wrote that?" said Julian Downward, formerly of Michael Waterfield's lab in London and now a postdoc at the Whitehead, with a workbench right across from René's. "Isn't it a bit unprecedented for somebody to tell a journal what they ought or ought not publish?"

"Yeah, well, now we're talking hardball science," said René.

The three of us then discussed that all-important topic, authorship order. The two big billings had gone to Steve and Ted. Steve was first author; Dryja was second. "For a while we were wondering how Bob was going to handle that," said René. "Steve was worried that Bob would end up being last author, but I guess Bob kept his promise to Ted. Ted's getting the credit." I was happy, but surprised, to see that Snezna was third author.

"And Snezna got credit, too," I said.

"Yes, how did that happen?" Julian asked René. "How did she go from being 'gratefully acknowledged' to being a full author?" Often, people who helped with scientific experiments but who didn't do enough to merit a position on the marquee are thanked in a paragraph at the end of a paper; the paragraph is in very small print. "Who wanted to put her in the 'gratefully acknowledged' section?" I asked.

"Bob," said René. "He thought that making a cDNA library wasn't a huge intellectual contribution, and he thought that too many authors would take away from the importance of Steve and Ted. He said that he thought those papers you see with three hundred authors listed look ridiculous and false."

"So how did it get changed?" asked Julian.

"Steve asked for it. He was really insistent about it. He told Bob that Snezna had stopped her own experiments to help him with the library, and that if it weren't for her, he would never have been able to do it so quickly. Bob said, Okay, if that's what you want. So if you ask me, Snezna got a pretty good deal out of her cDNA library — authorship on the biggest paper from the Weinberg lab in two or three years."

Steve came back to the bay. "I've got the final cover letter here," he said to René. "Do you want to read it?"

"Ooh, can I see it too?" I said sweetly. He grinned and clasped it against his chest. I put my hands together in supplication. "Please?"

Steve shook his head. "Uh-uh," he said. "You'll know what's in it —"

"I already know," I interrupted.

"Exactly," said Steve. "I knew you would. But that doesn't mean you should see a hard copy of it; it's too juicy."

The cover letter wasn't the only place where the Weinberg scientists gloated over their victory. "Have you taken a look at the height wall lately?" Steve asked me. I went to look at the wall along the Weinberg corridor where, tradition decrees, all young Whitehead scientists have their heights marked in ink. For a long time, the tallest mark belonged to Lewis Stout, a six-foot-seven postdoc in David Baltimore's lab. Lew was beaten when the allowable entries were extended to visiting scientists, and a six-foot-nine Dutch biologist happened to be passing through the lab. Before the cloning of the retinoblastoma gene, the shortest slash on the gauge was that of a four-foot-eleven technician. But now I saw that other names had been written in below. Web Cavanee squatted at a two-foot mark; Bill Benedict was below Cavanee, at about one-foot-eight; and Gallie sank to the bottom, a mere ten or twelve inches tall.

Bob Weinberg called again at 5:10 and asked to speak to Steve.

Cori was still in the computer room with Ted Dryja, going over the figure legends. At 5:30, the man from DHL Worldwide Express arrived to pick up the paper, but it still wasn't ready. The driver said that he had to get the package to a plane that was leaving at 6:30, and he reminded the frazzled scientists that it was rush hour in one of the most traffic-clogged cities of America. "It's a good thing I know the back roads to the airport," he growled.

At 5:45, the paper was dispatched, and the tension in the lab snapped like a severed banjo string. Steve printed out beautiful, professional-looking copies of his opus to give as presents to all the people who'd contributed to the project. I asked him for a copy, and he handed me one. It was still warm, like freshly baked bread. Steve slumped into a chair and expelled a breath beyond exhaustion. "I'm pretty happy with it," he said. "I think it's a good paper."

"Who actually wrote it?" I asked. "A lot of it sounds like Bob."

"Maybe eighty percent of it is Bob," said Steve. "Ted and I divided up the rest."

"Bob was sitting at the word processor most of last week, writing this thing," said René. "But that's why it reads so well. He's a really good writer."

"There are some parts of it that I argued against," said Steve. "And I even won a couple of arguments. I got him to take out the word *anti-oncogene*. That made me happy."

"Why were you against it?" I asked.

"What does it mean?" said Steve. "Anti-oncogene? It's a flashy word, but it's meaningless. And we don't know if it's an anti-oncogene. We don't know for sure that it regulates N-*myc* expression, or whatever. I don't like that kind of catchy term." He pointed out one or two other sentences that he thought he'd have to correct when the galleys were sent to him.

"Brenda Lee Galleys," said René. "I can see there are going to be a lot of puns about that one."

"So do you think the reviewers are going to flag you on a lot of points?" I asked.

"You know what Cori said?" Steve replied. "She thinks that they're not going to bother sending the paper out to reviewers. She thinks they'll accept it in four days" — the minimum turnaround time for *Nature* — "but that they'll hold the paper until September, when everybody is back from vacation and it will have the greatest effect. And as we all know, Cori is usually right." In other words, Cori Bargmann predicted that the editors at *Nature* would be so impressed with the paper that they wouldn't feel it necessary to get the opinion of outside reviewers.

But Cori Bargmann, who's usually right, for once was not.

That night, ten people went out for a celebratory dinner to Du Barry's, a median-priced, median-taste French restaurant in Boston. Snezna was among the guests, and she was back to her radiant self; she'd told me earlier that she was proud to be an author on the retinoblastoma paper. Toasts were made to Ted Dryja, to the Weinberg lab, to luck. Snezna proposed a toast to Wen-Hwa Lee. Nobody lifted a stem.

"Oh, come on," she said. "The poor guy practically walked for half of chromosome thirteen."

"To Wen-Hwa Lee," said Steve, and the wineglasses tinkled generously.

"I guess at this point," said Julian, so quietly that you almost couldn't hear him, "we can afford to hand out booby prizes."

But the Weinberg lab was rejoicing a little too soon.

The retinoblastoma paper was not accepted in four days. It was not accepted in fourteen days. By early August, Weinberg still hadn't received a definite yes from Britain. By mid August, he had his yes, but he wasn't at all pleased with the terms of acceptance. The *Nature* editors had coolly informed him that the paper did not have enough data in it to warrant being published as a major article; they would publish it as a Letter. "I thought it was the nuttiest thing I'd ever heard," said Mike Wigler. "You begin to wonder if those guys aren't letting their native eccentricities run away with them." What's more, *Nature* couldn't fit in the paper until mid October at the earliest.

"I was a bit angry with them," said Weinberg. "There were at least two other journals — *Science* and *Cell* — that were eager to have the paper, and I could easily have pulled it from *Nature* and given it to one of them." Weinberg opted against a last-minute retrieval, however, for fear of further slowing down its publication. So he and Steve grudgingly chopped the report to half its original length and returned it to London.

In September, Ted Dryja heard awful news: either Bill Benedict or Wen-Hwa Lee was on the verge of publishing a report about the retinoblastoma gene in a competing journal, possibly *Science*. Dryja desperately called Peter Newmark to plead for an earlier publication date, but Newmark replied that there was no space before mid to late October. After all the monomaniacal dedication, it looked as though Dryja and Friend would be beaten, anyway.

I told Steve and René that I would snoop around and try to learn what I could about Benedict and Lee. I was being selfish; I wanted a Weinberg win as much as anybody. "What are you going to tell them?" René scoffed. "'Hi. I want to know when you're publishing your paper on the retinoblastoma gene because I've been staying in the Weinberg lab and they're worried.'"

"Sure," I snapped back. "That's exactly what I'll say."

I called Benedict first. Playing innocent, I said that I was writing a story about retinoblastoma, and that I had heard his lab had cloned the gene. Could he tell me when the paper would be published so that I could include the information in my article?

"I have twenty-five or thirty people in my lab, and we've been working on retinoblastoma for three years now," said Benedict. "We're close to cloning the gene, but I don't think we'll be ready for two or three months. Have you ever heard of a guy named Ted Dryja? He's been working with Bob Weinberg's lab at the Whitehead, and they've supposedly cloned the gene. I think they've even submitted a paper on it."

"Oh, really?" I said. "So you think their report will be out before yours?"

"Yes, I think so," said Benedict. "There's somebody in Weinberg's lab named Steve Friend. He's only been working on retinoblastoma for a few months, but from what I've been able to gather, he's got the right gene. I spoke with Steve recently, in fact. I met him in Colorado last February, and we talked about retinoblastoma. I guess he feels that since I gave him a few pointers, he can tell me how things are going."

"Oh, really?" I said. "So you gave him the idea about how to clone the gene?"

"Uh, no, not exactly. But he's very nice. He's not paranoid, the way some people are. Another person you should talk to is Wen-Hwa Lee. I've heard he's close to getting the gene, too. He may even have it by now. He used to be a postdoc with me, but he has his own lab. So as you might imagine" — Benedict chuckled — "I don't hear too often from him anymore."

After I hung up, I immediately called Lee in La Jolla. I introduced myself to him as I had to Benedict and expressed curiosity about the publication date of his paper on the retinoblastoma gene.

"For one year and a half, I have been walking from the esterase-

D gene," said Lee. "We are very close. I think I have the right gene, but I can't be sure. We are not doctors, you know, so we have a hard time getting tumor samples. So I do not have the evidence, expression data, to be sure."

"So you haven't started writing up your paper yet?" I asked.

"Not yet. Almost," said Lee. "Have you talked to this guy in Boston, Ted Dryja? Or Steve Friend? He works with Bob Weinberg. I hear they are publishing their paper soon on the gene. I think it's the same gene we have. I can't be sure, but from the information I've heard, I think it is. But they're farther along than we are. The combination of those two groups is very big, very powerful. They're hard to beat. And they have had a lot of luck. We have been very methodical. Our approach is step by step. Theirs is luck, and they got there first. Nothing I can do about it, right? We're careful, we take more time, so they got there first. But maybe in the end, our approach will be better. Nobody can be lucky all the time, right?"

I hung up the phone and sighed with relief, secure that my story was not going to be scooped out from under me. Benedict and Lee had talked enviously of Dryja's luck and of Weinberg's invincibility, which winners of a race aren't likely to do no matter how crafty they are.

None of the Weinberg scientists knew how the rumors of competing publications got started, but they soon would be grateful that their competition was so doggedly close.

The Dryja-Friend retinoblastoma paper was indeed published first, in *Nature* in October of 1986. Despite its truncated format, the report was heralded widely as a genuine breakthrough. I'd never had such an easy time collecting supportive quotes. "I'm quite high on this," said Ray White of the University of Utah. "It's the next obvious step in oncogene research, and I think it would be really hard to overstate its importance."

"I take my hat off to these guys," said Web Cavenee. "You can call it luck, but they did the right experiment, an elegant experiment, and it worked. What more do you have to do before they stop calling you lucky and start calling you a good scientist?"

"I'm delighted this has happened," said Al Knudson. "Before, we could only concoct theories about what the retinoblastoma gene does. Now that we have the gene, we can get to work on the facts."

"They've started the ball rolling," said Bill Lewis of the University of Toronto, who was studying Wilm's tumor, another recessive cancer. "For somebody like me, it's very encouraging to see that these recessive genes can be cloned."

I liked Wigler's response best. When I asked him why he thought the cloning of the retinoblastoma gene was so exciting, he replied, "It's like getting through to the other side of the mirror."

But against the hymnody rang a few chords of dissent: loud, atonal chords that squawked across continental borders and could not be ignored. A week after Dryja's paper was published, I visited Brenda Gallie in Toronto to find out why she hadn't succeeded with Dryja's probe. A graduate student in her lab explained to me that the Gallie team was new to molecular biology, and that they hadn't used a sensitive enough assay when they initially screened the retinal cells. When they learned of Friend's success, the researchers repeated their Northern blots more carefully and detected expression of the probe in healthy fetal retinal cells, just as Steve had.

And then Gallie declared to me that the earlier failure didn't matter. She did not think Dryja and Friend had the right gene. She complained about all the data that didn't fit — the same expression data that had worried Weinberg months earlier and that had never been "fine-tuned" out of existence. She complained that the behavior of the gene didn't conform to the standard model of retinoblastoma. The Dryja-Friend gene was *always* turned off in eye cancer cells, she noted, when logic said that there should be abnormal messages of the gene at least part of the time. The gene was turned on when it shouldn't be — in tumors that had no relation to retinoblastoma. If this was a gene that modulated the maturation of eye and bone cells, she said, Friend's results made no sense. "There are just too many things about it that don't fit our model," she said. "But it's a good model, and it's served us well for a number of years now. I don't see why we should have to throw away the whole model just because Dryja and Weinberg have cloned a gene that they claim is the retinoblastoma gene.

"If I had to place my money on the gene right now," Gallie continued, "I'd bet against it. I think the gene they have is very close to the retinoblastoma gene. It's the closest thing that's been cloned so far, and it will be a great tool for helping us to clone the real gene. But I don't think this is it. I've requested a copy of the clone from Weinberg, and I imagine we'll get it soon. When we do, we'll be able to test it in our own system. We've got about seventy or eighty independent retinoblastoma cell lines here, so we're in pretty good shape to look at the gene in some detail."

Gallie received a clone from the Weinberg lab in the late fall of 1986, and soon afterward her criticisms grew sharper. She checked the DNA and RNA of fifty of her retinoblastoma samples, and her results convinced her that Dryja and Friend had the wrong gene. She could find no abnormal messages of the gene's RNA. She didn't spot a single instance of an internal deletion of the gene. The gene was turned on in adult eye cells — another contradiction of the retinoblastoma model. All the data were wrong, wrong, wrong, and she began to complain in public. Other scientists took notice, among them Ruth Sager, the Harvard biologist. When Weinberg talked about retinoblastoma at a conference in Utah, Sager stood up and passionately countered his results with a recitation of Gallie's anti-Dryja findings. "It was highly unusual, and not entirely pleasant," said Weinberg. "If one gets up and presents data, normally one talks about one's own data rather than acting as proxy for someone else. I was angry, but I attempted to keep my cool. It would have been ridiculous and counterproductive for me to lose my temper."

Panic swept from one side of the Longfellow Bridge to the other. Dryja tried to joke with Gallie that retinoblastoma must be a different disease in Canada, but his wit fell flat. He and Weinberg dreaded the possibility that they were wrong, that the retinoblastoma gene would turn out to be another metastasis-type fiasco. They needed to get the protein propagated by the retinoblastoma gene. They had to prove that the protein could revert cancer cells to normal eye cells. But such experiments were slow and difficult. Skepticism was waxing among their peers. "The rumor mill says that they have the wrong gene," a researcher in the Wigler lab told me in February of 1987. At the Whitehead Institute, spirits mirrored the winter blear.

The coming of spring, however, brought fresh breezes of proof. Steve Friend found another type of submicroscopic deletion in retinoblastoma cells that pointed at his gene like a bright blinking arrow. At last he had a deletion that began on the right side of chromosome 13 and moved partway into his gene, the kind of deletion that he and René had agreed would be compelling DNA evidence for the clone. And in a March issue of *Science,* Wen-Hwa Lee reported that he too had cloned the human retinoblastoma gene — the same gene that Friend had cloned — but Lee had the sort of confirming expression data the Weinberg lab lacked. Blotting the RNA of six retinoblastoma tumors, Lee detected abnormal messages. When the gene was disabled by something as small as a point mutation, said Lee, it still was typed into a short, useless message of

RNA. Lee's RNA results fitted more snugly into the accepted model of how the retinoblastoma gene should behave. And Lee also found deletions of the gene in some of his retinoblastoma tumors.

"I think that when you take my paper and Dryja's paper together," said Wen-Hwa Lee, "there can be little doubt that the gene is the true one."

Lee's paper did not squelch all reservations about the retinoblastoma gene. It didn't stop Gallie from attempting to clone the "true" retinoblastoma gene on her own. But by the fall of 1987, the scientific consensus had listed toward Dryja, Friend, and Lee. Said Al Knudson, the father of retinoblastoma genetics, "Scientists are supposed to be cautious and take a let's-wait-and-see attitude. But I don't have many qualms about the gene. I think they have the proper clone, and the thing I'm waiting to see is what the gene does."

Content that they were on the right path, the Lee, Weinberg, and Dryja labs focused their energies on the real problems: pulling out the protein synthesized by the retinoblastoma gene and learning the office of the protein in eye and bone cells. That question is not likely to be answered overnight. Deciphering the function of a protein, any protein, is arduous, as the mavens of the *src, ras,* and *myc* proteins know only too well. Yet scientists have their theories about the retinoblastoma protein. Lee suggested in his paper that the protein may be located in the nucleus, where it could muzzle the patter of growth-promoting oncogenes — if not the N-*myc* gene, he said, then some other rambunctious, cell-splitting oncogene. Weinberg proposed that the retinoblastoma protein may be, or may communicate with, a membrane receptor that embraces a growth-suppressing factor gliding by in the blood stream. Interferon, for example, is known to inhibit cell division; the retinoblastoma protein could be a docking port for a molecule much like interferon. In any case, the retinoblastoma protein provides the cell with a vital antigrowth signal, and no eye cell can afford to forgo both copies of the gene. "It will take some time to nail down the function, but it will happen," said Weinberg. "This is one area where there's room enough for all of us."

The cloning of the retinoblastoma gene marked the beginning of a new epoch in the molecular analysis of cancer. Heartened by the apparent success of Dryja's random-access method for squeezing out a splinter of the retinoblastoma gene, scientists are seeking other recessive cancer genes by jumping onto chromosomes and crawling around from left to right. They are closing in on recessive genes that have been implicated in kidney cancer, childhood brain cancer, and small-cell lung carcinoma, one of the most common and deadly of the lung cancers. In the summer of 1987, teams of scientists in Britain and Israel reported that they were creeping toward a recessive gene, located on chromosome 5, that may play a role in colon cancer. Biologists believe that in the multistep evolution of most, if not all, of the hundred cancers known to science, the demise of an anticancer gene must be one of the steps. By isolating the many recessive genes scattered across our twenty-three pairs of chromosomes, biologists may be better equipped to understand the chthonic dialogue between the proteins that promote and the proteins that repress the cleavage of one cell into two. They're more likely to solve what Jim Watson calls "the great dilemma of human cancer" if they have all the pieces of the puzzle before them.

Recessive oncogenes are no longer the "wave of the future." They are the commanding present, with all the concomitant high-pressure, competitive, overcrowded roisterousness of scientific fashion. "I've been working in recessive oncogenes for ten or eleven years now," said Susan Naylor, a scientist at the University of Texas who is working on a recessive gene for lung cancer. "I thought I could just move along at my own pace, unnoticed, unstressed. Nobody was paying much attention to our work. Then one day I wake up and I look around, and there's Bob Weinberg getting up on stage to talk about recessive oncogenes. I think, Whoa, I'd better get going, or my field is going to leave me behind in the dust."

Molecular biologists live in a beautiful crystal palace, where esoteric flowers of their own choosing leap into bloom and curl toward the light until the integrated artifice within obscures and supplants the natural disjunctions without. The solipsism and the compartmentalization are what make science so hypnotic. It's fun to think about DNA and messenger RNA, and whether a black hyphen on a blot is the right gene, and what the gene might do in fetal eye cells and teenage bone cells. It's fun to watch scientists compete, and it's pure joy to watch them triumph — although success often ranks as a great triumph only by the parameters biologists have set for themselves. Beyond the crystal walls of science, triumphs may not always look so grand. Transfection is not a cure for cancer; the identification of the *neu* lesion is not a cure for cancer; learning the function of *ras* would not be a cure for cancer. Yet the gradual and inexorable accretion of defined triumphs will, someday, arc above the abstract quandary of the darkling cell. The cell will be understood, if not absolutely, at least well enough so that its worst excesses can be subdued. Basic research and applied research will merge, are merging, as surely as the grudging cell is giving up its secrets.

There are many reasons that the cloning and analysis of the retinoblastoma gene is a more profound advance than many achievements in oncogene research. The feat led biologists through, as Wigler said, to the other side of the mirror. It helped round out the story of dominant oncogenes. And, as Web Cavenee pointed out to me, it's the first true human cancer gene to be found. "We can say, unequivocally, that the retinoblastoma gene is involved in these particular *human* cancers," he said. "We're not scaling up from mice and rats here. We're not forced to be cautious and say, Maybe this gene is sometimes involved in five percent or ten percent of human cancers. We know that it's a key component in retinoblastoma and osteogenic sarcoma — no ifs, ands, or buts about it."

Certainty in science is rare, and the certainty that retinoblastoma is a two-hit disease may prove overly simple. Retinal cells may have to undergo a series of progressive mutations after the initial loss of their retinoblastoma genes before they become fully cancerous. Enough scientists now believe the two-hit model so that, with the gene cloned and its protein nearly captured, they can't help themselves. They predict that retinoblastoma will be the first human cancer to bow beneath the might of molecular biology. Not immediately, of course. Not quickly enough to help a retinoblastoma infant born within the next five years. But soon. "If we're right, and the loss of this gene is one of the keys to forming cancer," said Dryja, "someday we may be able to design a treatment by making a drug that mimics what the protein of the gene is doing."

The convictions of David Abramson, one of the retinoblastoma experts at New York Hospital, reach further. "If you want to talk twenty-first-century stuff, let me play seer for a few minutes," he said. "I believe that within fifteen years, at the outside, we'll be able to stop retinoblastoma before it begins. I'm so sure of that that I've already given the drug a name. I call it retino-revert, or retino-prevent. The drug will be an analogue of the natural protein that is missing in retinoblastoma cells. So here's what I think will happen. We'll be able to diagnose a child prenatally and start giving this retino-revert to the mother to prevent retinoblastomas from growing as the fetus is developing. I know I'm going out on a limb with this one, but now that the gene's been cloned, people will be able to take me up on my challenge. Come back to me in 2001 and tell me if I wasn't right."

Dryja and Abramson are willing to talk wishfully of cures because they're doctors, and they don't want the retinoblastoma gene to remain a scientific curiosity for long. They want to help their patients. "It wouldn't bother me one bit," said Abramson, "if the disease were wiped off the face of the earth. I have plenty of other ways to earn a living."

Even basic researchers, when confronted with the clean equation between a segment of DNA and a severe childhood disease, acknowledge a shift in their thinking. "For most of us scientists, a cell is just a cell. It's something that you manipulate in a petri dish," said Cavenee. "But when you see one of these retinoblastoma kids, your perspective changes. That's what happened to me. I'd been working on retinoblastoma for a couple of years, and I'd never seen a patient. I'd seen retinoblastoma cells under a microscope, but big deal. They could have been rat embryo fibroblasts for all I cared. Then we went over to Sweden, because we were testing the possibility that we could use some of our markers for prenatal diagnosis. While we were there, I met this little boy with retinoblastoma who was about to be operated on, enucleated. He was beautiful, blond, about two years old — the same age as one of my own children. His mother kissed him goodbye, and they took him away for surgery.

"The next time I saw him was in postop. His whole head was bandaged: both of his eyes had been removed. When he woke up, he started to cry, and his mother tried to soothe him. He asked her, 'Mommy, when are they going to take these bandages away so that I can see?' I thought, See? See? Jesus Christ. It wasn't fair. He didn't deserve it. He didn't do things like this" — Cavenee lifted his pack of strong Canadian cigarettes from the table and threw it back down — "so forget about scientists' reputations and scientists' egos. Maybe going to the hospital should be a required part of a scientist's training. Once you've seen something like that bandaged little boy, you realize who you're working for. Any advances we may make, anything that comes from our research, isn't for us. It's for them."

Cavenee's admonition was directed at basic researchers, but I thought it was time to see whom Steve Friend, Ted Dryja, and the rest were working for. I'd learned about the retinoblastoma gene, and now I wanted to see what a wretched lack of the gene had wrought.

On Mondays and Thursdays, New York Hospital holds its renowned retinoblastoma clinic, headed by the best retinoblastoma specialist in the world, Dr. Robert Ellsworth. Parents bring their children from across the United States, Europe, the Middle East, occasionally from Asia and South America. When a family can't afford the air fare to New York, corporations pay the bill. All out-of-town families stay free of charge at the Ronald McDonald House, half a mile or so from the hospital. The initial treatment of the tumor, once it shows enough to bring the disease to a parent's attention, generally entails enucleation or whole-beam radiation and lasts an average of two weeks. Afterward, retinoblastoma children must see Ellsworth at least once every six months until they're six or seven years old and beyond the age of recurrence. Until then, new eye tumors can spring up unexpectedly and proliferate rapidly, and Ellsworth wants to catch the malignancies while they're small enough to be attacked with a blast of nitrous oxide, which kills the tumor tissue by freezing it. The less radiation he must use, the better; and the fewer eyes he must remove, the better still. "Our goal in treating bilateral retinoblastoma is to save vision in at least one eye, and most of the time we're able to do that," he said. "One good eye is all anybody needs to get through life." I was not to watch any grisly enucleations or beam-radiation procedures; I would see patients returning to Ellsworth for their semiannual checkups.

When I arrived at the clinic on Monday morning, the nine children in the sunny waiting room already had been given drops to dilate their pupils, and most were caterwauling in terror. Nobody was tempted by the toys, dolls, and stuffed animals that littered every surface. I saw two children who had empty eye sockets; they'd been unilaterally enucleated and had not yet been fitted with false eyes. I saw several other children who each had one plastic eye. Although the optometrist at the hospital who makes the false eyes labors lovingly to match his product to the size, shape, and color of the child's remaining eye (he himself has an artificial eye), the orbital and lid muscles of an enucleation victim never quite recover. False eyes look too wide, too protuberant, too immobile — too false.

When Ellsworth and his colleagues were ready, the patients were dragged one by one into the operating room. Inevitably, as the child realized where he was being led, he screamed, flailed, kicked with the fury of a trapped animal; he knew the operating room, and he'd learned to hate it. But once a child was pinned to the table and fitted with an anaesthesia mask, screams rapidly faded into the thick, measured breathing of unconsciousness.

"It's actually good when they scream like that," said Ellsworth. "It's not good for the parents or our ear drums, but they go under faster because they're pulling in more oxygen, and that helps the anaesthesia to take effect."

I soon understood why the children had to be knocked out for the exam. Ellsworth clamped open the patient's eyelid with a lid speculum and began poking his pen-shaped scleral indentor around the perimeter of the eyeball, deforming the orb as though he were fussing with a soft-boiled egg. Peering at a mirror on the side of Ellsworth's ophthalmoscope, I could watch what he was examining deep within the eye: pale tumors afloat on a shiny black sea. Magnified about eight times, the tumors looked as big as Brooklyn, though even the largest was only the size of a pinhead.

Every tumor had a unique profile. Some were scattered across the eye like birdseed; those were tumors that had broken apart as they grew. Other tumors were white and fluffy, like hair mousse: those were tumors that had been killed successfully with radiation, Ellsworth explained, and they would eventually shrink. New tumors, or old malignancies that still harbored active cancer cells, had a pinkish cast to them. "That's what we call fish flesh," he said. He found fish flesh tissue in the eyes of three patients. Each time, he aimed a thin tube into the corner of the child's eyeball and then pressed a foot pedal, releasing a waft of nitrous oxide onto the retina. As the gas infiltrated the eye, it froze the tumor and surrounding tissue into an opaque, cracked "iceball." As the gas dissipated moments later, the iceball rapidly melted, leaving a rash of dead cells in its wake.

Ellsworth worked with astonishing speed and deftness. He was so skilled that while he probed and prodded his patients' eyes, he was able to sustain a steady stream of randy banter with others in the room. He flirted with nurses and women doctors; he told racy and sometimes racist jokes; he commented vulgarly on the appearance of the tumors, offering chocolate sauce to anybody who wanted to spoon up one that looked especially creamy. To test my mettle, Ellsworth asked me to touch the base of a child's eye socket after he'd popped out the girl's plastic eye. The skin of the crater felt warm and damp, and projecting from the bottom was the tip of a thin metal rod that had been inserted in place of the girl's excised optic nerve. My stomach clenched, but I didn't say a word.

Yet for all his tasteless operating-room patter, Ellsworth is an extraordinary physician. He took the time out to confer at length with the parents, explaining what he had done that morning and what remained to be done in future visits. He answered their questions without condescension. Most impressive, he knew the personalities and pastimes of all his patients and their family members. He talked with one father about getting together for a round of golf; he asked a mother how her other, nonafflicted son was faring in college.

Among Ellsworth's favorite families was the Vanceks: three-year-old Michaelle (pronounced "Michelle"); her mother, Carmen; and her father, John. Michaelle seemed to be the nurses' pet, too. Tear-stained and angry though she was, she looked so adorably pretty that everybody took turns hugging her. They laughed at her spunk, at how she'd yelled at a nurse that she was "an ugly lady" and immediately repented, returning to say "I'm sorry I said you're an ugly lady. You're really a beautiful lady." They pointed out to me that Michaelle never relaxed her grip on her little red *manga* (Spanish for sleeve), the tatty remains of an old pajama top and Michaelle's equivalent of a security blanket; even while unconscious, she clung to her *manga*. By morning's end, I too was in love with Michaelle, so I approached her mother and asked whether I could visit the Vanceks to talk about their daughter and their ordeal. Carmen Vancek, a beautiful, nut-brown Ecuadorian immigrant, replied in broken English that she'd discuss it with her husband but that she thought he'd probably agree to see me.

Several days later, I stopped by the Vanceks' apartment on Manhattan's East Side. Carmen and I were alone for the first few hours. Michaelle was spending the morning at a city-run nursery school, and John, the superintendent of the building where the Vanceks lived, was trying to get a balky elevator repaired. We sat in a little dining area off the living room and drank coffee. In one corner was a china cabinet stocked with trinkets and mementoes from Ecuador: Ecuadorian dolls, Ecuadorian ashtrays, a dinner plate embossed with a formal photograph of Carmen in her white communion dress. The apartment is in a luxury building, but it's probably the least luxurious apartment available: the ceilings are low, the living room square and small; Carmen, John, Michaelle, and Michaelle's older brother all sleep in a single small bedroom. The Vanceks have very little money. Carmen cleans houses and sells off-price clothes to supplement John's income.

Carmen told me how she had first discovered that there was something wrong with Michaelle. When her daughter was about two years old, in 1985, Carmen saw what looked like a tiny diamond glittering in Michaelle's right pupil. Over the following months, the diamond seemed to grow larger. "Sometimes, in certain lights, we could see it," said Carmen. "Sometimes it was invisible." Michaelle's eye began to shed tears, constantly and copiously. Carmen was worried, but the first New York physician the Vanceks consulted assured them that Michaelle's eye was merely irritated.

The shiny white spot got bigger, and the tearing worsened. John's mother insisted to Carmen that Michaelle's eye did not look natural, and she suggested that the family contact another doctor. Carmen showed me snapshots of her daughter taken at the time when the elder Vancek spoke up. I was horrified. Michaelle's left eye was dark and lively, reflecting a small glint of the camera flash, as all eyes do when photographed indoors. But her right eye looked like the flash itself. It glared; it menaced; it almost leaped off the surface of the picture. The pupil was pearly bright and pearly cold. Michaelle Vancek's right eye obviously was dying.

The second physician took one look at Michaelle and referred the family to Bob Ellsworth. In addition to diagnosing the diamond in Michaelle's right eye as retinoblastoma, Ellsworth found another, smaller tumor in her left eye. Seeing that the girl had bilateral retinoblastoma, Ellsworth knew that she had a hereditary case; but because there was no history of the disorder in either Carmen's or John's family, Ellsworth assumed that one of the parents' sex cells had been mutated, and now Michaelle carried the retinoblastoma trait.

Given her relatively advanced age — Michaelle was almost two and a half years old by this time — she needed quick and aggressive therapy. Ellsworth first tried to treat both eyes with whole-beam radiation. But while the pinprick tumor in her left eye regressed under the withering dose of gamma rays, the bulkier right mass did not; in fact, after a few weeks of therapy, it was still growing. "So in January or February, Dr. Ellsworth said that the eye had to come out," said Carmen. She started to weep. "He was worried the cancer might spread. He said it was our decision, but that he thought it was necessary to do immediately. He said the tumor was getting bigger. So the next week he did it. He took the eye out."

Carmen was sobbing loudly now. She ran into the kitchen, returned with a box of Kleenex, and buried her face in a tissue. I tried to think of something sympathetic to say, but the best I could come up with was "Did she feel any pain after the operation?"

To my surprise, Carmen brightened at the question. "No, I don't think she really felt bad," she said. "She didn't act any different than she did before. She talked about the 'boo-boo' in her eye, but she didn't seem to have any pain.

"She was stronger than me. For the first few weeks, before she got the plastic eye, it was hard for me to look at her. Sometimes, when I saw the empty socket, I would cry, just a little. She'd say, 'Mommy, why are you crying? Do you have a boo-boo in your eye, too?' Or she'd go to her father and say, 'Daddy, you take care of Mommy.' And then she'd say to me, 'Mommy, you take care of Daddy. Then both of you can take care of me.' She is a very, very smart little girl." Ellsworth had told me that children with retinoblastoma often are brighter than average, perhaps because they spend so much time in hospitals and are exposed to the stimulating influence of many adults.

John Vancek came into the apartment, and Carmen left to fetch Michaelle from day school. An émigré from Yugoslavia, John Vancek has lived in the United States since 1957, and he speaks English with only a trace of an accent. He's more than twenty years older than his wife. He told me how he had met Carmen at a Christmas party in 1981. "She'd just arrived from Ecuador," he said. "She didn't speak a word of English, and I didn't speak a word of Spanish. We communicated in our own version of sign language. Six months later we were married."

John wanted me to write about retinoblastoma because, he said, when he and Carmen first found out about their daughter's cancer, they'd never heard of the disease. He hoped that if other parents in a similar situation were able to read something about retinoblastoma, they'd realize they weren't alone. "We didn't know anybody else in the world had this cancer until we went to the clinic," he said. "Then we met lots of other parents in our situation. I want people out there to know there are many children suffering from this cancer. Maybe that will help them to feel stronger."

"Do you think Michaelle is adjusting well to only having one eye?" I asked.

"She's adjusting incredibly well," he said. "Of course, she's too young to know what's happened to her. But even so, she just runs around like any other little girl. Maybe when she's really active she sometimes runs into the wall a little more often than she used to, but there's nothing serious.

"It's been much harder on us, especially on my wife. I think it's worse because Michaelle is a girl. We look at her and realize how much more beautiful she'd be if she had both eyes. But that's a small thing, a silly thing. The important thing is she's alive and healthy. Ellsworth tells us that her other eye looks very good. They've been able to cure the tumor in that eye, and he says she will see fine with it. She's got so much spirit that I know she'll do better than many children who don't have physical defects.

"But there's one thing I hope. I hope they invent a new treatment. It's too late for Michaelle, but maybe it won't be too late for others — like my grandchildren."

Carmen returned with Michaelle, who was dressed in a red jumpsuit and striped shirt. She's small for her age, and the asymmetrical largeness of her one plastic eye makes her look smaller still. Yet in a strange way, the artificial eye heightens the Mayan beauty she's inherited from her mother — the way some people are attractive not in spite of a bumpy nose or crooked mouth, but because of it. Michaelle was a little wary of me at first. "She probably remembers you from the clinic," said Carmen. "Maybe she thinks you're going to take her back there." But after a while, Michaelle crouched under the table where I was sitting and began playing a game of peek-a-boo with me. She used the table leg to hide behind, and I noticed that she peek-a-booed only with her left eye; casually and perhaps unconsciously, she had shifted her view of the world to accommodate her semiblindness, her loss of spatial sense and of most peripheral vision. I thought fleetingly of her DNA, of each molecule in each cell kinked out of whack by something that could be as small as a single nucleotide. I thought it was astonishing that scientists knew, in such detail, the sequence of events that had left this lovely girl to play games through one dark, tumor-stippled, but seeing eye. I stared at her and she stared at me, more curious now than wary.

A moment later, she dashed out from beneath the table, spread her arms as though she were an airplane, and began running in circles around the living room. I watched her run faster and faster, in wider and wider circles. I was waiting for her to soar.

Epilogue

IN FEBRUARY 1986, when Thaddeus Dryja traversed the Longfellow Bridge between Boston and Cambridge, bearing gifts for Steve Friend, the Weinberg lab was plunged into a preferred state of bedlam. Yet Bob Weinberg didn't change. He wouldn't allow prevailing obsessions to distract him from his sustaining obsession. Through the spring, summer, and fall, while others engaged in adrenergic workfests, he continued to build the addition to his house. Plank by plank, and Sunday by Sunday, it rose from the pitched and gravelly clearing of his New Hampshire kingdom. By mid autumn, it was essentially finished. I'd helped pour the concrete foundation for his annex; now I wanted to see the final product. So I called Weinberg and invited myself north for a couple of days. "Besides," I said to him over the phone, "it will give me an excuse to visit my relatives in Fitzwilliam."

One late Saturday afternoon in October I inched my car up the steep, rutted driveway to the house. Amy and the children were back in Boston; Bob was alone in Rindge with Kevin, a bluff, strawberry-blond New Hampshire handyman who'd been assisting him with the construction. Their weekend goal: to lay down the pine floorboards. I'd timed my arrival to correspond to the end of their working day, but as I emerged from my car and heard the whinny of an electric saw, I figured Bob was in one of his ambitious, tally-ho moods. I didn't care. The sight of the façade of the new structure was enough to keep me amused for a long time. The new house was identical with the old house. Mother cell and daughter cell squatted side by side, one a beaming clone of the other. Both presented Weinberg's idiosyncratic synthesis of bungalow, Elizabethan manor, and *Rathskeller.* Weinberg had told me that he'd designed the exterior of the addendum to match the original, but I hadn't expected such a precise recapitulation.

Eventually, Weinberg poked his head out the door and saw me. "You're here!" he said. "Aren't you cold, standing out there? Is that sweater warm enough?" I looked down sheepishly at my thin wool sweater; once again I hadn't dressed sensibly for a rustic outing. "Come in and sit by the wood stove," he said. "We're almost ready to call it a day." He was wearing his paint-spattered overalls and clunky work boots, and he looked radiant, with the docile radiance that only physical exertion can bring. I gave him a box of chocolates I'd brought from the Fitzwilliam Inn. He immediately tore off the cellophane wrapping and popped two candies into his mouth. "I think I'm entitled," he said. "I've been burning calories since seven-thirty this morning."

We went into the original house, and he urged me to relax by the stove for a few minutes while he and Kevin finished laying two more floorboards. I pulled up a chair, braced my feet against the wood stove, and tipped my head back to give me a broad view of the room. The old house is morosely comfortable, faithful to its façade; the walls are sheathed in dark wood panels, and heavy piers support the high ceiling. A Beethoven piano sonata played on the tape deck, and shavings of the setting sun drifted across the floor and furniture. I was on the verge of nodding off when I noticed that the heat from the wood stove was penetrating suspiciously close to my socks. I looked at the bottoms of my shoes and gasped: the edge of the stove had melted deep gulches through the rubber soles. Although the shoes were still wearable, I was glad Bob wasn't there to witness what I'd done. I jumped up and frantically started washing the dirty dishes in the sink.

Bob came into the kitchen. "Don't worry about those!" he cried, shutting off the faucet firmly. "Come, look at the new room." I followed him through the door that led from the familiar to the fresh, hoping that my softened, sticky soles weren't leaving traces of rubber on the floor.

Stepping into the new room was like abandoning the shadowed hush of a cathedral nave for the bridal buoyancy of a transept chapel. The interiors of the two houses are roughly the same size.

But where the original is close and woodsy and divided into kitchen, living room, and sleeping quarters, the annex is one white, open space, free of struts and beams. The walls are constructed with sufficient thickness to support the roof and to resist the oblique thrust of the ceiling, an elaborate dome sectioned by ribs into four petals. Slicing into the southern wall, twelve or fourteen feet above the floor, are two exquisite windows capped by round Renaissance arches. Rounded windows are difficult to build, and I pictured Bob perched high atop a ladder, slowly troweling on two plaster eyebrows, pleased with the perfectibility of details.

"This place is spectacular," I said. "The last time I was here, when you just had the frame up, I thought the whole thing looked too complicated. But all the parts fit."

"You saw the framework for, what, the walls?" he asked. "Before we had the roof put up?"

"No, I saw the framework for the roof, too. It was some sort of series of interlocking gables — two or three for every side. I thought it looked like a fancy Loire Valley château."

"Ah, yes. Well, we had to surmount a few engineering challenges to match the new house to the old, because the grade of the hill beneath it is steeper. But it's an interesting space, don't you think?"

"It's very interesting. And it's so soothing. You're going to have a great time coming up here on weekends to think. Why don't you just have a phone installed so that you never have to go back to the lab?"

"I used to have a phone up here. But I took it out. The temptation to call the lab at odd hours of the night was just too great."

I asked Weinberg where he had learned to be a professional carpenter.

"It's funny you should bring that up, because I was just discussing it with somebody the other day," he said. "For many years, I believed that my grandfather taught me everything I know. But I now realize that he wasn't a very good carpenter. His solution to any construction problem was simply to bang at it with a hammer. Now, that approach may suffice when you're throwing together a bookshelf, but for a house, it won't wash."

I looked again at the windows; they were as though drawn with a T-square and compass. I scrutinized the completed sections of the floor: not a warp or chink or gap to be found. Weinberg's commitment to quality reminded me of nostalgic advertisements for old-time Scotch or shoes or suits. He's the mythic master craftsman, who'd sacrifice sleep, meals, and eyesight to stitch a perfect lapel.

(And when I watched Bob and Kevin lay the remaining floorboards the next day, I noted that several times Kevin would have been happy to ignore or fudge around a flaw or to hammer an obstreperous plank into line, but Weinberg wouldn't hear of it. "Kevin, let's pull this one up and start from the beginning," he'd say. Or "Kevin, this edge needs sanding." Or "Kevin, you'll have to turn this board around so it's facing in the other direction." "Are you sure, Bob?" Kevin would say. "It looks to me like the right direction." "No, it's facing the wrong way. Can you turn it around, please?" Kevin the Stout never grumbled; he'd pull up a seven-foot board, lay it down, and pull it up again until Bob was satisfied.)

"If you do things right the first time around," said Weinberg, "you'll avoid disasters later on."

The last light of day was waning, and the white walls turned warm and silvery. Bob switched on the gaslight fixtures he'd bought especially for his new home; they looked like little replicas of Parisian street lamps, and they cast a broad, bronze glow just bright enough to read by. For a few moments, we stared at the lamps, transfixed by the light and the gentle hiss of burning kerosene. Then we joined Kevin in the kitchen, where he was preparing chicken cacciatore for dinner. After we ate, Bob broke out a bottle of peppermint schnapps. Weinberg rarely drinks, and after a small glass of the liqueur he became a bit lightheaded. "I should do this more often," he said. "It makes me voluble." I'd come to Rindge with a mental list of topics to cover, but Weinberg wanted to be voluble only about retinoblastoma.

"It just fell into our laps," he said. "And it's all ironic, given my earlier skepticism. I didn't think that Dryja had a chance of walking into the gene, but I'm glad I was wrong. I'm glad that he chose to collaborate with us, despite my ill-concealed doubts. I believe that the cloning of the retinoblastoma gene will prove to be of great significance. It will have a profound impact on cancer biology."

The retinoblastoma gene already had exerted its impact on Weinberg. "I'm having ideas again, one after another," he said. "My mind is working overtime these days. The creative juices are flowing. I'd been worried that I was all dried up, that along with the other accouterments of increasing age is a deterioration of brain function. But I'm coming up with new models again. I'm obsessed with ideas. I'm beginning to think about the synergism between dominant and recessive oncogenes. They're not profound ideas — my ideas usually aren't — but they're enough to keep me happy."

Bob Weinberg sizzled and crackled with paradigms. It wasn't the schnapps that loosened his tongue; it was his new love affair with recessive oncogenes. He cared about how the retinoblastoma protein worked, but he cared more about its place in the larger panorama of cancer. He told me about his latest models of multistep tumorigenesis. They sounded somewhat similar to his old *myc* and *ras* theories, but now they included loss of retinoblastoma-type genes. The deletion of antigrowth genes, he said, may well be as important to human cancer as the activation of heavyweight growth genes, like *ras*.

I asked him whether he still believed that the cancer cell was simple. He walked over to the wood stove, tossed in another log, and replied that he was more convinced than ever of the cell's fundamental simplicity.

"You know about my long-term hobby of genealogy," he said. "Well, I realized recently that it's very similar to my passion for breaking down the complexity of a cancer cell into a few basic steps. Trying to understand relationships of people in a complex welter of a community, and tracing them back to a small number of progenitors, is like trying to understand the relationships of genes within a cell.

"All of the work that has gone on in oncogenes will come full circle," he said. "It will come together. What's so exciting to me is that the retinoblastoma gene and other recessive oncogenes fit so well into the picture of multistep carcinogenesis. They're another piece of the puzzle. I'm convinced that the retinoblastoma-like genes and dominant oncogenes are going to be united into one grand conceptual framework in a couple of years. We'll be able to describe, in real physiological terms, what multistep carcinogenesis is about."

"Will you continue working with Dryja, or are the two of you going to go your separate ways?" I asked.

"I have made a solemn promise to Ted," Bob said, thrusting a poker into the stove. "I am committed to a long-term collaboration. We're together on this one. We will not abandon him." He returned to the table and poured himself a splash of liqueur. "Which brings me to another point. In some ways I feel that the retinoblastoma gene has been something of an unearned run for us. All the other advances that have been made in my lab, regardless of their relative merit, we got by the sweat of our brow. We really worked for them. Steve Friend did a very nice job cloning the gene, but it wasn't the equivalent of, say, transfection or finding the point mutation in the *ras* gene. Dryja did the work — and may I add that he was lucky on his end. He was lucky to have pulled out a good probe. But we were luckier still in collaborating with him. We hadn't been working on retinoblastoma for six years. We jumped aboard at the last minute."

"Yeah, people have been talking about how lucky you were," I said. "Sometimes they sound almost annoyed by your luck. They may not think it's fair that you keep coming up with one discovery after another."

"Well, what is fairness?" he said. "What is luck? There are all kinds of luck. You might say I was lucky to have been born with a certain native intelligence. That's luck, isn't it? And you might say that I was lucky to have been raised the way I was, and given the education I was given, and to have the cultural background that I do. Look at the history of German Jews. Look who's come from that great culture. Sigmund Freud. Felix Mendelssohn. Albert Einstein. Is it luck to have been born a German Jew?"

I had never noticed it before, but as Weinberg was talking, his head cocked slightly, his thick mustache still mussed from the meal, and his hair skewed every which way, I realized that he looked like the young Albert Einstein. The idea made me smile. "You *are* lucky, Bob," I said. "For whatever reason, you seem to be one of the luckiest scientists around."

When I left Rindge the next day, Weinberg stood on his doorstep and waved goodbye until I pulled out of sight. I thought about his models for human cancer. Even when he incorporated the latest results on recessive oncogenes, his models remained pure Bob Weinberg: clean and holistic. The loss of a gene that controlled growth. The mutation of a gene, such as *ras,* that encouraged growth. The models were too good to be true. Cells couldn't be that simple.

Then again, the model for Bob's house had looked like nonsense, a doll house confection. Each piece in outline seemed an arbitrary flourish. Yet when pegged together in the woods of New Hampshire, the rabble of parts assumed logic and meaning. From the Renaissance windows to the Parisian gaslights to the medieval dome, one element alluded to and interpreted the other, and every component deferred to the pure white sweep of the walls. By compiling a host of complexities, Weinberg had reduced them to simplicity. He had designed a calming retreat where he could ponder the cell, retrieve order from entropy, and know that he'd been right all along. There was no reason for despair.

Appendix

Major Publications Cited in the Text

Following is a selection of the major papers published by members of the Weinberg and Wigler labs during the years 1979 to 1986.

Shih, C., B.-Z. Shilo, M. Goldfarb, A. Dannenberg, and R. A. Weinberg. "Passages of Phenotypes of Chemically Transformed Cells via Transfection of DNA and Chromatin." *Proceedings of the National Academy of Sciences* 76 (1979): 5714–5718.

Goldfarb, M., K. Shimizu, M. Perucho, and M. Wigler. "Isolation and Preliminary Characterization of a Human Transforming Gene from T24 Bladder Carcinoma Cells." *Nature* 296 (1982): 404–409.

Parada, L. F., C. J. Tabin, C. Shih, and R. A. Weinberg. "Human EJ Bladder Carcinoma Oncogene Is Homologue of Harvey Sarcoma Virus *ras* Gene." *Nature* 297 (1982): 474–479.

Tabin, C. J., *et al.* "Mechanism of Activation of a Human Oncogene." *Nature* 300 (1982): 143–149.

Land, H., L. F. Parada, and R. A. Weinberg. "Tumorigenic Conversion of Primary Embryo Fibroblasts Requires At Least Two Cooperating Oncogenes." *Nature* 304 (1983): 596–602.

———. "Cellular Oncogenes and Multistep Carcinogenesis." *Science* 222 (1983): 771–778.

Toda, T., *et al.* "In Yeast, *RAS* Proteins Are Controlling Elements of Adenylate Cyclase." *Cell* 40 (1985): 27–36.

Bernstein, S. C., and R. A. Weinberg. "Expression of the Metastatic Phenotype in Cells Transfected with Human Metastatic Tumor DNA." *Proceedings of the National Academy of Sciences* 82 (1985): 1726–1730.

Bargmann, C. I., M.-C. Hung, and R. A. Weinberg. "Multiple Independent Activations of the *neu* Oncogene by a Point Mutation Altering the Transmembrane Domain of p185." *Cell* 45 (1986): 649–657.

Friend, S. G., *et al.* "A Human DNA Segment with Properties of the Gene That Predisposes to Retinoblastoma and Osteosarcoma." *Nature* 323 (1986): 643–646.

Index

Aaronson, Stuart, 320
Abbott, Berenice, 172
abl gene, 134, 325
Abramson, David, 331, 360–61
adenine (*a*), 33, 37–39
adenomas: and *ras* gene, 137, 138
adenoviruses: transfection, 56–57
adenylate cyclase pathway: *ras* mutation and, 279–82, 286, 297
AD-12 retinal cell line, 314–15
aging: cancer and, 142
AIDS virus, 141, 173, 194
alanine, 41
Alberts, Bruce, 173
alcohol consumption: cancer and, 141
alternate pathway theory, 282–302
Alu sequences, 101–2, 105, 108, 109, 184, 200, 201, 240
American Academy of Achievement, 261
American Association for the Advancement of Science, 11, 79
American Cancer Society, 11, 27, 128, 274; oncogenetic funding by, 134
Ames, Bruce: work on carcinogens and mutagens, 71
amino acids, 41
angiogenesis factor, 215–17
anticancer vaccine, 187; Bernards's antineuroblastoma vaccine research, 246, 258–59, 260, 268; prospects for retinoblastoma treatment, 360–61
anti-oncogenes. *See* recessive oncogene
Applied bioTechnology, 170
Armand Hammer Foundation research awards, 11
artifacts, 195–96, 197, 203
Asilomar regulations, 77–78, 102
authorship of research papers: rules of, 103, 129–30, 350–51
autoradiograph, 36, 172
Axel, Richard, 270, 299; and Spandidos controversy, 85–86

bacterium: RNA of, 40
baker's yeast, 276

Baltimore, David, 79, 145, 146, 194, 212, 235, 247, 276, 307; approval of book project by, 8; on role of young scientists, 9; on clinical applications of research, 12, 17; on normal cell processes, 15; at Salk Institute, 50; awarded Nobel Prize, 52, 165, 173; at Cancer Institute, 52–53, 54; discoveries in retroviruses, 52–54, 165; on Goldfarb's work, 68; and Spandidos controversy, 81; and polio virus research, 91; and oncogene transfection and cloning project, 99; and establishing of Whitehead Institute for Biomedical Research, 163–68; as director of new Whitehead Institute, 164–68, 172–74, 175–76, 248; molecular biology textbook of, 173; as public speaker, 176

Barbacid, Mariano, 163, 268, 300, 318, 346; on applied medical research, 12, 13; and T24 oncogene cloning, 109, 115; and point mutation research, 129, 136, 138, 143

Bargmann, Cornelia (Cori), 159, 171, 175, 176, 200, 211, 212, 227, 231, 307, 342, 346, 349–50, 352–53; and Weinberg house-building project, 2; on Weinberg, 44, 48; and *ras* mutation research, 123, 130; and *neu* gene project, 245–67, 315–19, 340; appearance and personality of, 246–47; reputation of, 247–48, 306; sequencing of *neu* nucleotides, 250, 252, 255–56; discovery of point mutation in transmembrane of *neu* gene, 315–19, 322

Bar-Sagi, Dafne, 295, 296

Bell, Eugene, 166

benchwork: lab directors and, 22, 23; routine nature of, 30–31; craftsmanship in, 31

Benedict, William: cloning of esterase-D gene, 336, 339, 342; and retinoblastoma project, 351, 353–54

Berenblum, Isaac, 142

Bernards, André: relations with brother, 188–89; and pink Cadillac, 306, 308, 325; and *abl* oncogene, 325

Bernards, René, 175, 177, 213, 227, 247, 248, 305, 306, 314, 315, 316, 318, 327, 346, 348–51; and Weinberg house-building project, 2–3, 47–48; research partnership with Bernstein, 185, 186, 189, 191–95, 198–99; and metastasis project, 185–87, 189–95; concurrent multiple projects of, 187; personality and background of, 187–89; sibling rivalry, 188–89; makes subclones, 192–93; reaction to Bernstein's artifact error, 194–95, 198–99; and antineuroblastoma vaccine, 246, 258–59, 260, 268; and retinoblastoma

project, 314, 315, 327, 346, 349–50
Bernstein, Shelly, 227, 265, 345; and metastasis project, 180–87, 189–202, 213; research partnership with Bernards, 185, 186, 189, 191–95, 198–99; artifact error, 194–95, 196, 197, 198–202
Biogen, 170
biology. *See* molecular biology
Birchmeier, Carmen, 274, 275, 292, 294–96, 297; and *Xenopus* oocyte-*ras* project, 280–81
Bishop, Michael, 151; on cancer research, 13; lab staff of, 25; work on Rous sarcoma virus, 65, 66, 67, 68; on *ras* gene mutation, 128–29
bladder oncogene: transfection of, 100, 102–3; cloning projects, 103–10, 111–16
Boas, Franz, 46
Bolger, Graeme, 311–12
bone cancer, 19, 312, 314, 326, 331; *see also* osteosarcoma
Borst, Piet, 188
Bos, Johannes: and *ras* oncogene research, 136–38
Boston University, 82
Bradley, Scott: and mutation theory research, 123, 128, 130, 131, 143
brain cancer, 19
breast cancer, 15, 19, 141, 162; and *ras* gene, 134–35, 136; and *neu* oncogene, 320–21
Bristol-Myers research awards, 11
Broach, James, 277, 281, 282
Broek, Daniel, 109, 273, 277, 283, 289, 291, 292–93, 294–96
Brown, Michael: Nobel Prize research of, 16
Burkitt's lymphoma, 15, 149

Cairns, John, 134
Cameron, Scott, 274
cancer: mortality rates from, 12, 21; incidence of, 13; funding of research, 14; remission as "definition" of cure, 14; advances in treatment, 14–15; as disease of lifestyle, 141, 214; as disease of aging, 142; *neu* gene in prognosis of, 321–22; *see also* anticancer vaccine; multistep tumorigenesis theory; oncogenes; oncogenetics; *and specific cancers,* e.g., retinoblastoma
Cancer Center: oncogenetic research at, *see* Weinberg lab; Weinberg transfers from, 168–69
cancer medical research: vs. molecular biological research, differing perspectives, 6, 11–15, 16; clinical applications of molecular biology research, 12–22; *see also* retinoblastoma
carcinogenic chemicals, 74
carcinogens: and mutagens, 71
Cavenee, Webster: and

Cavenee, Webster (*cont.*)
retinoblastoma research, 310–11, 315, 333, 334, 339, 351, 355, 360, 361
Cell, 79, 162, 176, 197, 353; Spandidos's report in, 79, 85
cells: molecular differences between normal and tumorigenic, 20; number of different genes in human, 36, 37; mutations and malignancy, 142–43
cellular division: Weinberg's model for, 4
Cepko, Constance, 2
cervical cancer, 141, 184
Chang, Esther, 119, 124, 125, 130
chemotherapy, 20
Cheng, King-Lan, 252, 253, 257–58
Chernobyl fallout, 142
chicken leukemia virus, 113, 149, 242
chicken lymphoma oncogene, 111
Children's Hospital, Jimmy Fund Clinic, 185, 189, 202
Chipperfield, Randy, 239, 306; and *ras* mutation project, 118, 119–20, 121, 123–24, 129–30, 139–40; death of, 140
chromosome *5*, 359
chromosome *11*, 310
chromosome *13: q14*, 310, 311, 312, 331–33, 336, 346; Dryja's work with, 334–35
chromosome *23*, 331
chromosomes: genes and, 37–39
clinical applications: of molecular biology research, 12–22
cloning of genes, 38, 50; and Asilomar regulations, 77–78; *see also* oncogene cloning
Cochran, Brent, 234
codons, 41
cold room, 30, 172
Cold Spring Harbor Laboratory, 8, 9; transfection discoveries at, 102; T24 cell cloning project, 108–9; point mutation research at, 130–31; summer meetings at, 271; description of, 271–72; staffing of, 271–72; alternate pathway project at, 282–302; Japanese researchers at, 283–84; arachidonic acid experiment at, 295–96; *see also* Wigler, Michael
collaborative work: authorship accreditation issue, 103–4, 129–30, 350–51; Shih-Shilo conflict, 104–7; Bernards-Bernstein, 185–99; Bargmann-Hung, 244–67; retinoblastoma project, 309–58
colon cancer, 15, 19, 359; oncogene cloning project, 104–7, 108; oncogene, 111; and *ras* gene, 134–35, 136–37, 138
controls: need for, 196–98
Cooper, Geoffrey, 107, 197; transfection work by, 68–69, 102; transfection report in *Nature*, 102; *ras*-oncogene homology report, 116; *ras*-

protein receptor report, 197
cooperation theory. *See myc-ras* gene project
coronary heart disease: treatment of, 16
Crick, Francis, 70
Croce, Carlo, 149, 150
croton oil, 142
Cuzin, François, 147
cyclic AMP, 279
cyr1-1 strain, 287–89
cytomegalovirus virus, 96
cytosine (*c*), 33, 37–39

Dannenberg, Ann, 94
deoxyribonucleic acid. *See* DNA
Dessain, Scott, 307, 343
Dhar, Ravi: and *ras* oncogene project, 124, 125–27, 129, 130, 131
Dhawan, Vivek, 207, 214, 324
diacyl glycerol, 204–5
diet: cancer and, 141
"dirt science," 210
Discover, 266
DMBA (dimethylbenzathracene), 93, 95, 142
DNA: technological advances in recombinant, 28–29; blot types, 29–30; as synonym for life, 32, 32–43; nucleotides of, 33, 37–38; immortality of, 33; human vs. chimpanzee, 33; seeing, 34–35; proteins and, 36, 37, 42; double helix, 38, 39, 40; and reverse trancriptase, 52–56; early models of, 70; mutation of, cancer and, 71; Asilomar regulations on recombinant, 77–78
DNA-protein-binding assay, 325
DNA-RNA transcription, 39–42, 54
DNA transfection project, 56–69; under Smotkin, 53–54, 55, 56–61; Smotkin proposes, 56–58; under Goldfarb, 62–65, 67, 68
dog-racing outing: by Whitehead Institute staff, 306–8, 333
Doolittle, Russell, 242
Dotto, Paolo, 234, 264–65
double-blind controls, 197; in oncogene transfection project, 94–95
double helix, 38, 39, 40
Down's syndrome, 331
Downward, Julian, 242, 350, 353
Dryja, Thaddeus: work with retinoblastoma, 327, 333–41, 349–50, 352–54, 355–56, 357–58, 360–61, 368, 371–72; and H3-8 probe, 337–38, 339
Duke University, 164
Dulbecco, Renato: Weinberg's work under, 50–51

E. coli: cloning oncogene into, 78, 97, 98, 99, 105
EC gene, 124
EJ gene, 100–101, 102, 103, 104, 105, 107, 108, 110, 111, 114, 116, 121, 148–49, 180; and *ras* gene, 114–16, 117–18, 124
Ellsworth, Robert, 331, 362–64

enhancers, 38, 42
epidermal growth factor receptor (EGF receptor), 242–43, 244, 245, 249
Epstein-Barr virus, 96
erb-b oncogene, 242–43, 244, 246
esophageal cancer, 142
esterase-D gene, 336, 339
experiments: elegant, 31; *see also* projects *and specific projects by name*
eye cancer, 15, 19; *see also* retinoblastoma

Feramisco, James, 295
Fetal membrane tissue malignancy, 15
Fink, Gerald, 174, 219, 269, 276
Finkelstein, Robert, 120, 208, 227, 231, 307
Finney, Michael, 2
Flechtheim, Alfred, 46
floor meetings, 176
Folkman, Judah, 215
fos gene, 134, 259, 312
Francke, Uta, 239
fraud. *See* research fraud
Friend, Stephen, 196, 333; reputation of, 306; personality and background of, 308–9; and retinoblastoma project, 308–9, 312–14, 338–46, 348–53, 357

Gallie, Brenda, 336, 338, 342, 351, 356–57, 358; objections to Weinberg lab conclusions on retinoblastoma, 356–57
Gallo, Robert, 194
Garte, Seymour, 157
gene library. *See* phage libraries
Genentech labs, 26, 139
General Motors Cancer Research Prize, 11, 23
genes: molecular principles of, 32–43; number of different in human cells, 36, 37; DNA, RNA, and protein, 36–42; stages of, 37–38; nucleotides, 37–38; and chromosomes, 37–39; promoters, 38; enhancers, 38; start site, 38; stop site, 38; viral transfection, 56–69; *see also* cloning of genes *and individual genes*
genetic polymorphism: research on, 101
Gianni, Massimo, 54, 61
Gilbert, Walter, 170
Gilman, Alfred, 212
Gilman, Laurie, 225–26, 234
Gilman, Michael, 178, 208, 216, 306, 314, 340; personality of, 217–19; relations with Weinberg, 218–20, 221–22, 228–31; training of, 221; and *myc-fos* gene project, 222–34, 237, 268, 325; family relations of, 225–26, 227, 228
glutamic acid, 317–18
Goldfarb, Mitchell, 346; and DNA transfection project, 62–65, 67, 68, 74, 213; and oncogene transfection project, 75–78, 91, 93; thesis on Harvey sarcoma virus, 76, 114; and regulations on oncogene

cloning, 77, 78; transfection work at Cold Spring Harbor Laboratory, 102; cloning T24 cell, 108–9, 115
Goldstein, Joseph, 16
Gordon Conference, New Hampshire (1979), 96
Gouvea, H. de, 328–29
G proteins, 212, 279, 290, 300
graduate students: in research labs, 25–26, 231; concern over thesis, 26; at Weinberg lab, 26–27; stipends for, 27; at Whitehead Institute, 176–77
Graham, Angus, 83, 90
Graham, Frank: and DNA transfection, 56, 57, 58, 63
grant awards: to Weinberg, 11; and demand for cancer cure, 14
grant proposals: Weinberg and, 23, 24
Graves, Michael, 3
Green, Cecil, 166
Green, Howard, 92
Green, Ida, 166
Grosschedl, Rudolf, 140, 239, 262
group meetings, 176
guanine (*g*), 33, 37–39
guanosine triphosphate (GTP), 208–9, 212, 279, 300
Gusella, James, 101
Guthrie, Woody, 188

Harford, Joe, 197
Harvey sarcoma retrovirus: and transfection project, 60, 62, 63, 65, 67–68, 73, 74, 76, 111, 114; and T24 clone, 115–16, 117–18
Hayward, William, 128, 157–58
Heidelberger, Charles, 73, 76, 93, 98
herpes virus, 96
Herskowitz, Ira, 286, 289–90
HER-2 gene, 320
Hodgkin's disease, 14–15
Hoffman, Robert, 90; on cancer research, 13
homology: between retroviruses and transfected oncogenes, 110–16, 118, 119, 159
homozygogosity, 332
Hood, Leroy (Lee), 25, 150
Housman, David, 101, 102
H3-8 probe, 337–38, 339, 342, 356
Hung, Mien-Chie: and *neu* gene project, 244–46, 248–67, 316, 319; problems with *neu* project, 252, 254, 256–58, 262–67, 305; personality and family relations of, 252–54, 257–58; preference for drosophila, 262, 264; on dog-race betting, 307
Hunter, Tony, 244, 311, 347
Huntington's chorea, 188

immortalization theory, 148–49, 153, 154, 156
initiators, 142
inositol trisphosphate, 204–205, 209
interferon, 166, 358
introns, 40, 42, 101

Jaenisch, Rudolf, 174
Jensen, Earl, 100, 148

Jensen tumor cells. *See* EJ gene
job seminars, 81

Kamen, Robert, 147, 148
Kataoka, Tohru: and yeast *ras* project, 277–78, 282, 284; grant proposal preparation by, 293–94, 296
Kennedy, Edward, Jr., 312, 331
kidney cells: transfection of human, 56, 58
King, Jonathan, 166
Kitchin, David, 331
Klagsbrun, Michael, 215–16
Knudson, Alfred, 329–30, 355, 358
Koehler, Georges J. F., 70
Komaroff, Anthony: on Weinberg, 44, 53, 90, 99–100, 127

lab secretaries: salaries for, 27
lab technicians: salaries for, 27
Lalande, Marc, 334
Land, Hartmut (Hucky), 145–48, 150, 151–54, 156, 159–63, 306, 307; personality of, 145–46; and rat fibroblast project, 145–47; and *myc-ras* gene project, 150–53, 154, 156, 158–60, 161–63, 180, 312
large-T antigen, 148
Latt, Samuel, 334
Leder, Philip, 149, 150, 155
Lee, Wen-Hwa: cloning of esterase-D gene, 336, 339; and retinoblastoma clone, 347, 348, 353, 354–55, 357–58
leucine, 41
leukemia oncogene: effort to clone, 102
leukemia virus, 15, 19; and transfection project, 57–60, 73; chicken, 113, 119, 242
Levinson, Arthur, 129, 139
Lewin, Benjamin, 89
Lewis, William, 355–56; and Spandidos controversy, 85, 86, 87, 88–89
Lodisch, Harvey, 174
Logan, Jonothan, 134
long terminal repeats (LTR's), 75
Loos, Adolf, 223
Low, Francis, 165, 167
Lowy, Douglas, 116, 124, 130
lung cancer, 19, 98, 141, 359; mortality rates, 15; and *ras* gene, 135
Luria, Salvador, 51, 52, 173; on transfection, 133

MacArthur award, 175, 181
MacDonald, Marcy, 188
Macmillan Ltd., 79
"magic bullets," 70
Maniatis, Thomas, 201
Massachusetts Institute of Technology (MIT): graduate students at Weinberg lab, 26; and creation of Whitehead Institute, 164–68, 170
MBO-I, 107
McGuire, William, 320
Medawar, Sir Peter, 221
medical research in cancer. *See* cancer medical research
mentor-protégé relationship, 28
messenger RNA, 40–41

metastasis gene project, 180–87, 189–202, 259; cloning of metastatic gene, 191–202; artifact error discovered, 194–202; *ras* gene identified, 199–202
3-methylcholanthrene, 76, 93, 95, 97
middle-T antigen, 148
Milstein, Cesar, 70
MIT Press, 79
molecular biologists: perspective on cancer, 6, 12–15, 16; training of, 9; numbers in oncogenetic specialty, 13; socioeconomic background of, 211
molecular biology: funding of, 11, 14, 23–24, 134; conference and forum circuit, 22–23; limits of knowledge in, 36–37
molecular biology laboratories: workings of, 7–10; lab costs, 23–24; structure, funding, and staffing, 24–30; technological advances, 28–29; *see also* Cold Spring Harbor Laboratory; Weinberg lab; Whitehead Institute for Biomedical Research
Moloney leukemia virus, 53
monoclonal antibodies, 70
Morgenstern, Jay, 146, 159, 193, 248
mouse leukemia virus, 54
mouse oncogene, 111; cloning assigned to Shilo, 97–98; *see also* oncogene cloning project
mouse retrovirus: and transfection project, 58
Mulligan, Richard, 174–75, 176, 201, 257
multistep tumorigenesis theory, 142–43, 148–49, 154–56, 163, 182, 372
Murray, Mark: and oncogene cloning project, 103, 104, 105, 106, 108
mutagens: and carcinogens, 71
myc-fos gene project, 222–34, 259, 325
myc gene, 113, 187, 198, 312–13; N-, 313–14
myc-ras gene project, 149–63, 178

National Academy of Sciences, 11, 194
National Cancer Institute, 14, 162; oncogenetic funding by, 134
National Institutes of Health: funding of cancer research, 14; graduate student stipends from, 27; and Asilomar regulations, 78, 93, 274
Nature, 79, 168, 176, 197; lab candidates' papers in, 25; Cooper's transfection report in, 102; Wigler's paper on cloning in, 109; Lowy's homology report in, 116; point mutation reports in, 128, 129, 131; *myc-ras* gene report in, 153; reports on *ras* and *myc* genes, 157; Cavenee's article on retinoblastoma in, 333, 334;

Nature (*cont.*)
and Friend and Dryja's report on retinoblastoma, 348, 350, 352–53, 355
Naylor, Susan, 359
negative controls, 197
nematodes, 16
neu gene project, 235–36, 237–68, 315–19, 359; division of labor in, 246, 249–50; mutation in transmembrane, 316–19, 322, 326
neuroblastoma, childhood, 30
neuroblastoma oncogene project. *See neu* gene project
Newmark, Peter, 348, 350, 353
New York Hospital: retinoblastoma clinic, 362–64
NIH 3T3 cells: and DNA transfection project, 58, 59, 60, 63, 64–65, 68–69
Nikawa, Junichi, 284
nitrosamine, 142
Nixon, Richard, 14
N-*myc* gene, 313–14, 352, 358
Nobel Prize: Weinberg as potential candidate for, 1–2, 139, 194, 229; awarded to Watson, 12; awarded to Brown and Goldstein, 16; awarded to Luria, 51; awarded to Baltimore, 52, 165, 173; awarded to Temin, 52; awarded to Rous, 65; awarded to Milstein and Koehler, 70; awarded to Gilbert, 170
Northern blot, 29–30
nucleic acid. *See* nucleotides
nucleotides: DNA, 37–38; RNA, 39

oncogene cloning: and Asilomar regulations, 77–78, 102; metastatic gene, 191–202 (*see also* metastasis gene project)
oncogene cloning project: assigned to Shilo, 97–110; competition between Wigler and Weinberg labs, 102; competition between Shih and Shilo, 103–7; Parada's discovery of homology with retroviruses, 112–16
oncogenes: characteristics of, 5; number identified, 22 (*see also individual oncogenes*); protein and, 42–43; discovery of *src*, 66–68; *ras*, 67–68 (*see also ras* gene *and ras* oncogene project); *see also* metastasis gene project; recessive oncogenes
oncogenetics: 1970s discoveries in, 4–5; and medical research, differing perspectives of, 6, 12–15, 16; clinical application of, 12–22; phenomenological approach to, 20–21; virology and, 49–50; funding increases, 134; *see also* molecular biologists
oncogene transfection, 71–95, 163; Spandidos's claims on, 79–81, 83–90; human and animal oncogenes discovered through, 111

oncogene transfection project: Shih and, 91–95, 97–110; double-blind tests in, 94–95; Weinberg announces results at Gordon Conference, 96–97; first effort to clone oncogene, 97–110, 111–16; Parada's discovery of homology, 112–16; landmark status of, 133
Orr-Weaver, Terry, 174
osteosarcoma: retinoblastoma gene and, 312, 314, 326, 331, 343, 360

pancreatic cancer, 184; and *ras* gene, 134–35, 138
Papageorge, Alex, 130
papilloma viruses, 141
Parada, Luis, 183, 227, 248; and Weinberg house-building project, 2; on graduate students, 26–27; on oncogene cloning project, 99; on Shilo's work, 108; appearance and character of, 110; homology discovery, 112–16, 118, 119, 159; on mutation theories, 122; and *myc* gene project, 149–50, 151, 154, 158–59, 161–62, 163; and point mutation in transmembrane of *neu* gene, 317
Paskind, Michael, 121
Pawius, Petrus, 328
peer review system, 17
Penman, Sheldon, 49, 166
Perucho, Manuel, 108; and *ras* oncogene research, 136–38
phage library, 105–6, 107, 108
pharmaceutical companies, 19–20
Phillips, Robert, 336, 338
phosphatidyl inositol (PI) pathway, 204–5, 207–215, 216, 217, 260
phospholipase C, 209
phospholipid pathway project, 204–5, 207–215, 216, 217, 260, 267
Pillai, Shiv, 214
pituitary cell: DNA of, 34
plant cells: oncogenics and, 34
platelet-derived growth factor (PDGF), 223, 224, 242
PNAS: Shih's transfection report in, 102; Cooper's homology report in, 116; Bernstein's report in, 180, 181, 185
point mutation, 126–29, 130–36, 139, 143, 157, 163, 180, 203, 275, 279, 310, 346; accreditation on Whitehead lab report, 129–30; in transmembrane of *neu* gene, 316–19; and retinoblastoma, 343–44
polio virus research, 91, 173
polymorphism: and *ras* oncogene, 132
polyoma virus, 147–48, 150
p185 project, 239–44
positive controls, 197
postdocs: in research labs, 25–28, 220–21, 231; training of, 27; salaries for, 27; at Weinberg lab, 27–28, 220–21; professional independence of, 27–28; at

postdocs (*cont.*)
Whitehead Institute, 176–77, 220–21
Powers, Scott, 273, 275, 282, 294, 299, 342, 346; and yeast *ras*, 275–76, 277, 283, 284, 286–87; and sup-H gene, 285–86, 288, 289–91, 300
projects: development of, 31; conclusion of, 31; *see also specific projects*
promoter mutation theory, 121–26, 131, 132
promoters, 38, 42, 142
protein kinase C, 204, 209
proteins, 37, 40–43; and DNA, 36, 41–42; transcriptional, 39; amino acids, 41; and oncogenes, 42–43; *p185*, 239–44
protocols: research, 31
proto-oncogene: use of term, vs. oncogene, 5, 67
P-3 room, 30

q14 region of chromosome *13*, 310, 311, 312, 331–33, 336

radiation therapy, 20
ras gene, 67–68, 75, 111, 113, 114, 183, 198, 311, 312; and EJ gene, 114–16, 117–18, 149, 180, 203; pervasiveness in cancer, 134–38; twelfth codon of, 135, 136, 143, 155, 180, 275, 310 (*see also* point mutation); search for function of, 138–39, 203–4; identified with metastasis gene, 199–202; phospholipid pathway project, 203–4, 208–9, 210; yeast and functioning of, 270–71, 275–79, 300–301; animal vs. yeast, 276–77; adenylate cyclase pathway, 279–82; and *Xenopus* oocyte project, 280–81; alternate pathway theory, 282–302; *see also ras* oncogene project
ras oncogene project, 116, 117–39; promoter mutation vs. structural mutation theories, 121–23, 131; point mutation, 126–29, 130–36, 139, 143, 157, 163, 180, 203, 275; second mutation, 131; *see also myc-ras* gene project
rat embryo fibroblasts (REF's) project, 144–47, 190
rat neuroblastoma, 98, 111, 239–40
Rauschenberg, Robert, 172
Rauscher, Frank, 134
Ray, Man, 172
Reagan, Ronald, 137
recessive oncogene, 309–11, 344, 359–60, 372–73; retinoblastoma as, 309–11, 332–33, 344–45
recombinant DNA. *See* DNA
research fraud, 83; Spandidos controversy, 79–81, 83–90
restriction enzymes, 118, 124–25
retinoblastoma: incidence of, 309; pathology of, 311–12, 328–30; mortality from,

328; summary of research on, 328–30; hereditary and genetic patterns, 328–33; sporadic, 330; two-hit theory of, 330–31; future prospects for treatment, 360–61; victims of, 361–67
retinoblastoma gene, 308, 309–15; as recessive oncogene, 309–11, 332–33; effort to clone, 311, 313–15, 326–27, 339–46, 347–53, 360–61; and osteosarcoma, 312, 314, 326, 331, 343, 360; internal deletions in, 343–44; Friend and Dryja's report on, 348–58; Lee's report on, 357–58
retroviral oncogenes, 111; homology with transfected oncogenes, 112–16, 118
retroviruses, 95, 111, 173, 174; reverse transcriptase, 52–54; Rous sarcoma virus, 65–66, 68
reverse transcriptase, 52–54, 165
Revlon, 166
ribosomes, 41, 42
Rifkind, Richard: on potential for cancer cures, 12, 15, 141
RNA, 35, 36, 39–43; nucleotides, 39–41; transcription, 39–42, 54; phases of, 40; messenger, 40–41; retroviruses, 52–53
RNA-DNA transcription: reverse transcriptase and, 52–56
Rogelj, Snezna, 238, 307, 324–25; and Weinberg house-building project, 3, 47–48; and phospholipid pathway project, 205, 207–215, 216, 217, 221, 222, 237, 259, 267; personality and appearance of, 205–7, 211; disaffection with cancer research, 214; and angiogenesis factor research, 216–17; and retinoblastoma project, 340, 345–46, 350–51
Rous, Peyton, 16, 65; awarded Nobel Prize, 65
Rous sarcoma virus, 65–66, 68
Ruley, Earl, 153–54

Saccharomyces cerevisiae, 276
Sager, Ruth, 143–44, 357
Salk Institute for Biological Studies, 50, 51
Samonella typhimurium, 71
sarcomas: rarity of, 98
Schechter, Alan, 148–49, 159, 211, 216, 234, 257, 265; and *ras* pathway, 203; personality of, 234–35, 238–39; and *neu* gene project, 239, 240, 241, 245, 246, 263–64, 319–20; and p185 project, 240–44
Science, 79, 353; Lee's retinoblastoma report in, 357–58
scientific journals, 79; *see also Cell; Nature; PNAS; Science*
scientific problems: "interesting" vs. important, 48
scientific research: dynamics of, 7; politics of, 7; *see also*

scientific research (*cont.*)
molecular biology;
oncogenetics
scientists: dynamics of relationships among, 7, 9; senior vs. junior, 9, 176–77; frustrations of, 305; *see also* graduate students; molecular biologists; postdocs
Scolnick, Edward, 282; and *ras* gene, 67, 68, 115, 116, 118–20, 130, 281
Sharp, Phillip: on cancer research, 17; on academic lab function, 28; on Spandidos controversy, 83; and oncogene transfection and cloning project, 99, 133; and *neu* oncogene, 257
Shih, Chiaho: character of, 91–92, 252; and oncogene transfection project, 91–95, 97–110, 213, 235, 239–40, 241, 311; transfects mouse lung cancer cells, 98, 99; attempts to transfer human carcinomas to mouse cells, 98–99; work on EJ gene, 100–101, 103, 106, 107–10, 112–16, 118, 121, 306; work with *Alu* sequences, 102; transfection report in *PNAS*, 102; competition with Shilo over oncogene cloning project, 103–7, 112
Shilo, Ben-Zion: and oncogene transfection project, 93, 97; and oncogene cloning project, 97–99, 100, 101–2, 112; competition with Shih over oncogene cloning project, 103–8, 112
Shimizu, Kenji: and T24 cell cloning, 108
Shubik, Philippe, 142
sickle cell anemia, 126–27
simian virus 40, 50, 53
Siminovitch, Louis: and Spandidos controversy, 81, 83–89
single-blind controls, 197
sis oncogene, 242, 246
skin cell: DNA of, 34
Slamon, David: and *neu* gene research, 319–23
smoking: cancer and, 141
Smotkin, David, 207; and DNA transfection, 53–54, 55, 56–61, 68, 69, 73, 213; departure from Cancer Center, 61–62
somatic hybrids: research with, 101
Southern, Edward, 29
Southern blot, 29–30
Spandidos, Demetrios: controversy over transfection research claims of, 79–81, 83–90
Sparkes, Robert: cloning of esterase-D gene, 335–36
Spector, Mark, 82, 83
src gene, 75, 111, 113, 134, 156, 160–61, 178, 198, 244, 249; discovery of, 66–68, 71, 133
Stanley, Pamela, 83–84
start site, 38
Steffen, David, 76, 112, 119, 120; on Chiao Shih, 92
Stehelin, Dominique, 66
Stella, Frank, 171

Stern, David, 32, 34, 177, 183, 194, 197–98, 218, 227, 231–32, 308, 316, 345; on Weinberg, 48, 206; and p185 project, 241–42; and *neu* gene project, 244–46, 253, 259, 268
Stern, Robert, 3
stop site, 38
Stout, Lewis, 351
structural mutation theory, 121–26; mix-match experiments on, 123–26, 128, 131; point mutation, 126–29, 130–33
Summerlin, William, 81–82
sup-H gene, 285–86, 288, 289–91, 300
suppressor mutation, 285–86
Symbolics, Inc., 170
synergysm theory. *See myc-ras* gene project

Tabin, Clifford, 158, 241; on benchwork, 30; on Shih-Weinberg conflict, 100; on Shih-Shilo competitiveness, 104; on Parada, 113; character of, 119–20; and *ras* oncogene project, 119–32, 139–40, 143, 144; and accreditation on point mutation report, 129–30, 139, 306; and rat fibroblast experiment, 144–45; and *myc-ras* gene project, 150–53; and *ras* pathway, 203
Technicon, 164, 166
Temin, Howard, 111; on clinical applications of research, 12; discovery of reverse transcriptase, 52, 53; awarded Nobel Prize, 52
testicular cancer, 15
thesis: graduate student concern over, 26
Thomas, Lewis, 168
thymine (*t*), 33, 37–39
Toda, Takashi: and alternate pathway project, 277–78, 284, 286–87, 291–92, 297; and *cyr1-1* strain, 287–89
transcriptional proteins, 39
transfection. *See* DNA transfection project; oncogene transfection project
transgenic mice, 262, 316
transmembrane domain mutation, 315–19, 322, 326
T24 cell cloning, 102, 108–9, 115; Wigler's and Barbacid's work on, 108–9, 115, 124; and Harvey sarcoma retrovirus, 115–16, 117–18, 124
tumor galls, 34
tumorigenic genes. *See* oncogenes
Twain, Mark, 299
tyrosine kinase, 243–44

Ullrich, Axel, 239, 320
University of Toronto, Medical Research Council, 88, 89
uracil, 39
U.S. Steel research awards, 11

valine, 317–18
Vancek family, 364–67
van der Eb, Alex: and DNA transfection, 56, 57, 58, 63

Varmus, Harold, 159, 160, 309–10; on cancer research, 13; work on Rous sarcoma virus, 65, 66, 67, 68
Verma, Inder, 346
vinyl chloride, 142
viral DNA transfection, 56–69
virology, 49–50, 68
Vogelstein, Bert: and *ras* oncogene research, 136–38

warm room, 30, 172
War on Cancer, 14
wart virus, 141
Waterfield, Michael, 242, 350
Watson, James D., 15, 70, 299; on clinical applications of research, 12; as director of Cold Spring Harbor Laboratory, 12, 270
Weinberg, Abraham, 46–47
Weinberg, Amy, 45, 143, 194; and house-building project, 3–4; marriage to Robert, 51
Weinberg, Aron, 2, 45, 46
Weinberg, Leah Rosa (Rosie), 2, 45, 48
Weinberg, Robert: as potential Nobel candidate, 1–2, 139, 194, 229; house-building project of, 1–4, 48, 99–100, 368–73; DNA models of, 4; faith in simplicity principle, 4, 18–19, 182; celebrity of, 6, 11, 158; approached as subject of book, 6–7; awards and grants received by, 11; on research vs. clinical application, 17–22; and conference circuit, 22; as lab director, 22–24, 47–48, 285; grant proposal writing by, 23, 24; and staffing of lab, 25–28, 158–59; luck as scientist, 43, 373; speaking style of, 44; photographs in office, 45–46; interest in genealogy, 46–47, 372; family and personal background of, 46–47, 49–52; as amateur polyglot, 47; relations with staff, 48; education and training of, 49–50; background in virology, 49–50; as postdoc, 50; aversion to benchwork, 50; and simian virus 40, 50, 53; early work at Cancer Center (MIT), 51, 52–54; first marriage of, 51–52; and DNA transfection project, 55, 56, 57–69; on staff turnover, 61–62; transfers interest from retroviruses, 68; and oncogene transfection theory, 71–95, 133; Longfellow Bridge brainstorm, 73–74; and Goldfarb's objections to oncogene transfection, 76–78; and Spandidos controversy, 79–81, 83, 90; on Baltimore's style vs. his own, 91; assigns oncogene transfection project to Shih, 91–95; announces results of oncogene transfection research at Gordon Conference, 96–97; assigns oncogene cloning to Shilo, 97–100; competition with

Cold Spring Harbor Laboratory, 102, 107; and Parada's homology discovery, 112–16; and *ras* oncogene research, 116, 117–19, 121–24, 127–31, 133, 138–39, 143; advocates promoter mutation theory, 121–23; and point mutation, 126, 128–29, 131, 133–34, 135, 139, 346; and rat fibroblast project, 144–47; multistep tumorigenesis theory of, 147–48, 154–55, 372; and *myc-ras* gene project, 149–63; transfers to Whitehead Institute, 163, 168–69, 178–79; and Applied bioTechnology, 170; and metastasis project, 182–84, 185, 186, 187, 189–90, 193–95, 199, 201, 202; elected to National Academy of Sciences, 194; and phospholipid pathway project, 203–4, 207–8, 212–13, 214, 215, 216–17; as teacher, 215; relations with Michael Gilman, 218–20, 221–22, 228–31; and *neu* gene project, 246, 248, 252, 254, 259–261, 264–68, 327; on Wigler, 269, 302; and retinoblastoma project, 308–9, 335, 339, 341, 343–44, 346, 350, 352–53, 371–72

Weinberg, Suzanne, 49

Weinberg lab: structure of, 24–30; as learning and research center, 28; DNA transfection project, 56–69; oncogene transfection project, 71–116; oncogene cloning project, 97–110; point mutation research, 116–36, 316–19, 343–44; *ras* oncogene project, 116–39; metastasis project, 180–202; phospholipid pathway project, 204–17; retinoblastoma gene project, 308–15, 326–61; *see also* Whitehead Institute for Biomedical Research *and individual scientists*

Weinstein, I. B., 160

Western blot, 29–30

Whitaker, Margaret, 166

White, Raymond, 177, 355

Whitehead, Edwin (Jack), 9, 164–68

Whitehead Institute for Biomedical Research, 1, 9; establishment of, 163–69; Baltimore as director, 164–68, 172–74; description of facilities, 170–72; working environment of, 172, 175–78; competitiveness of, 175–76; staffing of, 175–78, 248; group meetings at, 176; postdocs and graduate students at, 176–77; junior and senior scientists at, 176–77; as teaching institute, 215; dog-racing outing, 306–8, 333; *see also* Weinberg lab *and specific projects*

Wigler, Michael, 163, 196, 210, 221, 309, 326, 342,

Wigler, Michael (*cont.*) 353; as author's foil to Weinberg, 8, 9, 268; early transfection work, 72–73, 75; on research fraud, 82, 83; transfection work under, 102, 107; and cloning T24 cell, 108–9, 115; on homology oversight, 114; and point mutation research, 130–31, 143; on Weinberg, 132, 300; personality and background of, 269–71, 272–75, 298–300, 302–3; lab management style of, 272–75, 285, 292–94, 296–97; and alternate pathway theory, 282–302; inquires about author's project, 304–5; on retinoblastoma research, 356, 360
Wilm's tumor, 310
Wilson, Ruth, 34, 37, 308
Wynder, Ernest, 46

Xenopus oocyte: and *ras* gene, 280–81, 295

yeast genetics, 16, 219, 276–77, 283, 286
yeast *ras* gene, 270–71, 275–79, 300–301; alternate pathway theory, 282–302
Young, Dallan, 288
Yunis, Jorge, 71